BASICS OF ANALYTICAL CHEMISTRY AND CHEMICAL EQUILIBRIA

BASICS OF ANALYTICAL CHEMISTRY AND CHEMICAL EQUILIBRIA: A QUANTITATIVE APPROACH

By

BRIAN M. TISSUE
Virginia Tech
Department of Chemistry
Blacksburg, VA

Second Edition

For general information on our other products and services or for technical support, please contact our Customer Care Department within the United States at (800) 762–2974, outside the United States at (317) 572–3993 or fax (317) 572–4002.

Wiley also publishes its books in a variety of electronic formats. Some content that appears in print may not be available in electronic formats. For more information about Wiley products, visit our web site at www.wiley.com.

A catalogue record for this book is available from the Library of Congress

Paperback ISBN: 9781119707356; ePub ISBN: 9781119707387; ePDF ISBN: 9781119707349

Cover Images: Background: © kundoy/Getty Images; Figure: Courtesy of Brian M. Tissue
Cover Design: Wiley

Set in 11/13pt TimesNewRomanPSMT by Integra Software Services Pvt. Ltd, Pondicherry, India

SKY10062107_120823

CONTENTS

PREFACE

This text will introduce you to *analytical chemistry*: the science of making quantitative measurements. Quantifying the individual components in a complex sample is an exercise in problem solving. An effective and efficient analyst will have expertise in

- sampling, sample processing, and method validation;
- the chemistry that can occur in a sample before and during analysis;
- selecting an appropriate analytical method; and
- proper record keeping, data analysis, and reporting of results.

I do not attempt to be comprehensive in this text. Samples that require analysis are so diverse that it is not possible to describe every sample preparation protocol, separation method, and measurement technique. These details are contained in handbooks and method compilations, many of which are now accessible from online sources. This text emphasizes the fundamental chemical and physical concepts that underlie the analytical methods. With an understanding of the fundamental concepts, a scientist faced with a difficult analysis can apply the most appropriate techniques, identify when a particular problem cannot be solved with existing methods, and develop new analytical methods. The proficient analyst will also be alert to interferences and problems in analytical measurements and recognize when an "answer" might not be correct.

I organize the discussion of the core principles of analytical chemistry into three parts:

- Part I: Analytical concepts such as calibration and uncertainty, sample preparation, classical (wet-chemical) methods, and molecular UV/Vis spectroscopy
- Part II: Chemical equilibria involving acids, bases, complexes, and insoluble precipitates
- Part III: Electroanalytical methods, atomic spectrometry, molecular structure determination, and chromatographic separations

The analytical methods in Part I rely on reactions that go to completion. Part II is a detailed treatment of chemical equilibria—reactions in which reactants and products coexist. Chemical equilibria are critical to many aspects of chemical, biochemical, and environmental systems. Part III describes the most common instrumental methods of analysis, illustrating many of the tools of the trade for making quantitative measurements. Even if your future career veers away from science, you will find the problem-solving and graphical data analysis skills developed in this text to be useful.

Many of the topics in this text follow directly from first-year college chemistry. You will want access to a general chemistry text or online resource to refresh your memory of the underlying principles of physical processes and chemical species. The level of this text presumes that you know

Basic math	Algebra
	Exponential functions
	Calculating and plotting in a spreadsheet
Basic chemistry	Predicting properties based on the periodic table
	The nature of chemical compounds
	Stoichiometry and balancing reactions
Reaction types	Acid–base
	Complexation
	Precipitation
	Reduction and oxidation (redox)

The beginning of each chapter lists learning outcomes that serve as a brief outline to help categorize new material. After completing a chapter, make a concept map to help yourself see the big picture and underlying concepts. You will often encounter a repeat of concepts in the text. Making connections with prior material makes learning analytical concepts much easier. Treating every topic as something new becomes overwhelming. Each chapter contains sample calculations and practice exercises. I assume that your goal is success. Achieving success requires skills, and acquiring skills takes practice.

Variables and constants are italicized to not be confused with other text. As much as possible, I use the conventions and terminology in the *IUPAC Gold Book*.[1] You will find other symbols in other books and resources, so use the context to decipher the differences. This second edition adds substantial information on instrumental methods and manufacturers will often use different terms for similar instruments. Relevant spreadsheets and links to useful resources are available at https://www.achem.org.

Blacksburg, VA   BRIAN M. TISSUE
                                                                 August 2022

---

[1]See *IUPAC Compendium of Chemical Terminology*, "The Gold Book," https://goldbook.iupac. org; Accessed August 2022.

ABOUT THE COMPANION WEBSITE

This book is accompanied by a companion website:

www.wiley.com/go/tissue/analyticalchemistry2e

This website includes:

- Solutions to the end-of-chapter practice exercises.
- Spreadsheet templates and solution guides for the "You-Try-It" exercises.
- A glossary of analytical chemistry terms.
- General spreadsheets for common analytical calculations.

PART I

QUANTITATIVE ANALYSIS USING REACTIONS THAT GO TO "COMPLETION"

CHAPTER 1

MAKING MEASUREMENTS

Learning Outcomes

- Describe aspects of good laboratory practice (GLP).
- Use correct terms to describe analytical measurements and data.
- Calculate analyte concentration from measurement results.
- Use statistical formulas to express the precision of analytical measurements.
- Use calibration methods to obtain accurate results.

1.1 INTRODUCTION

There are few areas in our modern life in which the *quantity* of substances is not important. Industries and government agencies spend substantial resources to determine and monitor the safe levels of chemicals in foods, pharmaceuticals, and the environment. Setting permissible levels of contaminants is based on quantitative results from toxicological studies and raising or lowering a level has significant costs and consequences. Similarly, companies compete for sales by providing high-quality goods at the lowest price. Optimizing industrial processes depends on making decisions based on analytical measurements. Poor measurements or incorrect data interpretation will lead to poor decisions.

Basics of Analytical Chemistry and Chemical Equilibria: A Quantitative Approach, Second Edition. Brian M. Tissue.
© 2023 John Wiley & Sons, Inc. Published 2023 by John Wiley & Sons, Inc.
Companion Website: www.wiley.com/go/tissue/analyticalchemistry2e

You might not make many measurements yourself, but you probably rely on data and quantitative results to make decisions. You've probably read the ingredients or nutritional information on a product label to choose one product over another. I certainly want manufacturers to perform quality checks on the contents of the products that I buy. I'm also expecting an independent agency, say the FDA or USDA,[1] to check that there is not too much of a mineral, or contaminants such as Pb or rat poison, that could make the food unhealthy. Think about the last time that you had a medical checkup. Did the doctor determine your health by just looking at you? At the least you had a quantitative measure of pulse rate and blood pressure. Modern medicine relies on a variety of technological tools and clinical analyses. At some time, you might need to make a significant decision such as beginning daily doses of a cholesterol-lowering drug. We all hope that doctors and clinical technicians analyzing our samples were paying attention when they took an analytical chemistry course!

When you make a measurement, or you need to make a decision based on someone else's measurement, do you trust the value? This chapter introduces the terminology and statistical tools to describe and assess quantitative results. Some of the details will be new to you, but they all fit into a framework for collecting and reporting quantitative measurements. Table 1.1 begins building our vocabulary of measurement science and data-handling concepts by defining general terms. Many of these terms are used rather loosely in the scientific and manufacturer literature. You might need to dig into the details to know exactly what is meant when a procedure refers to the sample, the signal, etc. It is also common that a given term will have a different definition for different instruments or techniques. Resolving power is a measure of selectivity in mass spectrometry. However, resolving power and the related term resolution have different meanings when discussing a spectrum, a microscope image, or the separation of components in a mixture.

In quantitative analysis, we want a measurement or detector signal that we can relate to an analyte concentration. Doing so can be quite involved, and Chapter 2 discusses various sample preparation methods to isolate an analyte from interferences so that it can be measured. What mechanisms are available to detect an analyte? I can think of only three general strategies to detect and quantitate an analyte:

- measuring a physical property,
- using electromagnetic radiation, called spectroscopy, and
- measuring an electric charge or current.

---

[1]US Food and Drug Administration and US Department of Agriculture. See "FDA Fundamentals," https://www.fda.gov/about-fda/fda-basics/fda-fundamentals; accessed August 2022. Other countries and trade blocs have similar agencies.

TABLE 1.1 Measurement Terms

Term	Definition
Sample	*v.* To collect one or more samples.
Sample	*n.* Substance of interest. Assumed to be representative of remaining substance that is not collected. May refer to an unprocessed field sample or to a laboratory sample that has undergone one or more sample preparation steps. Test portion is the preferred term for laboratory samples.
Unknown	*n.* A sample, the source of which is usually known. Calling a sample an "unknown" indicates that the identity of the substance or the analyte concentration in the sample is unknown and to be determined.
Test portion	*n.* A portion of a collected sample that is processed and measured.
Test solution	*n.* Analogous to test portion but specific for a liquid solution.
Analyte	*n.* The chemical species to be identified or quantitated. It might exist as a pure substance or as one constituent of a multicomponent sample.
Qualitative analysis	*n.* Making measurements to determine the identity, structure, or physical properties of a substance.
Quantitative analysis	*n.* Making measurements to determine the amount of an analyte in a sample.
Detector	*n.* Device that responds to the presence of analyte, usually generating an electrical output.
Signal	*n.* The detector output that is displayed or recorded.
Sensitivity	*n.* The change in detector signal versus change in analyte concentration.
Selectivity	*n.* The discrimination of an analyte versus other components in the sample.

TABLE 1.2 Measurement Strategies

Physical Property	Spectroscopy	Electric Charge or Current
Mass or volume	Absorption	Electrical conductivity
Density	Emission	Electrical potential, for example, pH meter
Refractive index	Scattering	Voltammetry (reduction and oxidation current)
Freezing point depression		Mass spectrometry (ion current)
Thermal conductivity		

Table 1.2 lists some examples in each of these categories. We will discuss most of these methods, so do not worry if they are unfamiliar. Chapter 3 discusses classical methods that rely on physical measurements and Part III of the text introduces instrumental methods based on electrochemistry, spectroscopy, and mass spectrometry.

Although I list only three general detection strategies, each of these general categories encompass a multitude of specific analytical techniques. For example, spectroscopic methods have been developed to use most of the electromagnetic spectrum, including X-ray, ultraviolet (UV), visible (Vis), infrared, and radio waves. The different regions of the electromagnetic spectrum interact with matter differently and provide different types of information. This text concentrates on quantitative methods for analytes in aqueous solution. There are numerous other spectroscopic techniques to make quantitative measurements of solids and to determine physical properties of materials.

These general categories vary in sensitivity and selectivity. Measurements based on a physical property are usually less sensitive than spectroscopic or charge-based instrumental methods. The methods based on physical methods are useful when analyte concentrations are relatively high and when preparing standards to calibrate instrumental methods. When coupled with a separation column, detectors based on physical methods are the most universal and capable of detecting all analytes in the sample. Spectroscopic and electroanalytical methods can be extremely sensitive and selective for specific analytes. Selecting from one of these three general strategies for a given analytical problem depends on the nature of the analyte, the expected concentration, and the sample matrix.

New analytical methods and instruments are developed continuously. The breadth of research and development in analytical chemistry is too extensive to convey through just a few examples. For an overview of current research topics in analytical sciences, browse the Technical or Preliminary Programs of upcoming analytical chemistry conferences.[2]

1.1.1 Concentration Units

Most of the quantitative methods that we will discuss are aimed toward determining the concentration of an analyte in a sample. *Concentration* is the quantity of one substance, the analyte, divided by the total quantity of all substances in the sample. Concentration is different from an amount, in moles, mass, or volume, and we often interconvert between the two. Table 1.3 lists the SI base units that are used to derive other units.[3] In addition to these units, we drop or add the prefixes listed in Table 1.4 to indicate numerical factors. Common units that we work with in addition to the mole and kg are mmol (millimole),

TABLE 1.3 SI Base Units

Unit of	Name	Symbol
Length	Meter	m
Mass	Kilogram	kg
Time	Second	s
Electric current	Ampere	A
Thermodynamic temperature	Kelvin	K
Amount of substance	Mole	mol
Luminous intensity	Candela	cd

---

[2]EAS (Eastern Analytical Symposium), https://eas.org; FACSS (Federation of Analytical Chemistry and Spectroscopy Societies), https://facss.org; or Pittcon (Pittsburgh Conference on Analytical Chemistry and Applied Spectroscopy), https://pittcon.org; URLs accessed August 2022.
[3]SI is the abbreviation for "The International System of Units." For more information see https://physics.nist.gov/cuu/Units/index.html; accessed August 2022.

mg (milligram), and g (gram). The purpose of the prefixes is to simply express results in convenient values rather than using scientific notation. From the SI base units, we derive other useful units, and some are listed in Table 1.5.

The SI unit for an amount of a substance is the mole. A mole is a fixed number of something. If you have one mole of fish, you have 6.022×10^{23} fish (that's a lot of fish). The mole is a large number because atoms and molecules are incredibly small. The mole is determined from the number of atoms in 12 g of carbon-12 (^{12}C). We cannot count individual atoms or molecules easily, so the physical means of measuring an amount is to determine a mass and convert it to moles. The *unified atomic mass unit* or atomic mass constant is 1/12 of the mass of a carbon-12 atom and has units of u (amu is common in older literature). For large biomolecules, the equivalent unit of Dalton (Da) is common. The atomic weight or standard atomic weight of an element is the relative mass to the unified atomic mass unit. Atomic weights are therefore relative, unitless quantities. Since most elements have multiple isotopes, the atomic weight is the weighted average based on the abundance of the different isotopes.

We will neglect cases where the atomic weight of a specific sample varies from published values, but it can be significant for some light elements and radioactive materials that are enriched in one isotope. We will also not show conversion of unitless atomic weights to proportionality factors when converting between amount and mass. For our calculations, we will use atomic or molecular *formula weights* with units of gram per mole (g/mol). As a final practical note, mass and weight are not the same thing. Mass is an intrinsic property of a substance, but weight is a force that is dependent on mass and gravity. Modern analytical balances can be calibrated to read mass accurately, so we will follow common usage and assume that any weights that we measure are equal to mass.

TABLE 1.4   Common Numerical Prefixes

Prefix	Name	Factor	Prefix	Name	Factor
p	Pico	10^{-12}	k	Kilo	10^{3}
n	Nano	10^{-9}	M	Mega	10^{6}
μ	Micro	10^{-6}	G	Giga	10^{9}
m	Milli	10^{-3}	T	Tera	10^{12}

TABLE 1.5   SI Derived Units

Unit of	Name	Symbol
Area	Square meter	m^2
Volume	Cubic meter	m^3
Volume	Liter (=0.001 m^3)	l
Concentration	Kilogram per cubic meter	kg/m^3
Concentration	Moles per liter	M (mol/l)
Density	Mass per volume	ρ (g/ml)
Relative density (specific gravity)	Ratio of a density to a reference	d (unitless)

Different types of analytical methods will measure concentration, volume, or mass. Similarly, different types of detectors respond to analytes in different ways, that is, respond to either the analyte concentration or the analyte amount. Part of the common language that we need are conventions for describing amounts and concentration in different orders of magnitudes. Table 1.6 lists common units for describing analytical results.[4]

The reason to use different concentration units and numerical prefixes is convenience (see Example 1.1). We could quote every concentration as a percentage. For example, the EPA[5] lead limit in drinking water of 15 ppb would be written as 0.0000015%. This expression is not only awkward, it is susceptible to typographical errors in calculations. Some units will be accompanied by explanatory text, for example a percentage might be listed as a "weight percentage" or "percentage v/v" (volume per volume). For ppm, ppb, and ppt units, the "per l" definitions are equivalent to "per kg" only for dilute aqueous solution, that is, at a solution density of 1.0 kg/l.

There are many other units for specialized measurements. Water hardness is usually quoted as the equivalent amount of calcium carbonate in mg/l or grains/gallon (1 grain/gallon = 17.1 mg/l). The salinity of salt water is expressed in parts per thousand or as a ratio to a reference solution. When measured versus a reference, units are expressed as a practical salinity scale, PSS, where 1 PSS unit is equal to approximately 1 g sea salt per kg of seawater. There are numerous other cases where some standard protocol is adopted so that measurements made by different analysts in different locations can be compared directly.

Air sampling can require different approaches. The EPA limits for particulate matter in indoor air, $PM_{2.5}$, are 15.0 $\mu g/m^3$ for the annual average limit and 35 $\mu g/m^3$ for the 24-h limit.[6] As this example illustrates, quoting a single quantity might not be sufficient to describe the concentration or criteria for an analyte in a real sample or situation.

TABLE 1.6 Common Concentration Units

Name	Symbol	Description
Molality	m	Moles solute/kilogram solvent (mol/kg)
Molarity	M	Moles solute/liter solution (mol/l)
Mole fraction	X	Moles solute/moles total (unitless)
Percentage	%	Parts per hundred, often as weight percent
Parts per thousand	‰	
Parts per million	ppm	(mg/kg or mg/l)
Parts per billion	ppb	(μg/kg or μg/l)
Parts per trillion	ppt	(ng/kg or ng/l)

[4]Note that parts per thousand is sometimes abbreviated as ppt in older literature. It should not be confused with parts per trillion.
[5]US Environmental Protection Agency, https://www.epa.gov; accessed August 2022.
[6]Solid and liquid particles less than 2.5 μm in diameter suspended in air. US Environmental Protection Agency, National Ambient Air Quality Standards, https://www.epa.gov/environmental-topics/air-topics; accessed August 2022.

Example 1.1 Unit Conversion.   The EPA maximum contaminant limit (MCL) for benzene, C_6H_6, in drinking water is 0.005 mg/l. What is this limit written in units of ppm, ppb, M, and wt%?

The question does not specify the density of the water. Given that drinking water is purified, we will take it to be 1.0 g/ml or 1.0 kg/l. As we have only one significant figure in 0.005 mg/l, this assumption does not affect this calculation.[7] Given a density of 1.0 kg/l, the units mg/l and ppm are the same:

$$\frac{0.005\,\text{mg}\,C_6H_6}{1.0\,l\,\text{water}}\left(\frac{1.0\,l}{1.0\,\text{kg}}\right)=0.005\,\text{ppm}\,C_6H_6$$

Converting ppm to ppb is accomplished by multiplying by 1000:

$$0.005\,\text{ppm}\,C_6H_6\left(\frac{1000\,\text{ppb}}{1\,\text{ppm}}\right)=5\,\text{ppb}\,C_6H_6$$

To convert to molarity, M, we use the formula weight of benzene, which is 78.11 g/mol. Converting 0.005 mg to mol is

$$\frac{5\times10^{-6}\text{g}\,C_6H_6}{78.11\,\text{g/mol}}=6.4\times10^{-8}\,\text{mol}\,C_6H_6$$

Now dividing by 1.0 l to obtain concentration, c, gives

$$c_{C_6H_6}=\frac{6.4\times10^{-8}\,\text{mol}}{1.0\,l}=6.4\times10^{-8}\,\text{M}$$

To find the weight percent, wt%, we need grams of benzene per 100 g of solution. The MCL of 0.005 mg/l for a water density of 1.0 g/ml is 0.005 mg/1000 g. I multiply both numerator and denominator by 0.1 to get the units that I want in the denominator:

$$\frac{5\times10^{-6}\text{g}\,C_6H_6}{1000\,\text{g water}}\left(\frac{0.1}{0.1}\right)=\frac{5\times10^{-7}\text{g}\,C_6H_6}{100\,\text{g water}}$$

$$\frac{5\times10^{-7}\text{g}\,C_6H_6}{100\,\text{g water}}\times100\%=5\times10^{-7}\%\,C_6H_6$$

As you can see, 0.005 mg/l or 5 ppb is simply more convenient for describing this limit than other concentration units.

The following spreadsheet calculation is the first You-Try-It exercise in the text. An Excel spreadsheet file containing this exercise and a step-by-step guide are available at: https://www.achem.org.

[7]Pure water at 25°C has a density of 0.997 g/ml.

I place these exercises at appropriate locations in each chapter, and I recommend that you practice the just-completed topics before moving to new material. If you are new to or want more practice using spreadsheets, work through the spreadsheet-help.pdf document that is also available on the achem.org website.

You-Try-It 1.A

The conversions worksheet in you-try-it-01.xlsx contains two tables of measurement results. Convert these results to other common units.

1.2   GLP AND A FEW OTHER IMPORTANT ACRONYMS

1.2.1   Good Laboratory Practice (GLP)

GLP (Good Laboratory Practice) refers to specific regulations by which laboratories must conduct, verify, and maintain their procedures, results, and records. These regulations, effective in the United States in 1979, were a response to the unreliability of data that were submitted to government agencies to certify that agricultural chemicals, food additives, drugs, and cosmetics were safe and effective. The current GLP regulations are found in the Code of Federal Regulations as[8]

- Title 21, Part 58: US Food and Drug Administration,
- Title 40, Part 160: US Environmental Protection Agency pertaining to the Federal Insecticide, Fungicide, and Rodenticide Act (FIFRA),
- Title 40, Part 792: US Environmental Protection Agency pertaining to the Toxic Substances Control Act (TSCA).

The different regulations are similar in their overall structure and purpose, but they are tailored to specific types of chemicals and laboratories. The details of the Code of Federal Regulations are not important to us, but they are useful to recognize the origin of regulations that we will discuss. Reference to sections of the federal code use an abbreviation based on title, part, and section number. The following paragraph illustrates the coding with a small excerpt of the Code of Federal Regulations, 21CFR58.83, where the .83 in the title refers to this specific section.

```
--------------------------------------------------------------
[Code of Federal Regulations]
[Title 21, Volume 1]
[Revised as of April 1, 2006]
From the U.S. Government Printing Office via GPO Access
[CITE: 21CFR58]
[Page 308]
```

[8]Available from US Government Printing Office: https://www.ecfr.gov; accessed August 2022.

```
              TITLE 21--FOOD AND DRUGS

  CHAPTER I--FOOD AND DRUG ADMINISTRATION, DEPARTMENT OF
                HEALTH AND HUMAN SERVICES
  PART 58_GOOD LABORATORY PRACTICE FOR NONCLINICAL LABORATORY
                        STUDIES
  --Table of Contents

          Subpart E_Testing Facilities Operation

  Sec. 58.83 Reagents and solutions.
     All reagents and solutions in laboratory areas shall be
  labeled to indicate identity, titer or concentration,
  storage requirements, and expiration date. Deteriorated or
  outdated reagents and solutions shall not be used.
  ----------------------------------------------------------------
```

Besides codifying good science and common sense as federal law, the GLP regulations stipulate a framework for personnel responsibilities, analytical methods, and record keeping. These tasks include

1. Personnel and management responsibilities for individual laboratory workers, Study Director, and a Quality Assurance unit (QAU).
2. Written protocols, study plans, and standard operating procedures (SOP) for individual steps or instruments.
3. Record keeping, final written reports, and retention of records.

With the formalized GLP regulations, quality assurance (QA) now indicates an auditing role, while quality control (QC) refers to instrument calibration and method validation (discussed in Section 1.5). My point in this discussion is not that we must memorize government regulations but that all workers in an analytical laboratory have certain roles. Laboratory workers must follow documented and reproducible procedures and create an audit trail of all work. Following GLP ensures the reliability of analytical measurements and maintains the credibility of reported results. An analyst should develop "GLP" habits-of-mind to become adept at measurement science and to work safely in the laboratory.

1.2.2 Standard Operating Procedure (SOP)

An SOP is a document containing the instructions for a specific analytical procedure or instrument. SOPs can serve as a user's guide, reference source, and safety manual. As per GLP regulations, they are reviewed and approved by other members of the laboratory management team. The following example shows a simple SOP. Other SOPs might be the size of a novella, depending on the complexity of a task, an instrument, or a hazard.

━━━━━━━━━━━━━Standard Operating Procedure ━━━━━━━━━━━━

Title: Use of Ocean Optics FL-400 Flame-Resistant Fiber Probe

| **SOP No:** AC-21 | **Author:** B. Tissue | **Version:** 05/27/06 |

1. Purpose

To ensure correct and safe usage of the FL-400 Flame-Resistant Fiber Probe when recording flame emission spectra.

2. Scope

This SOP provides operating procedures for the Ocean Optics FL-400 Flame-Resistant Fiber Probe for recording flame emission spectra. Consult other SOPs or your instructor for sample handling and use of the flame source and spectrometer.

3. Precautions

- Do not touch the tip when hot.
- Handle the probe with care. It contains a glass fiber and should not be dropped or bent.
- Let tip cool for at least 10 seconds before placing in sample or cleaning solutions. Do not place a hot probe tip in a cool solution as the glass fiber could crack.

4. Specific Procedures

4.A. Set-up: The flame-resistant fiber probe should be connected to a 1-m optical patch cable using the stainless steel SMA splice bushing stored with the probe. Do not attempt to use the probe connected directly to the USB2000® spectrometer.

4.B. Use: To load a test portion onto the flame loop (1) dip into a solution and allow to dry or (2) wet the loop with dilute HCl and dip into solid sample. Insert only the loop into a flame and not the whole tip. A separate sample loop or splint may also be used.

4.C. Cleaning: The flame loop and fiber tip should be cleaned with distilled water. A mild detergent and ultrasound is acceptable if necessary.

5. Related SOPs and References

   a. SOP No. AC-20 Use of Ocean Optics USB2000® spectrometer.
   b. Ocean Optics web site: http://www.oceanoptics.com.

| Reviewed: *J.R. Morris 5/29/06* | Approved: *M. Anderson 6/7/06* |

SOPs provide a handy reference to ensure proper use of a procedure or instrument, which is necessary to obtain reproducible results. SOPs, and other procedural documentation, can serve as traceable legal documents in patent and criminal legal proceedings. As an example, try searching the internet for a forensic procedures manual.[9] Failure of an analyst to follow the approved procedures can result in measurements being inadmissible in court cases.

1.2.3   Still More Acronyms

There are numerous other regulatory practices that are similar to GLP. Some examples are

Good Clinical Practice (GCP). Regulations governing clinical trials.

Good Manufacturing Practice (GMP). Regulations governing the pharmaceutical industry. Also cGMP denotes current Good Manufacturing Practice.[10]

Organisation for Economic Co-operation and Development (OECD). An international organization concerned chiefly with economic and social issues. They also publish guidelines on science and technology including chemical safety, chemical testing, and GLP.[11]

International Union of Pure and Applied Chemistry (IUPAC). A nongovernmental agency that recommends standardization of chemical nomenclature, terminology, and chemical and physical data.

The point of this list is not that you should memorize acronyms. The purpose is that you can recognize the purpose and origin of regulations when you see them in the context of your work or study. Many of these regulatory structures are available online, and a search will often find all that you need to know about the details of the regulations. The chances are high that you will deal with scientific regulations sometime during your work life, possibly in more than one laboratory role as analyst, supervisor, or auditor.

1.2.4   Safety Data Sheet

I finish this section with a reference document that is very important for anyone working with laboratory chemicals. A safety data sheet (SDS) provides chemical and toxicological information for a specific chemical. A laboratory worker can use this information to select proper equipment, personal protective

---

[9]See for example the *Toxicology Procedures Manual* available at: https://www.dfs.virginia.gov/documentation-publications/manuals; accessed August 2022.

[10]GCP and GMP are both administered by the Food and Drug Administration in the United States.

[11]"OECD Principles on Good Laboratory Practice" and "Guidelines for Testing of Chemicals" can be found by searching at https://www.oecd.org; accessed August 2022.

equipment (PPE), and work practices to work with a chemical safely. A lab worker should read the SDS information *before* using a substance, that is, before spilling, inhaling, or igniting anything. It is very difficult to read an SDS when you are dizzy, burned, or unconscious. Some of the more useful sections of an SDS include

- Section 2, Hazard(s) identification
- Section 4, First-aid measures
- Section 8, Exposure controls/personal protection
- Section 10, Stability and reactivity (including incompatible materials)

SDS sheets must be readily available to lab workers. Medical personnel can refuse to treat a case of chemical exposure without an SDS for the substance involved. If you do not have an SDS, most chemical suppliers provide them online. The following excerpt illustrates the type of information that is available for hydrochloric acid.

Section 4: First-aid measures

General advice: First responders need to protect themselves. Show this safety data sheet to responders or medical personnel.
After inhalation: Move to fresh air. Seek medical attention if irritation persists.
After skin contact: Take off contaminated clothing. Rinse skin with water/shower for 15 min. Seek medical attention.
After eye contact: Rinse with plenty of water. Remove contact lenses. Seek medical attention.
After swallowing: Have victim rinse mouth and drink water. Do not induce vomiting. Do not try to neutralize. Seek medical attention.

1.3 PRECISION AND RANDOM ERROR

A good lab practice to improve the credibility of any type of measurement is to simply repeat it more than once. We call measurements of multiple test portions from the same sample *replicate measurements*. Placing a test portion in an instrument and recording the signal multiple times is *signal averaging*, which is useful if the signal fluctuates. Signal averaging is not the same as making replicate measurements. Signal averaging will not identify the bias if a test portion was diluted by a factor of 10 when the experimental procedure called for a dilution by 5.

After dividing a sample into several test portions, each test portion is treated and measured using the identical procedure. If an analytical procedure requires 10 steps, the 10 steps are done on each test portion. Doing replicate measurements can identify gross errors such as omitting one step in a procedure, one-time instrument glitches, or writing a value incorrectly. Even when making individual measurements, an analytical method will specify remeasuring a standard or sample at some frequency. The frequency could be twice daily, one duplicate measurement for each batch of samples, or once every 10 measurements. The frequency depends on the stability that is expected for a given method or instrument. Making duplicate measurements is especially useful to identify drift. *Drift* is the gradual change in instrument response over time. It introduces a bias in measurements that gets worse with time. Observing drift in a duplicate measurement can indicate that a method or instrument requires recalibration.

The repeatability in making replicate measurements is called *precision*.[12] We use the term *repeatability* when we are talking about replicate measurements made on the same sample and performed under identical conditions. When we compare measurements of the same sample by different analysts or different methods, we use the term *reproducibility*. The calculation of precision is the same for repeatability and reproducibility, the difference is the source of the measurements and the purpose of the result. Quantitative measures of precision include standard deviation, standard error, and confidence limits (CL). These measures quantitate the variation or spread in the individual measurements due to random fluctuations, which we call *random errors*. The distribution of random fluctuations follows a Gaussian-shape "bell curve," and precision is a measure of the width of this distribution. Graphically, we display the precision by placing error bars about a data point.

The *accuracy* of a measurement is how close an experimental result comes to the true value. The difference between a measurement and the true value is called the *systematic error* or *bias*. We determine the bias in a measurement by measuring samples of known composition, which we call *standards*. We will discuss accuracy and systematic errors fully in Section 1.5 on calibration, which is the process of measuring standards to be able to obtain accurate results for unknowns.

We can never know with 100% certainty if we have determined the true concentration of an analyte in a real sample. The accuracy of a measurement depends on the care in validating sample processing procedures, calibrating the measurement, and making sure that standards match the samples. The best that we can do, by following GLP, is to

- use reproducible methods;
- maintain equipment, reagents, and instruments in good working condition;
- verify the accuracy of our analytical abilities with standards of known concentrations;
- use a calibration procedure that is appropriate for the unknown sample.

---

[12]Imprecision, or the lack of precision, is probably a better term to describe the repeatability of measurements, but precision is the more common term.

As an example, the chemical treatment of a municipal water supply is based on the measurement of a pollutant being less than the EPA action level. Analysis of randomly collected water samples shows that it contains 14 ppb Pb, below the EPA action level of 15 ppb.[13] If you drink water from this system, you will probably have a few questions about the result. First, you might ask the analyst how confident she is of the accuracy of the result. Did the analyst follow GLP and check the accuracy of the analytical methods by measuring a known standard? Next, what is the spread in the measurement, that is, what is the precision of the reported result? Is the range of measured values 14.0 ± 0.1 ppb or is it 14 ± 5 ppb? You might be more concerned if the upper value of the error bars exceeds the EPA action level.

The terms *accuracy* and *precision* are often used interchangeably, but they are not the same. Precision does not tell us anything about the accuracy of a result due to systematic errors in the measurement procedure. Figure 1.1 illustrates the difference between accuracy and precision using the results of arrows shot at targets by three different archers. The archer on the left was very steady. She was precise, but she hit high on the target with every shot. She possibly aligned her sight incorrectly, estimated the distance to the target incorrectly, or made some other systematic error. Unfortunately, it is not always possible to identify the source of a systematic error without performing more experiments. The center archer hit the middle of the target repeatedly. We describe her shooting as being accurate and precise. The archer on the right was rather shaky and tended to hit far from the center of the target. However, if you average the positions of the four arrows on the right target, you will find that it falls on the bull's-eye. We describe his shooting as accurate but with poor precision.

Now let us think about what these archers might do when they try again. The center archer did very well, and she will not want to make any changes. The archer on the left might adjust her sight to try to be more accurate. If she was shooting from 20 m on her first try, should she now adjust her sight and shoot from 25 m? No, she wants to make one change and keep everything else

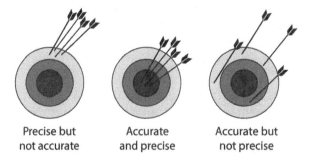

Precise but   Accurate   Accurate but
not accurate   and precise   not precise

Figure 1.1 The results from three archers.

[13]The EPA action level is the concentration of a contaminant that requires a response such as public notification, exposure monitoring, or remediation.

the same. The same goes for analytical measurements. If you want repeatable results, keep all conditions the same. The archer on the right just needs more practice to get steadier. As in any human endeavor, achieving high accuracy and precision requires practice and care.

1.3.1 Standard Deviation

As you can see from the archery example, shooting only one arrow might not provide a reliable measure of how well an archer can shoot. Based on only one shot, the archers on the left and right might appear very close in proficiency. The four shots show us that the archer on the left is much steadier than the archer on the right, but with some systematic error.

On making a series of repetitive measurements, we report the result as the *mean* or average, plus-or-minus, ±, some measure of precision. We can quantify the repeatability of replicate measurements using the difference of each individual measurement from the mean. We call this difference for each measurement the *deviation* or *residual*, d_i:

$$d_i = x_i - \bar{x} \tag{1.1}$$

where the mean, $\bar{x}$, is

$$\bar{x} = \frac{\sum\limits_{i=1}^{n} x_i}{n} \tag{1.2}$$

The summation symbol, $\sum_{i=1}^{n}$, indicates that we are to add up all n numbers in the data set, where the i is an index to refer to each individual measurement. I will often drop the $i = 1$ and n notation for clarity. Unless noted otherwise, take a summation in the statistical formulas for all data points, i.e., from $i = 1$ to $i = n$.

As the mean value is an average of all measurements, the deviations will be both positive and negative. Averaging the deviations produces zero, which is not a realistic measure of repeatability. A common measure of precision is the standard deviation or more specifically the *sample standard deviation*, s:

$$s = \sqrt{\frac{\sum (x_i - \bar{x})^2}{n-1}} \tag{1.3}$$

where n is the number of data points and $n - 1$ is called the *degrees of freedom*.[14] By squaring each deviation before summing them, we eliminate the

---

[14]The degrees of freedom is the number of measurements minus the number of results obtained from the measurements. When calculating the mean of a data set, it is $n - 1$. On obtaining the slope and y-intercept of a line, it is $n - 2$. Taylor, J. R. *An Introduction to Error Analysis: The Study of Uncertainties in Physical Measurements*, 2nd ed.; University Science Books: Sausalito, CA, 1996.

problem of an average deviation going to zero. Taking the square root of the summed squares gets the standard deviation back to the scale of the mean. The standard deviation provides a "typical" deviation for a series of measurements. We will see later that this typical deviation encompasses approximately 68% of all measurements in a set of data.

The equation for s is used to describe the precision of a relatively small number of replicate measurements. For a large number of measurements, say 20 or more for a reliable procedure, we can use the true or *population standard deviation, σ*:

$$\sigma = \sqrt{\frac{\sum(x_i - \mu)^2}{n}} \tag{1.4}$$

where μ is taken to be the true value. For a large number of measurements reporting s or σ will not be very different. Given that there are numerous measures of precision, you should specify the measure that you use when reporting the repeatability in a result.

The calculation of the mean and standard deviation can be accomplished using the built-in functions of a calculator or spreadsheet. These tools will often calculate both sample standard deviation, s, and population standard deviation, σ, so be sure you know what your calculator or program returns. The population standard deviation, σ, is only valid for a large number of measurements. Even though the numerical difference between s and σ might be small, you will almost always want the sample standard deviation, s, when working with analytical results.

Table 1.7 lists the results of four replicate titrations to determine the acetic acid concentration in a vinegar sample. These results are plotted graphically in Figure 1.2. The filled diamonds are the individual measurements for each student and the open squares, shifted to the right for clarity, are the mean of each data set. The vertical lines with horizontal caps are error bars showing the sample standard deviation, s, of each data set. The dotted line shows the true value, which was known for this sample.

These titration results are similar to the archery results in Figure 1.1. Student A was precise, but a systematic error made his mean result lower than

TABLE 1.7   Titration Results

	Concentration, M		
	Student A	Student B	Student C
	0.824	0.848	0.817
	0.849	0.861	0.869
	0.834	0.872	0.860
	0.839	0.865	0.901
Mean	0.837	0.862	0.862
Std dev	0.010	0.010	0.035

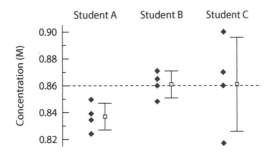

Figure 1.2 Results of four replicate titrations.

the true value. Students B and C were both accurate, their means were very close to the true value. Compared to Student B, Student C's measurements were less precise. The error bars on her results were larger than for the other two students, leading to greater uncertainty in her reported result. Given a choice, I will take a large error bar on an accurate result in place of a very repeatable wrong answer.

1.3.2 Relative Standard Deviation

The *relative standard deviation (RSD)*, s_r, is a unitless fraction that is calculated by dividing the standard deviation, s, by the mean of the measurement[15]:

$$s_r = \frac{s}{\bar{x}} \tag{1.5}$$

The RSD normalizes the standard deviation to the magnitude of the mean value, making it easier to compare the relative precision in different sets of measurements. We often express it as a percentage for convenience. IUPAC recommends the symbol $s_r(\%)$ for percentage relative standard deviation, but we will use the more common abbreviation of %-RSD. The expression for %-RSD is

$$\text{\%-RSD} = \left(\frac{s}{\bar{x}}\right) \times 100\% \tag{1.6}$$

If your measurements are also expressed as a percentage, be sure to specify if you are reporting the RSD or %-RSD to avoid confusion.

Example 1.2 illustrates the usefulness of using the %-RSD to compare the precision in different data sets when the means vary in magnitude. The table in the calculation shows measurement results obtained from four students for four different samples. Each student reports an equivalent standard deviation. Because the samples are different, quoting equivalent standard deviations does

---

[15]The relative standard deviation is also called the "coefficient of variation" in some disciplines.

not provide a true representation of each student's repeatability. Recalculating the precision as a relative percentage provides values that are easier to compare.

Example 1.2 Percentage Relative Standard Deviation.   The table lists the results from four students who each measured a different sample using the identical analytical procedure. Let us see which measurement result was most precise.

Student	Mean, ppm	Std Dev, ppm
A	1.03	0.01
B	10.03	0.01
C	3.58	0.01
D	0.11	0.01

Each measurement has the same standard deviation, but each measurement was not made with equal precision. Calculating the standard deviation as %-RSD for Student A gives

$$\%\text{-RSD}_A = \left(\frac{0.01\ \text{ppm}}{1.03\ \text{ppm}}\right) \times 100\% = 1\%$$

Repeating this calculation for the other students and adding the results to the table shows clearly which measurements are most and which are least precise. The result from Student B was measured to 0.1%, much better than the ones the other students achieved. We cannot say anything about the accuracy of the results based only on this data, but Student B certainly had a more repeatable technique in performing the measurement.

Student	Mean, ppm	Std Dev, ppm	%-RSD, %
A	1.03	0.01	1
B	10.03	0.01	0.1
C	3.58	0.01	0.3
D	0.11	0.01	9

1.3.3   Other Measures of Precision

The quantitative measures of precision described above, s and %-RSD, are the ones that we will use most often to describe analytical results. The %-RSD is most useful to compare the precision of different measurements that vary in magnitude or that use different units. The practical significance of s (or σ) is that 68% of the measurements will fall between $\pm 1s$. The 68% range assumes that the deviations from the mean are due only to random variations in the measurement. The 68% range of the standard deviation is not always the best measure to state the degree of confidence that we have in the repeatability of a measurement or in the spread in results from multiple samples. Several other quantities—variance, standard error, and CL—are described here for completeness. Each of these measures of precision derive from the standard deviation and are useful for different purposes.

1.3.3.1 Variance   The *variance* is simply the square of the standard deviation:

$$v = s^2 = \frac{\sum_{i=1}^{n}(x_i - \bar{x})^2}{n-1} \tag{1.7}$$

The advantage of working with variance is that variances from independent sources of variation may be summed to obtain a total variance for a measurement. The topic of the propagation of uncertainty is beyond the needs of this text, but it is treated in several other sources on practical statistics.[16]

1.3.3.2 Standard Error   The *standard error*, s_m, also called the *standard deviation of the mean* (*SDOM*), is the standard deviation divided by the square root of *n*:

$$s_m = \frac{s}{\sqrt{n}} \tag{1.8}$$

By dividing by $\sqrt{n}$, the s_m provides a better reflection of the amount of data collected than other measures of precision. Taylor considers it the "best" measure of the uncertainty in a set of multiple measurements.[17] A significant difference between the standard error and the standard deviation is that the standard error will decrease with increasing number of measurements, although slowly for large *n*.

1.3.3.3 Confidence Limits   For a set of measurements in which the deviations of the individual measurements vary only due to random fluctuations, we expect a normal or "bell curve" distribution. The mean ± standard deviation of a normal distribution will encompass approximately 68% of the individual measurements. Two standard deviations, ±2*s*, will encompass approximately 95% of the individual measurements. These values are the ideal case for a very

---

[16]Taylor, J. R. *An Introduction to Error Analysis: The Study of Uncertainties in Physical Measurements*, 2nd ed.; University Science Books: Sausalito, CA, 1996; also useful are Bevington P.; Keith Robinson, D. K. *Data Reduction and Error Analysis for the Physical Sciences*, 3rd ed.; McGraw-Hill: New York, 2002 and Garland, C. W.; Nibler, J. W.; Shoemaker, D. P. *Experiments in Physical Chemistry*, 7th ed.; McGraw-Hill: New York, 2003.
[17]Taylor, J. R. *An Introduction to Error Analysis: The Study of Uncertainties in Physical Measurements*, 2nd ed.; University Science Books: Sausalito, CA, 1996.

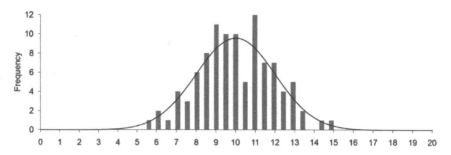

Figure 1.3 Histogram and normal distribution.

large number of measurements. Figure 1.3 shows a histogram of one hundred data points ($n = 100$) that have a mean of 10.0 and a sample standard deviation of 2.0. "Frequency" on the y-axis is the number of data points within a span of 0.5; for example, the bar at 11 shows that 12 data points fall between 10.75 and 11.25. The solid line is a normal distribution for this same mean and standard deviation.[18] You can see that there are noticeable differences between the histogram and the normal distribution even for $n = 100$. The normal distribution curve does match the overall shape of the scatter in the data points. In this example data, 68 values fall between 8 and 12 ($\pm 1\ s$) and 97 values fall between 6 and 14 ($\pm 2\ s$), as expected for a normal distribution.

There are frequent occasions when we must compare analytical results to other values. This type of task is known as *hypothesis testing*, and several common cases in chemical analysis are based on the following.

- Is an experimental procedure operating correctly, that is, is a measured result statistically equivalent to a known value?
- Is a measured result above or below some specified level?
- Is the difference between two sets of experimental measurements statistically significant?

To answer these questions, we use a more definitive description for the imprecision of repetitive measurements.

Consider again our example of lead in drinking water, for which the EPA has set an action level of 15 ppb Pb. Suppose that measurements from a random sampling of water in homes show a mean and standard deviation of 14 ± 1 ppb. For this municipal water supply, 68% of the measurements fall between 13 and 15 ppb. 16% of the water samples contain less than 13 ppb Pb, which is not an issue for the quality and safety of the drinking water. However, 16% of the water samples contain Pb greater than the 15 ppb EPA action level. In this case, reporting the standard deviation might not be the best measure of precision on which to base water treatment decisions. If I lived in this locality,

[18]The histogram data was generated in Excel using Random Number Generation in the Data Analysis Toolpak and the curve was generated with the NORMDIST function.

I might not want to take the risk that my tap water is in the 16% of cases with high Pb levels. So the question is: To what level should the average lead level be reduced?

In such cases, we introduce the "level of confidence" to specify the probability of a value being within or outside of a given distribution of measurements. CL is a statistical measure of the precision for replicate measurements that can be calculated for different levels of confidence. It is calculated from the standard deviation, s, using

$$CL = \frac{t \times s}{\sqrt{n}} \qquad (1.9)$$

where n is the number of measurements and t is a critical t-value taken from Table 1.8. These values are for a two-tailed test, meaning that the true value could be larger or smaller than the predicted distribution. For a 95% confidence level, you are allowing for a 2.5% probability of the true value being greater than your CI or a 2.5% probability of the true value being less than your CI.[19]

$$CI = (\bar{x} - CL) \text{ to } (\bar{x} + CL) \qquad (1.10)$$

TABLE 1.8 Table of Critical t-Values

	t-Values for Different Levels of Confidence				
$n - 1$	80.0% (0.20)	90.0% (0.10)	95.0% (0.05)	99.0% (0.01)	99.9% (0.001)
1	3.078	6.314	12.706	63.657	636.600
2	1.886	2.920	4.303	9.925	31.590
3	1.638	2.353	3.182	5.841	12.920
4	1.533	2.132	2.776	4.604	8.610
5	1.476	2.015	2.571	4.032	6.869
6	1.440	1.943	2.447	3.707	5.959
7	1.415	1.895	2.365	3.500	5.408
8	1.397	1.860	2.306	3.355	5.041
9	1.383	1.833	2.262	3.250	4.781
10	1.372	1.812	2.228	3.169	4.587
15	1.341	1.753	2.131	2.947	4.073
20	1.325	1.725	2.086	2.845	3.850
25	1.316	1.708	2.068	2.787	3.725
30	1.310	1.697	2.068	2.750	3.646
50	1.299	1.676	2.068	2.678	3.496
100	1.290	1.660	2.068	2.626	3.391
∞	1.310	1.645	2.068	2.576	3.300

[19]*NIST/SEMATECH e-Handbook of Statistical Methods*; https://www.itl.nist.gov/div898/handbook/eda/section3/eda3672.htm; accessed August 2022.

where CI, confidence interval, is the span of values that encompasses the true value with the stated level of confidence (see Example 1.3).

Table 1.8 also introduces the α notation, the values in parentheses, which is called the *significance level*. Referring to a significance of $\alpha = 0.05$ or a confidence level of 0.95 (or 95%) is the same for our purposes, and the symbol for confidence level is $1 - \alpha$.

Returning to the lead in the drinking water example, if the 14.0 ppb mean was the result of six measurements, the 99% CL is

$$CL = \frac{2.78 \times 1.0\,\text{ppb}}{\sqrt{5}} = 1.2\,\text{ppb} \tag{1.11}$$

If the mean was the result of 21 measurements, the 99% CL is

$$CL = \frac{2.09 \times 1.0\,\text{ppb}}{\sqrt{20}} = 0.5\,\text{ppb} \tag{1.12}$$

In these two cases, the number of measurements affects our level of confidence. With only six water samples the range of 14.0 ± 1.2 ppb encompasses the EPA action level, but for 21 measurements the result of 14.0 ± 0.5 ppb is safely below the action level.

Example 1.3 CI Calculation.   What is the 95-% CI for the following four trials of an iron analysis of a soil?

```
Weight Percent of Fe in Soil
Test Portion    Result
     1           3.44% Fe
     2           3.11% Fe
     3           2.98% Fe
     4           3.27% Fe
```

The sum of the four data points is 12.80, the mean is 3.20, and the standard deviation is 0.20 (determined from a spreadsheet).

Using the expression for CL and a t value from Table 1.8 for $n - 1$ of 3:

$$CL = \frac{(3.182)(0.20)}{\sqrt{4}} = 0.32$$

The CI at 95% confidence is $(3.2 \pm 0.3)\%$.

You-Try-It 1.B
Replicate measurement data is in the **precision** worksheet in **you-try-it-01.xlsx**. Open this worksheet and type formulas to find the mean,

standard deviation, and %-RSD of the experimental data. Change column widths and formatting to create a worksheet that will be suitable to cut and paste into a report.

The data shown in the example above is in the confidence-interval worksheet in you-try-it-01.xlsx. Open this worksheet and repeat the calculation for the 90% and 99% level of confidence. As a check, run the Descriptive Statistics in the Data Analysis Toolpak.

1.4 DISCARDING A SUSPECTED OUTLIER

If one value in a set of data appears very different from the rest, it might be possible to discard that value. The basis to discard a value is that some unrecognized error causes that value to not be representative of the sample. There are a number of criteria for removing an apparent outlier from a set of measurements. This section will illustrate Dixon's Q test and Peirce's Criterion. Note that performing a test is not necessary to discard data that you know is erroneous. For example, if you were making a UV/Vis absorption measurement and one solution was turbid due to a precipitate being disturbed from the bottom of a reaction tube, you may make a note of your observation in your notebook and discard the data value for that one experimental trial. Similarly, it is not necessary to discard a suspected outlier, and the conservative approach is to report all data with the larger standard deviation. If you do remove a point from a set of data, you must acknowledge the change to your data and specify your criterion for discarding the data.

1.4.1 Dixon's Q Test

To use Dixon's Q test as the criterion to discard a suspected outlier, calculate Q using the following formula:

$$Q = \frac{|\text{suspected outlier} - \text{closest value}|}{|\text{highest value} - \text{lowest value}|} \tag{1.13}$$

If Q is larger than the critical value, Q_c, for n repetitive measurements (see Table 1.9), then the outlier may be discarded. This expression and the table of Q_c is only valid for discarding one data point from a series of measurements (see Example 1.4). A more complete list at different levels of confidence can be found in the literature.[20]

---

[20]Adapted with permission from Rorabacher, D. B. "Statistical treatment for rejection of deviant values: critical values of Dixon's 'Q' parameter and related subrange ratios at the 95% confidence level," *Anal. Chem.* **1991**, *63*, 139–146. Copyright 1991 American Chemical Society.

TABLE 1.9   Critical Values of Dixon's
Q Parameter (Q_c)

n	95% ($\alpha = 0.05$)	99% ($\alpha = 0.01$)
3	0.970	0.994
4	0.829	0.926
5	0.710	0.821
6	0.625	0.740
7	0.568	0.680
8	0.526	0.634
9	0.493	0.598
10	0.466	0.568
15	0.384	0.475
20	0.342	0.425
25	0.317	0.393
30	0.298	0.372

Example 1.4 Q-Test Calculation.   Consider the following five potentiometric measurements of Mg^{2+} in five test portions of water from a fish aquarium. Ca^{2+} must be precipitated before measurement, leading to a source of error.

```
Potentiometric measurements for Mg
Test Portion    Measurement, mV
      1             39.8
      2             36.5
      3             39.9
      4             39.2
      5             39.6
```

The value of 36.5 mV appears lower than the rest of the values in the set. Can this value be discarded?

To answer this question, we find Q and compare it to Q_c:

$$Q = \frac{|36.5 - 39.2|}{|39.9 - 36.5|} = 0.794$$

At the 95% confidence level, which is usually suitable for working with analytical data sets, $Q_c = 0.710$ for five data points. Our calculated Q is larger than Q_c, so the suspected outlier may be discarded. Recalculating the mean after discarding the outlier gives a much smaller standard deviation 39.6 ± 0.3 mV, compared to an original value of 39.0 ± 1.4 mV.

1.4.2   Peirce's Criterion

Another approach to determine if data points are outliers and may be removed from a data set is to ask the question:

What is the probability of a suspected outlier occurring in a normal distribution given the number of measured values, n?

TABLE 1.10   Values of R for Peirce's Criterion

	Number of Suspected Outliers			
n	1	2	3	4
3	1.196			
4	1.383	1.078		
5	1.509	1.200		
6	1.610	1.299	1.099	
7	1.693	1.382	1.187	1.022
8	1.763	1.453	1.261	1.109
9	1.824	1.515	1.324	1.178
10	1.878	1.570	1.380	1.237
11	1.925	1.619	1.430	1.289
12	1.969	1.663	1.475	1.336
13	2.007	1.704	1.516	1.379
14	2.043	1.741	1.554	1.417
15	2.076	1.775	1.589	1.453
20	2.209	1.914	1.732	1.599
25	2.307	2.019	1.840	1.709
50	2.592	2.326	2.158	2.035

If the probability is very low, then the suspected outlier(s) may be discarded. Using the data in the previous example, the mean and standard deviation of the data are 39.0 ± 1.4 mV. The suspected outlier, 36.5 mV, is nearly two standard deviations from the mean. A value this distance from the mean should arise in approximately 5% of measurements. We would expect a value such as 36.5 mV to appear in a set of more than 20 measurements, but not when we have only 5 measurements.

Two similar methods that use this approach are Chauvenet's criterion and Peirce's criterion. I will describe Peirce's criterion because it is easier to implement using a published table of critical values (Table 1.10).[21] This method compares the deviation of the suspected outlier, that is, the difference from the mean value, to the probability of a data point being that different from the mean assuming a normal distribution of n measurements. Peirce's criterion can test multiple outliers, but we will use it here to test only one suspect value.

The procedure to use Peirce's criterion is to first calculate the absolute value of the deviation from the mean for a suspected outlier:

$$|d_i| = |x_i - \bar{x}|$$

where x_i is the suspected outlier and $\bar{x}$ is the mean. This deviation is then compared to the maximum deviation, d_{max}, that is expected for a random distribution given the number of data values. d_{max} is found from

$$d_{max} = sR$$

---

[21]Adapted with permission from Ross, S. M. "Peirce's criterion for the elimination of suspect experimental data," *J. Eng. Technol.* **Fall 2003**.

where s is the sample standard deviation and R is taken from a table of values. If the deviation of the suspected outlier is larger than d_{max}, it is improbable that the outlier occurred owing to a random fluctuation. The suspected outlier may be discarded if this condition is met:

$$|d_i| > d_{max}$$

You-Try-It 1.C

The above data is replicated in the outlier worksheet in you-try-it-01.xlsx. Use this worksheet to test for the suspected outlier using Dixon's Q test at the 99% CL and using Peirce's criterion. Use the = MIN(range) and = MAX(range) functions to find the potential outliers in the data set. Check that you get the same result as the Q-test calculation above.

1.5   CALIBRATION

The preceding section described the statistical tools to describe the repeatability of replicate measurements. *Calibration* is the process of measuring a known quantity to determine the relationship between the measurement signal and the analyte amount or concentration. Calibration allows the analyst to estimate the accuracy of a measurement, procedure, or instrument.

Calibration is critical to GLP and method development and validation. *Method development* is the process to determine the experimental conditions for sample collection, preparation, and measurement that produce accurate and repeatable results. Very often this work has been done by someone else and you are using one or more standard methods. When a standard method is not available, analysts usually start with a method for an analyte in a similar type of sample and modify it to obtain good results. *Method validation* is the process to ensure that you are obtaining accurate and repeatable results. To quote 21 CFR 211.165 (e) [22]:

> The accuracy, sensitivity, specificity, and reproducibility of test methods employed by the firm shall be established and documented. Such validation and documentation may be accomplished in accordance with Section 211.194(a)(2).

where section 211.194(a)(2) refers to proper record keeping of analytical methods or citation to standard methods.

Periodic calibration is necessary to maintain methods and instruments in proper working order and to identify and eliminate systematic errors. The

[22]Part 211—Current Good Manufacturing Practice for Finished Pharmaceuticals.

frequency and extent of calibration procedures will be method dependent and should be specified in SOPs for a given procedure or instrument. Systematic errors might result from a bias due to unidentified components in a sample, an uncontrolled instrument factor, or a persistent human error. An example of an instrumental bias is an incorrectly calibrated pH meter that always reads a pH value that is 0.5 units lower than the true value. An example of a method error is the partial loss of a volatile metal analyte during an ashing step to remove organic material. An example of human bias is a student who records titration endpoints beyond the correct endpoint owing to color blindness. Systematic errors can be identified and corrected by analyzing standards that closely match the real sample. As Figure 1.4 illustrates, if something seems odd, it probably is odd, and it might introduce a bias.

There are several common calibration procedures for analytical measurements:

- a simple proportionality (one-point calibration),
- using a calibration or working curve,
- using an internal standard,
- using the standard-addition method.

All of these calibration methods require one or more standards of known composition to calibrate the measurement. We will look at the key aspects of these different calibration approaches and then discuss the proper use and preparation of standards in the next section.

1.5.1 Linear Proportionality

Most of the analytical methods that we will discuss assume that the measured signal is directly proportional to an analyte amount or concentration. The generic expression of a direct or linear proportionality is

$$y = a_1 x \tag{1.14}$$

where y is the measured signal, a_1 is the proportionality factor, and x is the analyte amount or concentration. I will continue discussing these topics in

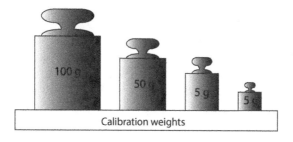

Figure 1.4 Source of a systematic error.

terms of concentration, but these concepts apply equally to detection systems that are sensitive to analyte amount. The proportionality factor, a_1, is called the *sensitivity* of the measurement. The larger the sensitivity, the larger the signal for a given analyte concentration. The basic concept of calibration is to measure the signal produced by a standard of known composition to determine the sensitivity, a_1. Knowing a_1, we can determine the analyte concentration in a test portion by measuring the signal from the test portion.

Given a linear proportionality, we expect a signal of zero in the absence of analyte, that is, $y = 0$ when $x = 0$.[23] It is good practice to always measure the signal produced by a standard that contains no analyte, which we call the *blank* or 0.0 concentration standard. It is not uncommon for the blank to have a nonzero signal, which can arise from an offset of the recording device or a substance in the blank that produces a signal. The calculation in Example 1.5 illustrates the use of a nonzero blank measurement in an analytical measurement.

Example 1.5 Calibration Calculation.   Total dissolved solids (TDS) in water and soils can be estimated from the electrical conductivity of solution. Solution conductivity is proportional to the number and types of ions in solution. It is expressed as siemens per unit distance (S/cm), where the unit S, siemens, is the measure of conductance. You add 50 ml of tap water to 10 g of dried and crushed soil from an irrigated field and shake for 5 min. After filtering you measure the conductivity of the water. For a standard you use 5.00 g of high purity KCl (f.w. 74.551 g/mol) in 1.00 l of tap water. Performing the measurement on solutions of a blank (water only), standard, and soil sample gives the following solution conductivities. What is the TDS in the soil sample?

Solution	Concentration, g/l	Measured Conductivity, mS/cm
Blank	0.00	0.05
Standard	5.00	9.26
Soil sample	?	5.50

This calculation is a one-point calibration. You expect 0.0 mS/cm for the blank, but there is some residual salt in the tap water. To correct the standard and unknown measurements, subtract the nonzero blank from each value. The corrected measurements are 9.21 mS/cm for the standard and 5.45 mS/cm for the soil sample. There are two approaches to work with a proportionality: determine the sensitivity, a_1, or set up a simple ratio. The two calculations are equivalent, so use whichever one is most intuitive for you. I will show the calculation first by finding the sensitivity and then using

---

[23]The potentiometric methods discussed in Chapter 8 are one notable exception.

it to find the unknown concentration. I find a_1 using the standard signal and concentration:

$$9.21\,\text{mS/cm} = a_1(5.00\,\text{g/l})$$

$$a_1 = 1.84\frac{\text{mS/cm}}{\text{g/l}}$$

We could also plot the blank and standard data points and determine the unknown from the equation of the line. We will use this approach when we have more calibration points to generate a calibration curve. Now using a_1 and the unknown measurement to find the unknown concentration:

$$5.45\,\text{mS/cm} = 1.84\frac{\text{mS/cm}}{\text{g/l}}x$$

$$x = 2.96\,\text{g/l}$$

Repeating the calculation in one step by setting up a ratio gives the same result:

$$\frac{x}{5.45\,\text{mS/cm}} = \frac{5.00\,\text{g/l}}{9.21\,\text{mS/cm}}$$

$$x = 2.96\,\text{g/l}$$

Note that we have assumed that the ions in our standard (K^+ and Cl^-) match the ions in our soil sample. If such is not the case, a different standard might be needed.

In the previous calculation, the standard concentration was higher than the sample concentration and the measurement was assumed to be directly proportional, or linear, between zero and the standard concentration. To illustrate a poor implementation of a linear proportionality, let us say that you want to weigh yourself on a bathroom scale. Not knowing the accuracy of the scale, you calibrate it with the best weight standard that you can find. Weighing a box of twenty-four 355-ml cans of diet cola from your kitchen gives a reading of 9.5 kg. You were expecting a weight of 24 × 355 g = 8.5 kg, so this calibration suggests that the scale is not accurate. Does this measurement mean that the scale is reading high by 1.0 kg each time? That is a simple calibration, subtract 1.0 kg from all measurements. But maybe your calibration experiment indicates that the scale is reading high by 12%. As the scale shows 0.0 kg with nothing on it, you presume that the scale is reading off by the fixed percentage. From this assumption you can generate a calibration plot, the dashed line in Figure 1.5. I label this line a one-point calibration even though I am assuming that the line goes through the origin.

On weighing yourself you obtain a reading of 60.0 kg. Using your calibration curve, you determine that your true weight is 53.5 kg. This measurement seems low to you, so you ask a friend, who knows that he weighs 80.0 kg, to step on the scale. The scale displays 80.0 kg. Plotting this calibration point on

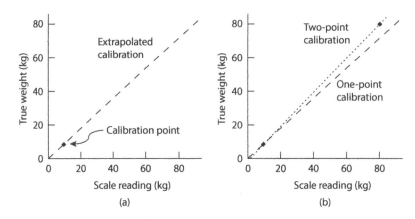

Figure 1.5 (a) One-point and (b) two-point calibration curves.

the curve give the two-point calibration curve shown by the dotted line in Figure 1.5b. Using this calibration line, your reading of 60.0 kg is equal to a true weight of 59.0 kg.

This example illustrates a number of concepts. First, between approximately 5 and 15 kg, either calibration curve appears to provide accurate results. The measurement using the one-point calibration has a large uncertainty due to extrapolating a calibration at a low weight to the higher weight range. It is good practice to use a standard that is similar to the unknown. An even better approach is to calibrate your measurement at multiple points throughout the expected measurement range. This example also shows that a calibration line is only as good as the standard(s) used to make it. In the example above, using an incorrect weight for the box of soda cans would introduce a bias into all measurements.

1.5.2 Measurement Range

The previous discussion assumed that a linear proportionality was valid over the range of measurements. This assumption must be validated by measuring standards. For analytes in solutions, we validate a measurement with standards at concentrations from zero to higher than what we expect for the unknown test solutions. Doing so avoids the pitfalls of extrapolating a calibration curve. It also allows us to determine the minimum and maximum signals, which leads to the minimum and maximum amounts of analyte that a given method can measure.

All quantitative methods will have a finite range of measurable signals and analyte concentrations. The range from the minimum to the maximum signal or concentration is called the *measurement range*. The minimum detectable concentration is discussed in the next section on detection limits. The high end of the measurement range will be limited by chemical characteristics of the analyte or physical limitations of the measurement method. At some point, the concentration of an analyte in solution might reach a solubility limit and be present as a second phase that is not detected. An example of a physical limitation is the ability of an instrument to detect a low level of light. A light-absorbing analyte can be quantitated by measuring the amount of light that passes through a sample. If the analyte concentration becomes high enough to make the sample opaque, it is not possible to measure even higher concentrations.

The *linear range* of an analytical method or instrument is the range for which the signal responds linearly to analyte concentration. Figure 1.6 shows a plot of fluorescence intensity versus concentration of riboflavin standards. Fluorescence is light emission that we expect to be directly proportional to analyte concentration. We see that the plot is linear until approximately 1.2 ppm riboflavin. The sensitivity of this measurement is the slope of the calibration line, but only in this linear region. Although this data appears to show a limited measurement range, fluorescence is very sensitive. The curve could be extended several orders of magnitude to lower concentration with diluted standards.

The short horizontal line in Figure 1.6 shows where the signal does not change for increasing concentration. We clearly cannot make an accurate measurement in this region. I estimate the maximum measurable concentration from the plot to be ≈1.8 ppm riboflavin. For concentrations between 1.2 and 1.8 ppm, the signal does change with increasing concentration. However, the sensitivity is concentration dependent and the linear calibration curve does not apply. Quantitating an unknown solution in this region can be done if a nonlinear calibration function is found that fits the calibration data.

In most cases, it is desirable to dilute a test portion and remeasure it in the linear range. For some analyses, such as a chromatographic separation that might take 60–90 min, it is accepted to use the nonlinear range of the detector rather than to repeat a long analysis.[24] Flame atomic absorption spectrometry

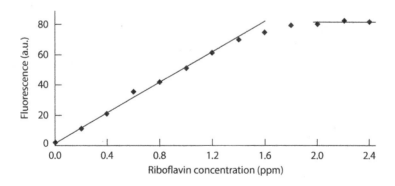

Figure 1.6   Linear range of a calibration curve.

[24]See Ettre, L. S. "Nomenclature for chromatography," *Pure Appl. Chem.* **1993**, *65*, 819–872; http://publications.iupac.org/pac/1993/pdf/6504x0819.pdf; accessed August2022.

is another method where nonlinear calibration functions are common and using a higher-order polynomial calibration function is accepted practice.

1.5.3 Detection Limits

Any measurement device will have random fluctuations in the signal that it produces. We call these fluctuations the *noise*. At low analyte amounts or concentrations, the noise in a detection mechanism can obscure the signal due to the analyte. For example, the readout on a four-place analytical balance will often bounce between 0.0000 and 0.0001 g due to vibrations affecting the weighing mechanism. Given these fluctuations, it is not possible to weigh less than 0.0001 g of a substance. When the signal from an analyte is comparable to the level of noise, we say that it is below the *limit of detection* (LOD) or not detectable.

The most appropriate method to describe and quantify the noise depends on the nature of the signal and the detector.[25] We will use the standard deviation of multiple measurements of a blank as a generic approach to quantify noise.[26] We call the average value of replicate blank measurements the *baseline* or *background*. Baseline and noise are not the same thing. A baseline value can be subtracted from a measurement signal. It is not possible to subtract the noise.

The LOD, or c_L, for an analysis is generally accepted as the concentration that gives a signal that is greater than the baseline by three times the level of the noise. We describe this condition as a signal-to-noise ratio of 3. The signal-to-noise ratio is often written as S/N or SNR. Based on a normal distribution, the probability of a measurement being three times the standard deviation, $3s$, from the mean is approximately 1 in 1000.[27] If an unknown measurement is made more than once, the probability is very low that a signal of $3s$ is due to random fluctuation.

Another quantity that is sometimes reported is the *limit of quantitation*, LOQ or c_Q. The LOQ is more stringent than the LOD and requires a signal level that is greater than the baseline by ten times the noise, SNR = 10. The difference between LOD and LOQ is that signal levels below the LOQ are reported as the analyte is detected. A quantitative result is only reported if the signal is at or above the LOQ. Calculating the LOD and LOQ are illustrated in Example 1.6.

---

[25]Common approaches include peak to peak, root mean square (RMS), and standard deviation.
[26]The US EPA recommends a more sophisticated approach to determine a method detection limit (MDL). For an example see *EPA Method 200.8 Determination of trace elements in waters and wastes by inductively coupled plasma-mass spectrometry* in *Selected Analytical Methods for Environmental Remediation and Recovery (SAM)*: Available at https://www.epa.gov/esam/ selected-analytical-methods-environmental-remediation-and-recovery-sam; accessed August 2022.
[27]$3s$ encompasses 99.73% of replicate measurements. The following definition for the LOQ also assumes that the noise has a normal distribution.

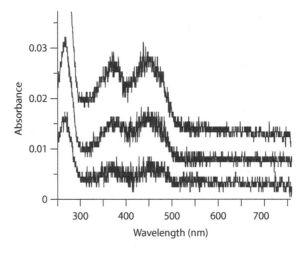

Figure 1.7 Absorbance spectra of riboflavin.

TABLE 1.11 **Riboflavin Spectral Data**

Spectrum	Concentration (μM)	Noise	266 nm		440 nm	
			Signal	S/N	Signal	S/N
Top	1.2	0.001	0.0392	40	0.0138	14
Middle	0.6	0.001	0.0234	23	0.0080	8
Bottom	0.3	0.001	0.0129	13	0.0028	3

Figure 1.7 shows UV/Vis absorption spectra of riboflavin in water at different concentrations. The three spectra are offset vertically for clarity. Absorbance is a unitless measurement related to the attenuation of light as it passes through a sample. Chapter 4 discusses spectroscopy in depth. The important point for us here is that absorbance, A, is directly proportional to absorber concentration, c.

Given no blank measurement, we take measurements at wavelengths where there are no peaks to determine the baseline and the noise. Averaging absorbance values near 590–650 nm gives a standard deviation of ±0.001. The signal for the peaks at 266 and 440 nm are the maximum values minus the baseline. Table 1.11 lists the noise and peak signals at 266 and 440 nm. Dividing the signal by the noise gives the S/N of each peak for each spectrum. In the absence of any interferences, the larger peak at 266 nm provides greater sensitivity. An unknown sample similar to any of these spectra could be quantitated as the S/N is greater than 10. If the test portion contains an interference that overlaps with the 266 nm peak, the 440 nm peak can be used for quantitation. Only an unknown with a signal of 0.01 or higher can be quantitated. An unknown with a signal of 0.003–0.01, similar to the middle and bottom spectra, could be listed as analyte detected.

Example 1.6 LOD and LOQ.   The measurement of five separate blanks gives an average and standard deviation of 0.10 ± 0.05 ppm. Calculate the LOD and LOQ.

As the LOD is three times the noise, it is:

$$0.10 + 3(0.05) = 0.25\,\text{ppm}$$

Similarly, the LOQ is

$$0.10 + 10(0.05) = 0.60\,\text{ppm}$$

As the blank value, 0.10 ppm, is not zero, we add it to both the LOD and the LOQ.

You-Try-It 1.D
The LOD-LOQ worksheet in you-try-it-01.xlsx contains two data sets. The exercise is to calculate the LOD and LOQ for each data set.

Now that we know how to find both the high and low limits of a measurement, there is one last descriptor that is useful to describe measurement range. The *dynamic range* of an analytical method or instrument is the ratio of the maximum to the minimum measurable signal. Since it is a ratio, it is unitless. For example, if an instrument has a minimum detectable signal of 0.001 and a maximum of 1.0, the dynamic range is 1.0/0.001 or 10^3.[28] Typical dynamic ranges for analytical measurements are 10^2–10^7 or higher. As an example, different types of gas chromatography detectors have dynamic ranges of 10^4–10^7. A larger dynamic range is advantageous to be able to measure a wider range of analyte concentrations without needing to preconcentrate or dilute the test portions. It is a useful specification to help decide if a given instrument is appropriate for an analytical task. In all cases, validation experiments are needed for an analyst to know the measurement range, and other limitations, of a given analytical procedure.

1.5.4 Linear Regression of Calibration Data

A *calibration curve* uses multiple standards to calibrate a measurement over an extended range. The advantage of a multipoint calibration curve is that any deviation from linearity will alert you to the limited linear range of the measurement (see Figure 1.6). A practical advantage of generating a calibration curve is that a single set of standards serves to calibrate the measurement of multiple samples. This approach is much more time efficient than performing a separate calibration for each individual test portion.

[28]This example is typical for miniature UV/Vis absorption spectrophotometers.

A calibration or working curve is prepared by making measurements of a series of standards to obtain the analytical signal versus analyte concentration. The relationship between the measurement, y_i, and analyte concentration, x_i, in the linear portion of the calibration curve is described by:

$$y_i = a_0 + a_1 x_i \qquad (1.15)$$

where a_0 and a_1 are the y-intercept and slope, respectively. The i subscript refers to an x_i, y_i data pair. I use this notation rather than the more common $y = mx + b$ because it is easily extended to higher order cases such as:

$$y_i = a_0 + a_1 x_i + a_2 x_i^2$$

The equation of the line is called the *calibration function* and the slope of this function is the sensitivity, a_1. This result is the same as for the simple proportionality discussed previously. The LOD, LOQ, sensitivity, and linear range can be different for each specific analyte that is measured by a given method. In general, you cannot use a calibration curve of one elemental or molecular species for other elements or molecules.

Given a set of calibration data, we obtain the best fit to the linear part of the data using linear regression. *Linear regression* uses the method of least squares to determine the best linear equation to describe a set of x_i, y_i data pairs.[29] The method of least squares minimizes the sum of the square of the deviations or residuals. In this context a deviation is the difference between a measured data point and the point on the proposed "best-fit" line. The best-fit coefficients a_0 and a_1 will be those that result in a calculated line that most closely matches the experimental data points. Quantitatively, this line will have the smallest total deviation between the experimental y values and the calculated y values. Thus, finding this best line, that is, the best a_0 and a_1 values, is done by minimizing the sum of residuals between the experimental and calculated y values. As was done for the standard deviation of a distribution, the residuals are squared so that positive and negative values do not cancel when summed (thus the name "least squares"). The quantity, χ^2, to be minimized is:

$$\chi^2 = \sum_{i=1}^{n} \frac{\left\{ y_i - (a_0 + a_1 x_i) \right\}^2}{s_y^2} \qquad (1.16)$$

where the width parameter, s_y, is:

$$s_y = \sqrt{\frac{1}{n-2} \sum \left\{ y_i - (a_0 + a_1 x_i) \right\}^2} \qquad (1.17)$$

---

[29]The equations here follow Taylor, J. R. *An Introduction to Error Analysis: The Study of Uncertainties in Physical Measurements*, 2nd ed.; University Science Books: Sausalito, CA, 1996 and de Levie, R. *Advanced Excel for Scientific Data Analysis*, 2nd ed.; Oxford Univ Press: New York, 2008.

Minimizing χ^2 is done by taking the partial derivatives with respect to a_0 and a_1 and solving the two resulting expressions. The results are as follows. The summations are taken from $i = 1$ to n (omitted from the summation symbol for clarity). The y-intercept is given by:

$$a_0 = \frac{\sum x_i^2 \sum y_i - \sum x_i \sum x_i y_i}{D} \tag{1.18}$$

and the slope is given by:

$$a_1 = \frac{n \sum x_i y_i - \sum x_i \sum y_i}{D} \tag{1.19}$$

where D is:

$$D = n \sum x_i^2 - \left(\sum x_i \right)^2 \tag{1.20}$$

The standard deviations for a_0 and a_1, respectively, are

$$s_0 = s_y \sqrt{\frac{\sum x_i^2}{D}} \tag{1.21}$$

$$s_1 = s_y \sqrt{\frac{n}{D}} \tag{1.22}$$

These equations can be typed in spreadsheet formulas, but most graphing calculators and spreadsheets provide linear regression results via built-in functions.

Figure 1.8 replots the linear portion of the riboflavin fluorescence data in Figure 1.6. This data was plotted in a spreadsheet, which determined the best fit to

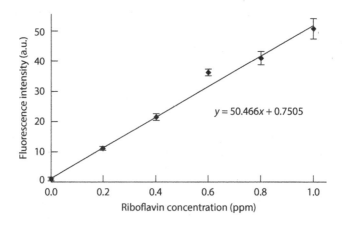

Figure 1.8 Example of a calibration curve.

the data points and displays the calibration function on the chart. The concentration of an analyte in an unknown sample is determined by measuring the unknown under identical conditions and comparing the measurement to the calibration data. In practice, we insert the unknown measurement, y_{unk}, into the best-fit linear equation for the calibration data to obtain the unknown concentration, x_{unk}. The calculation in Example 1.7 illustrates the use of a calibration curve.

Example 1.7 Calibration Curve. Use the calibration curve in Figure 1.8 to find the concentration of a sample that has a fluorescence intensity of 23.5. This sample was handled in an identical manner to the standards.

You are given a measurement for an unknown. Use the equation for the calibration curve (given in the figure) to find concentration. The equation on the chart does not show units, but from the units on the x- and y-axes, the slope has units of ppm^{-1}.

$$y = (50.466\,ppm^{-1})x + 0.7505$$
$$23.5 = (50.466\,ppm^{-1})x + 0.7505$$
$$x = \frac{23.5 - 0.75}{50.466\,ppm^{-1}}$$
$$x = 0.451\,ppm$$

As a check for a calculational error, place your finger at 23.5 on the y-axis of the calibration curve and trace over to the calibration line. Now trace straight down. When you do this, you will end up between 0.4 and 0.5, showing that the calculated result is reasonable.

You-Try-It 1.E
The calibration worksheet in you-try-it-01.xlsx contains measurements of a standard and two unknown solutions of acetylsalicylic acid (ASA or aspirin) using light absorption. The exercise is to use the measurement of the standard solution to determine the ASA concentration of the unknown solutions using a simple proportionality and using a calibration curve.

1.5.5 Interferences and Matrix Effects

An *interference* is a substance in a test solution that causes the measured amount of analyte to be higher or lower than the true value. An interference might be present in the original sample matrix or it might be introduced during sample processing or measurement. *Carryover* is contamination of analyte in a solution or instrument from a prior measurement. I've seen carryover in my student's results from several different types of instruments. It usually occurs due to overloading an instrument or failing to rinse glassware properly. Remeasuring a blank and finding analyte is a sign that you have a carryover problem to fix.

An interference might act by binding to an analyte, and reducing the analytical signal, or by producing an analytical response, making it appear that more analyte is present. Consider a spectroscopic measurement based on the amount of light absorbed by the analyte. If your standards are clear solutions, but your sample test portion is turbid (cloudy), will the calibration be accurate? The turbidity will scatter some light, appearing the same to the detector as analyte absorption. The measurement will be higher than the true value due to the scattering. This situation is easy to recognize, and can be corrected by filtering, but many types of chemical interferences will not be obvious visually.

The impact due to interferences that are present in a sample is often called a *matrix effect*. Matrix effects introduce bias in measurements. They can be difficult to manage because the extent of the bias depends on the details of the sample components, which can be variable in different samples. Much of Chapter 2 will describe methods to remove interferences and reduce matrix effects. An alternate approach to removing interferences is to add a reagent that equalizes matrix effects in all test solutions. The total concentration of salt will affect measurements made with an ion-selective electrode. To equalize sample variability, all standards and samples are adjusted to have a high salt concentration. The effect of the salt on the analytical response is still present, but it is the same in all standard and sample test solutions.

The disadvantage of using a calibration curve to quantitate unknowns is that a bias can be introduced in the results if

- experimental conditions differ in sample preparation or measurement of standard and sample test solutions
- matrix effects vary between standard and sample test solutions

A bias due to either of these causes will lead to inaccurate results. The first case can be improved by using an internal standard, where the internal standard and analytes are processed and measured together. In the second case, a bias will occur even when we maintain identical measurement conditions for standards and unknowns. One solution is to use standards that closely match the matrix of unknown samples, and Section 1.6 describes examples of this "matrix matching" approach. When sample matrix effects are highly variable, it is not possible to generate a calibration function for all samples. In this case, the standard-addition method provides a calibration procedure that compensates for matrix effects and can give accurate measurements.

1.5.6   Internal Standard

Unlike the "external" standards that are measured to generate a calibration curve, an *internal standard* is added directly to the sample test portion. Any addition to a test portion is often called a *spike*. This internal standard is measured simultaneously with the analytes and an analyte is quantitated based on the ratio of the analyte and internal standard signals. The internal standard is added in a known amount that is comparable to the concentration that is expected for the analytes. Since the internal standard is present at a single concentration, validation experiments must determine that all signals are in a linear range. Likewise, the relative sensitivity of a method for the internal standard and each individual analyte must be determined beforehand.

There are several requirements for a good internal standard. Since it is added to the test solution, it must not interfere or overlap with measurement of the analytes. Second, it must be a substance that is not present naturally in the samples to be measured. Third, it should be chemically similar to the analytes. This last point is especially important if the internal standard goes through sample preparation steps with the analytes.

Finding a chemically similar internal standard that does not interfere in an analyte measurement can be difficult for some methods. In molecular spectroscopy, absorption and fluorescence signals tend to be broad bands. It can be difficult to find an internal standard that does not overlap in the spectral measurement. In atomic spectrometry and many separation-based methods, the narrow lines allow the signal of an internal standard and analytes to be measured separately. In analysis of organic compounds using mass spectrometry, a deuterated organic compound with one or more H atoms replaced by D, increases the molecular mass by one or more and provides a very similar internal standard.

For methods that can measure multiple signals without interference, internal standard calibration can be more accurate because

- the simultaneous measurement reduces bias due to analyte loss or baseline drift during the measurement
- the internal standard and target analyte experience similar matrix effects
- the internal standard and target analyte experience similar losses during sample preparation

These advantages depend on the internal standard being a good surrogate for the target analyte. Only if the internal standard is similar to the analytes will sample processing and matrix effects also be similar. For example, you would not choose a nonpolar molecule as an internal standard for an organic acid if a sample purification step required the acidic analyte to be present as the anion. You would choose a similar organic acid, although not one that could be present in a sample.

1.5.7 Standard Addition

Matrix effects or interferences can cause the analytical response for an analyte in a real sample to be different than the same analyte concentration in an external standard. In these cases, using a calibration curve requires standards that closely match the composition of the sample. This *matrix matching* can produce accurate measurements because any bias is the same in both standards and unknowns. Certified reference materials, such as NIST standard reference materials (SRMs), are available commercially in numerous matrices for this reason.[30] The matrix-matching approach works well for repetitive analyses where the sample matrix does not change very much. For diverse and one-of-a-kind samples, replicating the composition of an unknown is time consuming and often impossible. An alternative calibration procedure for such cases is the standard-addition or known-addition method.

In the *standard-addition method*, a known amount of the analyte is added to the sample test portion to provide an "internal" calibration to the measurement. This process is similar to using an internal standard, but the spike is the actual analyte rather than a surrogate. The advantage of the standard-addition method is that interferences or matrix effects will affect the spike to exactly the same extent as they affect the unknown amount of analyte. Equalizing matrix effects between the analyte and spike eliminates bias in a measurement. The disadvantage of the standard-addition method is that it requires two or more measurements for each test portion, making it more time consuming.

In a simple implementation of the standard-addition method, an unknown sample is measured by some method and the signal is recorded. A known amount of the analyte, the spike, is added to the test solution and it is analyzed again. The spiked test solution will show a larger analytical signal than the original sample because of the additional amount of analyte. Figure 1.9 shows a plot of the two measurements. The difference in analytical signal, ΔS, between the spiked and unspiked test portions is due to the standard addition, Δc.

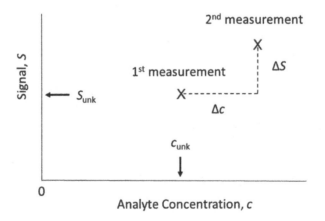

Figure 1.9 Plot illustrating calibration using standard addition.

[30]National Institute of Standards and Technology, https://www.nist.gov; accessed August 2022.

The ratio $\Delta c / \Delta S$ is known and provides the calibration to determine the analyte concentration, c_{unk}, in the sample.

$$\frac{c_{unk}}{S_{unk}} = \frac{\Delta c}{\Delta S} \qquad (1.23)$$

Example 1.8 shows two approaches to use this relationship to determine c_{unk}.

Example 1.8 Standard Addition. A 100-ml solution containing Cu^{2+} is treated with reagents to produce a blue color. Measurement by UV/Vis absorption spectroscopy gives a reading of $A = 0.444$. For this problem it is sufficient to know that the absorption reading (unitless) is directly proportional to concentration. This test portion is now spiked with 0.033 mmol of Cu^{2+} with a negligible change in the solution volume. Remeasuring gives an absorption reading of $A = 0.527$. What is the concentration of Cu^{2+} in the 100-ml solution?

The concentration of the spike, 0.033 mmol/100 ml, produces a reading of $0.527 - 0.444$ or 0.083:

$$c_{spike} = \frac{0.033\,\text{mmol}}{100\,\text{ml}} = 3.3 \times 10^{-4}\,\text{M}$$

Given that $A \propto c$,

$$0.083 = (\text{constant})\ 3.3 \times 10^{-4}\,\text{M}$$

We can determine the constant or set up a simple ratio with the unknown measurement since the proportionality constant is the same before and after adding the spike. Writing a ratio, we find the Cu^{2+} concentration in the original solution from:

$$\frac{0.083}{3.3 \times 10^{-4}\ \text{M}} = \frac{0.444}{c_{unknown}}$$
$$c_{unknown} = 1.8\,\text{mM}$$

The advantage of this calibration method is that any interference, for example, some fraction of the Cu^{2+} being complexed by interfering substances, affects the spike in the same way as it affects the unknown concentration that we are trying to measure.

You-Try-It 1.F
The standard-addition worksheet in you-try-it-01.xlsx contains a series of measurements for an unknown with multiple standard additions. The exercise is to determine the unknown from one addition and from the line generated by all of the data.

1.6   MAINTAINING ACCURATE RESULTS

1.6.1   Analytical Standards

The calibration methods described above rely on good standards. A standard, or certified reference material (CRM), is a material of known purity, of known analyte concentration, or of a verified physical property. We are most concerned with concentration standards. Standards provide the reference to calibrate analytical instruments and determine unknown concentrations. A given pure standard material is often diluted to prepare a series of standards of different concentrations, for example, to generate a calibration curve. In doing so, the solvent and laboratory equipment becomes integral to maintaining the purity of the reference material. There are several accepted definitions to describe standards:[31]

- Primary Standard: a commercially available substance of purity 100 ± 0.02% (Purity 99.98% or more)
- Working Standard: a commercially available substance of purity 100 ± 0.05% (Purity 99.95% or more)
- Secondary Standard: a substance of lower purity which can be standardized against a primary grade standard

The purity of commercially available chemicals is sufficient to meet the level of a primary standard, so we will not consider working standards further in this text.

A *primary standard* is a reagent that is extremely pure, stable, has no waters of hydration, and has a high formula weight. There is no general way to predict substances that will be good primary standards, but lists of them are tabulated in handbooks. Table 1.12 presents a few examples for illustration.

To illustrate the certification procedure, the description from the NIST Certificate of Analysis for the sodium carbonate, SRM reads

> This material was assayed by automated coulometric back-titration [Pratt, K.W.; Automated, High Precision Coulometric Acidimetry. Part II. Strong and Weak Acids and Bases; *Anal. Chem. Acta*, **1994**, *289*, 135.], to a strong acid endpoint, of weighed, dried, sodium carbonate samples after addition of excess coulometrically standardized hydrochloric acid and elimination of the product carbon dioxide. The certified mass fraction represents the result of eight titrations of samples from four randomly selected bottles from the entire lot of SRM 351.

To ensure that a reliable purity is reported, eight measurements were performed after random sampling from the packaged materials. This protocol, for

---

[31]Analytical Chemistry Section of the International Union of Pure and Applied Chemistry. "Sodium Carbonate as a Primary Standard in Acid-Base Titrimetry," *Analyst* **1965**, *90*, 251–255.

TABLE 1.12   Examples of Primary Standards

Material[a]	Formula Weight, g/mol	Typical Purity, %[b]	NIST SRM[c]
Na_2CO_3	105.988	99.9796 ± 0.0090	351
Tris	121.135	99.924 ± 0.036)	723d
KHP	204.221	99.9911 ± 0.0054	84k
Benzoic acid	122.121	99.9958 ± 0.0027	350a
$K_2Cr_2O_7$	294.185	99.984 ± 0.010	136e

[a]Tris is tris-(hydroxymethyl)aminomethane ($C_4H_{11}NO_3$) and KHP is potassium hydrogen phthalate ($KHC_8H_4O_4$).
[b]The uncertainties are for the 95% level of confidence.
[c]For more information see the NIST SRM website: https://www.nist.gov/srm; accessed August 2022.

a material with well-known purification procedures, is the level of analysis that is necessary for a material to be used and sold as a primary standard. Routine samples might not require this level of analysis, but analysis of poorly characterized samples can require significant method development to obtain reliable sample purification and analysis procedures. As another example, the potassium dichromate SRM was analyzed by three different methods, coulometry, titration, and gravimetry (these methods are discussed in Chapter 3). Luckily, many standards and CRMs for common analytical measurements are available from commercial suppliers and government agencies. In addition to NIST, the following examples illustrate the breadth of offerings that are available:

- Sigma-Aldrich, Calibration, Qualification & Validation
- US Geological Survey Geochemical Reference Materials and Certificates
- New Brunswick Laboratory (nuclear reference materials)
- Natural Resources Canada—Certified Reference Materials (CRMs for mineral, metallurgical, earth science, and environmental industries)
- Certified Reference Materials New Zealand (CRMs for chemical and spectroscopic analysis)

Do you wonder why a high formula weight is desirable for a primary standard? A higher formula weight allows you to obtain more significant figures when weighing a given molar amount of the primary standard. Analytical balances (Figure 1.10) often display mass to 0.0001 g (0.1 mg), so a larger mass of standard can provide more significant figures. As an example, consider using Cr metal versus $K_2Cr_2O_7$ to prepare 100.0 ml of a 0.01000 M standard solution of chromium. Using Cr metal requires 0.0520 g and using $K_2Cr_2O_7$ requires 0.1471 g. The $K_2Cr_2O_7$ provides an additional significant figure compared to that obtained when using Cr metal, provided the $K_2Cr_2O_7$ and Cr metal have equal purity.

A *secondary standard* is a standard that is prepared in the laboratory or by a third party for a specific analysis. It is usually standardized against a

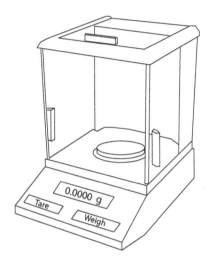

Figure 1.10   Analytical balance.

primary standard. The National Institute of Standards and Technology (NIST), and private companies, sell secondary standards for analytical calibration. NIST is probably the most well known and provides a wide variety of SRMs for validating and calibrating analytical methods. A key aspect of an SRM is that you can often obtain a standard that matches your sample matrix. Table 1.13 lists examples to illustrate the range of materials that are available.

In addition to the purity of standard materials, the purity of reagents for sample dilution or other steps in sample processing also play an important role. The most common laboratory reagent, water, is usually available as tap, distilled, and deionized water. If diluting a test portion for analysis of metal ions, you would need to use deionized water. If cleaning glassware, you would dissolve your soap in tap water. After cleaning you might rinse the soap out

TABLE 1.13   Examples of Reference Materials

Chemical Composition	Physical Properties	Engineering Materials
Elements in iron, steels, and other alloys	Si electrical resistivity	Particle size
Sulfur in fossil fuels	Radioactivity	Magnetic storage media
Polychlorinated biphenyls (PCBs) in oils	Strength and melt flow of polyethylene pipe	Surface flammability
Elements in foods and beverages (e.g., milk powder, wheat flour)		

with tap water, followed by rinsing with distilled or deionized water. Common grades of other reagents include

- AR or ACS grade: analytical reagent grade, varies for different materials, but typical purity is 99% or higher.
- USP: United States Pharmacopeia, materials specified as food and pharmaceutical grade.
- HPLC grade: high-purity solvent for chromatography.
- spectrophotometric grade: high-purity solvent with very low absorbance in the UV spectral region.

Manufacturers often have their own trademarked names for these grades. The purity required for a given application will often be specified in analytical procedures.

1.6.2 Volumetric Glassware

Maintaining the validity of high-quality standards requires proper handling to avoid errors from contamination or reducing the accuracy of the concentration during weighing and solution preparation. The tools for accurate handling include the analytical balance, volumetric glassware, and proper laboratory techniques.

Working with standard solutions requires volumetric glassware (Figure 1.11), the most common varieties being

- volumetric pipette (or pipet),
- burette (or buret),
- volumetric flask.

Volumetric glassware will have a tolerance etched on the surface and usually a label such as TD (to deliver) or TC (to contain). "TD" means that you let the liquid drain and you do not blow out any residual liquid. "TC" is found on volumetric flasks and some pipettes, which are designed for preparing but not delivering a calibrated amount.

Volumetric glassware is available as Class A and Class B, with the manufacturing tolerance for Class B being twice as large as for Class A. The stated tolerance of Class A volumetric glassware is often on the order of 1 part per 1000, for example, 50.00 ± 0.05 ml. The tolerance is slightly better for glassware volumes larger than 100.0 ml, but slightly lower for smaller volumes, for example, a 5-ml Class A pipette is listed as ± 0.01 ml. There will be small differences between the stated volume and the true volume because of temperature and manufacturing variability. Achieving the highest accuracy with volumetric equipment requires calibration by weighing the volume that the glassware delivers to determine its true capacity. Example 1.9 shows a typical result.

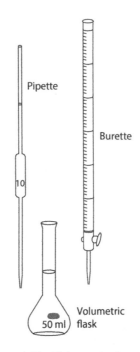

Pipette

Burette

Volumetric
50 ml) flask

Figure 1.11   Volumetric glassware.

Graduated cylinders and graduated pipettes, Mohr and serological, are also useful in the analytical laboratory. These pieces of glassware are not as accurate as volumetric glassware, but they are useful when preparing solutions where only approximate amounts are required. Their accuracy is on the order of 1–10%, depending on the amount of liquid and the size of the glassware (see Example 1.10). They are not suitable for preparing standard solutions, but they are used to quickly mix reagents for buffer solutions or adding reagents that are present in excess. You will also encounter mechanical pipettors of various styles that use disposable pipette tips to deliver solutions. The accuracy of these instruments can often be less than 1%, making them suitable for preparing some standard solutions, especially when small volumes (microliters) are involved. Many of these pipettors will use multiple parallel tips to complete sample procedures in multiple tubes or well plates quickly.

Example 1.9 Glassware Calibration.   Calculate the true volume delivered by a 25.0-ml volumetric pipette that delivers 25.013 g of distilled water. The laboratory temperature is 22°C and the density of water is 0.99777 g/ml.

Using dimensional analysis, we see that we must divide the mass by density to obtain the volume:

$$\frac{25.013\,g}{0.99777\,g/ml} = 25.069\,ml$$

This calibration now gives us a more accurate value for this pipette. To be more certain of the value, we would repeat the calibration two or three times to get an average value.

Example 1.10 Glassware Use.   Calculate the relative error in delivering 98 and 8 ml of solution using a 100-ml graduated cylinder. The readability and thus the uncertainty of the graduated cylinder is ± 1 ml.

The uncertainty in delivering 98 ml is 98 ± 1 ml or

$$\frac{1\,\text{ml}}{98\,\text{ml}} \times 100\% = 1\%$$

The uncertainty in delivering 8 ml is 8 ± 1 ml or

$$\frac{1\,\text{ml}}{8\,\text{ml}} \times 100\% = 12\%$$

The uncertainty in delivering a volume is dependent on the relative amount of the liquid compared to the size of the glassware. Delivering 8 ml of solution will be more certain using a 10-ml graduated pipette rather than a 100-ml graduated cylinder.

Chapter 1. What Was the Point?   The main point of this chapter was to introduce and clarify the terminology of analytical chemistry. Many terms in science have very specific meanings, with accuracy and precision being prime examples. This chapter presented an overview of the work environment and the issues in making analytical measurements. The remainder of the text concentrates on individual steps and specific concepts relevant to making measurements. As we go through the following chapters, think about how these topics fit into the "big picture" of making a measurement and reporting a result.

PRACTICE EXERCISES

1. You have a graduated cylinder (2.5 cm i.d.), burette (1.00 cm i.d.), and pipette (2.0 mm i.d.), where i.d. is the inner diameter of the glassware. List these volume measurement devices in the order of most precise to least precise. Explain your rationale for the order of your list.

2. Determine the imprecision as %-RSD for a tenfold dilution of a 0.0500 M Co^{2+} standard solution when delivering 10 ml of the standard solution into a 100-ml volumetric flask using each of the following pieces of glassware. You may assume that any error introduced in filling the volumetric flask is less than the imprecision in the 10 ml volume.

    (a) a 10.0-ml volumetric pipette that has a stated accuracy of 0.02 ml

    (b) a 10-ml Mohr pipette that has a stated accuracy of 0.1 ml

    (c) a 25-ml graduated cylinder that has a stated accuracy of 1 ml

3. The Mars Climate Orbiter missed its planned orbit in 1999 because some numerical values in a navigational program, calculated to many significant figures, were not converted to their metric equivalent. Describe the accuracy and the precision of these calculations.

4. SOPs for calibration of pH meters often specify either a one-point calibration with a $pH = 7$ buffer solution or a two-point calibration using buffer solutions of $pH = 4$ and 10. Discuss the relative merit of these two procedures.

5. Plot the data for the two peaks in Table 1.11 and determine the sensitivity at each wavelength. Determine the slope assuming the intercept is zero and also allowing it to vary. Is there any difference in these two cases for either peak?

6. The volume of a 2.3291-g rock was determined in four repetitive measurements to be 1.020, 1.031, 1.014, and 1.025 ml. What is the density of the rock?

7. High purity sodium chloride (NaCl) and sodium oxalate ($Na_2C_2O_4$) are both suitable to use as primary standards. In the following calculations of standard solutions, assume that all volumetric glassware has an accuracy of 1 part per thousand, for example, a 100.0-ml flask will deliver 100.0 ± 0.1 ml; an analytical balance will measure to 0.0001 g; and the chemicals are 99.99% pure.

    (a) Calculate the uncertainty as %-RSD in making 100.0 ml of 0.150 M standard solutions using these two forms of sodium.

    (b) What is the primary source of uncertainty if you prepare 50.00 ml of 0.0100 M standard solution using NaCl?

    (c) What is the primary source of uncertainty if you prepare 1.000 l of 0.500 M standard solution using NaCl?

    (d) What is the minimum volume of standard solution that you should prepare if you wish to make a solution of 0.100 mM sodium using NaCl and maintain an uncertainty limited by the glassware and not the mass measurement?

8. A 10-ml pipette delivers 10.021 g of water at room temperature. If the density of water at room temperature is 0.99707 g/ml, what is the error, both in terms of ml and as a percentage, in assuming that the pipette delivers exactly 10.00 ml?

9. Consider the possible interferences that you might encounter in determining the amount of Pb in each of the following samples. Describe what blanks and spikes will be necessary to obtain an accurate measurement for each. The sample must ultimately be in a 0.1 M nitric acid solution for analysis by atomic absorption spectrometry.

(a) drinking water (this one should be easy).

(b) chicken livers (organic components might complex some of the metal).

(c) a geologic mineral (must be digested in a flux at high temperature—possible loss of analyte owing to volatilization).

10. For the Pb analyses in the previous exercise, discuss the relative pros and cons of using a calibration curve versus standard addition to calibrate the measurement.

Calculational Practice

For the following spreadsheet exercises, title the top of the worksheet with the exercise number, your name, and date. Refer to the you-try-it and spreadsheet-help files as necessary. If you must do extensive revisions of your spreadsheet formulas, try developing more flexible ways to write the formulas.

11. (a) Enter the following data set in one column in a spreadsheet: 78.93, 78.77, 79.09, 78.52. Find the mean, standard deviation, %-RSD, standard error, and 95-% CI.

(b) Add 78.8 as an additional data point to part (a) and recalculate the requested quantities.

12. Find the mean and standard deviation for the following two data sets and plot the data as bar graphs.

(a) three data points: 98.2, 97.5, 99.4

(b) eight data points: 95.6, 94.1, 95.9, 95.7, 94.6, 94.2, 94.2, 95.3

(c) You suspect that the 99.4 measurement in the first data set and the 95.9 measurement in the second data set might be outliers. Perform the appropriate statistical test to determine if these data points can be removed from the calculations for the data sets.

(d) If you needed to do extensive revision of your spreadsheet formulas for either data set, copy the data to a new worksheet and redo the formulas to be more robust for changes.

13. Copy the following data (Data Set 1) into a worksheet.

Measurement	Voltage (V)
1	0.453
2	0.444
3	0.457
4	0.448
5	0.451
6	0.495
7	0.447

(a) Write your own formulas to find the mean and standard deviation of the data set. Compare your mean and standard deviation to the results returned by any built-in functions in your spreadsheet software.

(b) Plot (insert chart) Data Set 1.

(c) Copy the data and formulas from Data Set 1 to a new location or new worksheet. The sixth measurement in the data set is much higher than the other data. Show a Q-test calculation to determine if this data point can be discarded. If it can be discarded, delete the outlier and revise your formulas if necessary to recalculate the average and standard deviation.

14. Enter the following data (Data Set 2) into the worksheet.

Concentration (ppb)	Analytical Signal (V)
0.00	0.002
5.00	0.259
10.00	0.489
25.00	1.284
50.00	2.407
100.00	4.903

(a) Make a plot of Data Set 2. Do a linear regression using a trendline to find the slope and intercept of the best line that fits this data.

(b) Write your own formulas to find the slope and intercept, including their standard deviations. Compare to the results returned by the trendline.

(c) A sample of unknown concentration measured by the same procedure of the standards in Data Set 2 gives a measurement of 0.999 V. Find the concentration of the unknown, including an estimate of uncertainty.

CHAPTER 2

SAMPLE PREPARATION, EXTRACTIONS, AND CHROMATOGRAPHY

Learning Outcomes

- Choose appropriate sampling plans and control samples.
- Interpret the steps in sample preparation and processing protocols.
- Describe the nature of different solvents, electrolytes, and solutions.
- Predict the miscibility or solubility of a solute in different solvents.
- Calculate the efficiency of a liquid–liquid extraction.
- Choose appropriate stationary phases for a given class of solute in solid-phase extraction (SPE) and column chromatography.

2.1 SAMPLING AND CONTROL SAMPLES

2.1.1 Sampling Plans

Although this section discusses sampling only briefly, be aware that it is a critical aspect to the credibility of analytical results. You can find complete books written on the subject. The purpose of a sampling plan is to obtain samples for analysis that are representative of a "population." The population might be a geographic area, an aquatic system, a rail car of material, or anything else of interest. As noted in Chapter 1, we expect manufacturers and government

Basics of Analytical Chemistry and Chemical Equilibria: A Quantitative Approach, Second Edition. Brian M. Tissue.
© 2023 John Wiley & Sons, Inc. Published 2023 by John Wiley & Sons, Inc.
Companion Website: www.wiley.com/go/tissue/analyticalchemistry2e

inspectors to monitor product quality. For packaged food, a "Nutrition Facts" label shows the amount of fats, carbohydrates, protein, vitamins, and minerals in an item. Even for fresh food, we rely on biological and chemical testing to ensure that food is safe to consume. We immediately realize that not every single piece of food is tested. If it were, we'd have no food to consume. Ensuring a safe food supply depends on validated testing procedures, regular inspections, and enough sampling to monitor food production.

Sampling is the critical first step in performing an analysis. Often, someone else does the sampling, and the analyst in the laboratory receives a box of test tubes or bottles of samples. Table 2.1 groups sampling plans into three broad categories, although there is overlap and variation in these general approaches. The following discussion uses an example of environmental sampling, but sampling plans are also needed in product quality control and many other types of studies. Choosing a particular sampling approach will often depend on the purpose of the sampling.

Judgmental sampling is based on site observation or prior knowledge of a site. It is an obvious choice if, for example, you see a rusted chemical drum in a field or dead fish in a river. It is often done in a stratified manner where more samples are taken close to the location where the analyte is expected to be prevalent and fewer samples are collected at farther distances. Figure 2.1 shows a map with a stratified sampling plan. The purpose of this sampling is to determine if emissions from building 156 is a source of chemical contamination in the surrounding soil. Many samples, indicated by the **X**s on the map, are taken near the suspected source and fewer are collected farther away. The reason to use judgmental sampling is to minimize the number of samples that must be analyzed. It is suitable because it aims to answer a specific question: "Is Building 156 a source of a given substance?" Judgmental sampling does not provide much information about the spread of a target analyte. You would not use the results from this analysis to make broad statements about the concentration of an analyte for a whole campus, town, or state.

The systematic approach is useful to try to find hot spots of contamination. A hot spot is a localized high concentration of an analyte, leading to a source of contamination. Samples are taken on a periodic path or sampling grid. It is easy to implement in the field, and it can provide the initial survey of an area

TABLE 2.1 Sampling Plans

Method	Description
Judgmental sampling	Using your judgment or knowledge of a site to choose sampling locations
Systematic sampling	Collecting samples following a periodic path or a systematic grid. Often used to locate *hot spots*
Random sampling	Selecting sampling sites at random. Reduces bias in sampling but can result in a large number of samples for analysis

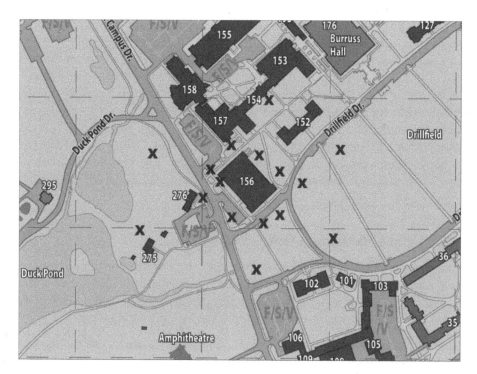

Figure 2.1 Map of a stratified sampling plan. *Source:* Virginia Tech campus map, Aug 2006. Available at http://www.vt.edu. Accessed 2013 Feb 13, used with permission Virginia Polytechnic Institute and State University.

that is suspected to need remediation. Systematic or periodic sampling is common in quality control of a product, including raw material, intermediates, and final product. Sampling might be done by the manufacturer, government inspectors, or border agents monitoring imported products. An established approach for material that is expected to be uniform is to sample $\sqrt{N}+1$ for N containers of product. If 100 barrels of precursor entered a manufacturing facility, 11 barrels would be opened and sampled to check the purity or other properties. If lots of material are expected to be nonuniform or from a new source, more extensive sampling plans are implemented.[1]

The random approach is similar to the systematic approach except that the sampling locations must be selected completely at random. Random samples are selected by using a random number generator to generate random locations on a grid or random individuals from a population. The advantage of this approach is that it eliminates bias from the sampling plan and will thus be the most credible when reporting results. The disadvantage is that random sampling can generate a large number of samples for analysis.

To understand the difference between random and other sampling plans, think about buying a house that is next to an industrial park. You have some concerns

---

[1] *WHO guidelines for sampling of pharmaceutical products and related materials*, World Health Organization, Technical Report Series, No. 929, Annex 4, 2005.

about contamination from the industrial site, so you ask for soil samples to analyze. Is your confidence that the property is free of contamination going to be greater or less if the samples were collected completely at random on the property, compared to being selected by the landowner in some other manner?

This discussion has only given examples in two dimensions. Another consideration for soil samples and standing bodies of water is the depth at which to sample. For streams and rivers, samples might be taken at low- and high-water conditions due to recent rain or snow melt. Analyte concentration can vary seasonally and even daily, especially for atmospheric studies. Developing a sampling plan must factor in these types of parameters so that the measured results will be representative of the system being sampled.

2.1.2 Sample Handling, Blanks, and Spikes

Another aspect of sampling is to ensure that the samples, once collected, are stored and preserved properly. Some analytical procedures will have very stringent requirements for sample handling and preservation. For illustration, two general methods for preserving drinking water samples are as follows:

- Solution samples for metal analysis can be treated with HNO_3 and stored in plastic for 6 months.
- Solution samples for organic analysis can be treated with HCl and $Na_2S_2O_3$ and stored in glass for 14 days (if kept at 4°C).

In the first example, the water is acidified with HNO_3 to prevent precipitation of metal ions as hydroxides. In the second case, the water is acidified to inhibit bacterial growth and $Na_2S_2O_3$ is added to remove chlorine disinfectant, which could degrade the analytes. Refrigeration further reduces analyte degradation and bacterial growth. In both cases, the container is chosen to reduce the chance of contaminating the sample with similar analytes, i.e., metal ions leaching from glass or organic compounds leaching from plastic.

The method of storage, preservation, and subsequent sample tracking are all issues that must be documented and included in a final report. Several types of control samples are collected when sampling the unknowns to validate that samples are handled properly and that analytical methods are giving accurate results. Table 2.2 lists the most common controls and their preparation. As in the standard-addition method, a spike is an intentional addition of analyte to a control sample.

The blanks serve two purposes—they provide a measure of the random variation or noise in a procedure or measurement and they can identify false positives. The noise measurement is necessary to predict the LOD and LOQ. A *false positive* is a measured result showing the presence of an analyte when none was actually present in the original sample. A field blank that shows the presence of the analyte in the measurement indicates an erroneous technique or contamination somewhere in the sample collection, processing, or measurement steps. Obviously, the collection and storage containers should be

TABLE 2.2   Control Samples

Control Sample	Preparation
Field blank	A blank prepared in the field that goes through all sample processing and analysis procedures
Spiked field blank	A spiked blank prepared in the field that goes through all sample processing and analysis procedures
Equipment blank	Blanks prepared in laboratory
Laboratory control standards	Standards or certified reference materials prepared in laboratory

clean before collecting samples. Preventing contamination can be difficult for certain analytes at ultratrace levels. As an example, sampling natural waters to measure organic pollutants must prevent exposure to engine fuel vapors during collection and transportation. The equipment blank prepared in the laboratory can help identify the source of contamination or problems. If a field blank shows a false positive but the equipment blank does not, the problem is probably contamination during sample collection or storage. If both field and equipment blanks show false positives, the problem might be in-lab contamination or analyte carryover in an instrument. These problems might be correctable so that the sample test portions can be remeasured after cleanup of the laboratory or instrument contamination.

The spiked field blanks can identify false negatives and provide a measure of analyte recovery after sample processing steps. A *false negative* is a measurement showing no analyte when an analyte was present in the original sample. False negatives can arise due to loss of analyte during storage, sample preparation, or measurement procedures. Since the amount of the spike is known, the measurement of the spike will show if analyte recovery after sample preparation is *quantitative*, that is, 100%. If loss is due to poor recovery in sample processing steps, the spike measurement can be used to correct results. Validation experiments are necessary to ensure that analyte recovery is the same for all samples and not variable.

Measuring the laboratory control standards, usually called the *standards*, provides the calibration for unknown samples and validates the linear range of the analytical procedures performed in the laboratory. As for the blanks, comparing laboratory standards and spiked field blanks can help identify if losses are occurring in the laboratory or during collection and storage of the samples. A false negative will also arise if the analytical method has insufficient sensitivity for the target analyte. Using a range of standard concentrations for the laboratory samples will provide a measure of the detection limit for the method. The spiked field blank will provide an additional check if losses during sample preparation result in a drop in the analyte concentration below a detectable limit.

2.1.3 Control Charts

In addition to SOPs that detail the correct use of protocols and instruments, method performance and the accuracy of results must be tracked over time. These procedures include a rigorous baseline measurement of a standard with remeasurement after appropriate periods of time. These control measurements might be performed daily, weekly, monthly, or based on a set number of instrument hours or measurements. The results of these procedures are plotted in *control charts* and are a key part of statistical process control (SPC). Identifying a method or an instrument as being *out of control*, meaning that it gives an erroneous measurement, can indicate the need for maintenance, calibration, repair, or user training. SPC is beyond the scope of this chapter, but it provides extensive procedures to ensure the reliability of laboratory measurements. Figure 2.2 shows an example of a control chart for an instrument that is checked on a monthly basis. The dotted line is the mean of the measurements, and the dashed lines show $\pm 3s$. In this particular example, the instrument measurements remain "in control," that is, within three standard deviations of the mean.

All of the topics we have discussed in this chapter and Chapter 1 give us the tools to follow GLP. Although most of this chapter explains individual measurement and chemical concepts, we can summarize the big picture of analysis as follows:

- Designing, following, and documenting an appropriate sampling plan
- Following SOPs for sample processing and measurement methods
- Using control samples to validate results
- Maintaining and documenting that laboratory instruments remain in control
- Reporting and archiving all sample information, measurement results, and data analysis

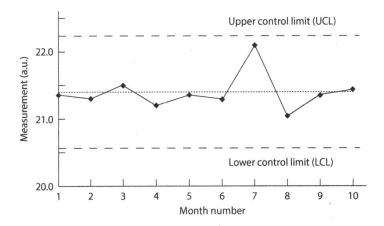

Figure 2.2   Example of a control chart.

2.2 SAMPLE PREPARATION

2.2.1 Role of Sample Preparation

Before discussing specific sample preparation and processing methods, let us start with a broad overview of the steps involved in making an analytical measurement. We will concentrate on the important aspect of describing the reliability of a measurement when reporting an analytical result. Table 2.3 lists potential sources of errors in the different steps of an analytical procedure. Errors introduced during any of the steps in collecting and processing a sample, or in making and reporting the measurement, will affect the final result. These potential errors, affecting the accuracy of a reported result, can include both systematic and gross errors. We rely on the field blanks and spikes to help catch both types of errors in the sample handling steps. The laboratory control samples serve the same role for the measurement step. The last step in the list in Table 2.3, reporting results, is the only one that is correctable without repeating experimental steps. Making errors in the sample handling or measurement steps requires repeating the procedures. If samples are no longer available or are out of date, resampling is necessary. Human patients are particularly unhappy when they must resupply samples because of an analyst's error or instrument malfunction.

For environmental and biological sampling, nonuniform analyte distribution will often be the largest factor contributing to the uncertainty in an analytical measurement. Systematic errors in sampling can be much larger and unpredictable than random errors, and there is no simple statistical tool to account for such heterogeneity. As an example, analysis of natural water systems can vary seasonally and depend on recent weather conditions. Analytical measurements must often include an evaluation of the factors and conditions that affect the reported results. Testing for diabetes is often done by measuring blood glucose after 8 h of fasting. Making a measurement for a patient who does not have the discipline to skip his morning donut will obviously affect the measurement, and such an occurrence might not be known to the analyst.

TABLE 2.3 **Error Sources in Analytical Procedures**

Measurement Step	Potential Errors
Sampling	Insufficient field sampling
	Poorly implemented sampling
Sample handling	Errors in sample preparation procedure
	Sample contamination
	Loss of analyte
Analytical measurement	Errors in method or instrument calibration
	Errors in instrument use
Reporting result	Errors in data treatment
	Errors in reporting results

After the uncertainty introduced by analyte heterogeneity and sampling issues, the variable composition of real samples is probably the second greatest source of uncertainty in analytical results. These variations in sample composition cause errors due the presence, or variable concentration, of interferences. We referred to these types of problems in Chapter 1 as *matrix effects*, because of their dependence on the sample composition. This factor is less severe for measurements of very similar or simple samples, but it might be significant depending on the accuracy desired for a measurement.

The goal of many sample processing procedures is to remove interferences. Some steps will be tailored for specific interferences. For example, the details of an organic extraction step for PCBs in fish samples (Figure 2.3) are slightly different depending on the fat content of the fish samples. When the difference between an unknown sample and the expected composition is not obvious, validation experiments are necessary to confirm that an analytical method is suitable for the sample. The use of blanks and spikes, as discussed in Section 2.1, is necessary to identify problems in the sample preparation and analysis steps. Using the standard-addition method in the measurement step is one remedy for accurate calibration of samples of variable composition. The disadvantage of this approach is the greater number of measurements needed for each test portion.

Adapting an analytical procedure to a very different type of sample can require a large time investment in method development to obtain accurate and reproducible results. Table 2.4 lists examples of NIST standard reference materials (SRMs) that are available to calibrate Pb measurements. The point of discussing this list is not to memorize SRM numbers but to see the wide variety of sample matrices that an analyst might encounter. Sample preparation methods will be quite different for these different sample types. The

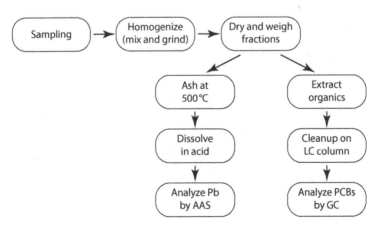

Figure 2.3 Abbreviated sample preparation for analysis of Pb and PCBs. Abbreviations: AAS, atomic absorption spectrometry; LC, liquid chromatography; GC, gas chromatography.

TABLE 2.4   NIST SRMs for Lead Analysis

SRM Number	SRM Name
SRM 87a	Aluminum-silicon alloy
SRM 617	Trace elements in glass
SRM 955c	Lead in caprine (goat) blood
SRM 966	Toxic metals in bovine blood
SRM 1131	Solder (60Pb-40Sn)
SRM 1632c	Trace elements in coal (bituminous)
SRM 1643e	Trace elements in water
SRM 2580	Powdered paint nominal 4% lead
SRM 2586	Trace elements in soil (contains lead from paint)
SRM 2976	Mussel tissue (trace elements and methylmercury) freeze dried
SRM 8436	Durum wheat flour
SRM 8603	Chinese lead ore

concentration of Pb in these different sample matrices covers a wide range, which can also affect the choice of suitable processing and measurement methods. The good news is that a wide variety of certified reference materials are available, so standards that match specific types of sample matrices can be obtained.

Given the wide range of analytical problems and sample types that you might encounter, this chapter can only illustrate a few sample preparation procedures. What I do hope that you take away from this chapter is the ability to read an analytical protocol and determine the purpose of each step. Being able to identify the chemistry of processing steps will let you fix problems when procedures do not work as planned.

2.2.2   Direct Analysis

There are some analytical techniques that do not require sample preparation, at least for certain samples. Measurement of pH with a pH electrode is one that comes to mind immediately. If you want to know the pH of a water sample, you simply dip your pH electrode into the water and record the readout. Of course, the accuracy of this measurement will depend on how well you calibrated your pH meter with standards. This direct analysis works well because the pH electrode is very selective for only H^+, at least in its normal working range of pH ≤ 10. Other electrodes have been developed for a wide variety of ions found in aqueous solution, for example, NO_3^-, NH_4^+, and Cl^-. These electrodes can also give a direct reading of a water sample, if interfering species are not present. When interferences are present, they must be eliminated by masking or sample processing steps to remove them from the test portion. *Masking* is the addition of a reagent to bind the interfering substance (examples are presented in Chapter 7).

The following list provides some examples of analytical methods that can use samples directly with no or minimal sample processing. Minimal sample preparation might be adding a few drops of reagent to a solution or grinding a test portion to a powder to introduce it into an instrument.

- H^+ (pH) and other ions in water using ion-selective electrodes
- Metal ions in drinking water using atomic spectrometry
- Chlorine in swimming pool water using a colorimeter
- Polymer identification using Fourier-transform infrared (FTIR) absorption spectroscopy
- Chemical residue detection by ion mobility spectrometry (airport scanners)
- Elemental surface analysis using X-ray photoelectron spectroscopy (XPS)
- Elemental analysis of solids using X-ray fluorescence
- Fragrances and other volatile organic compounds by thermal desorption gas chromatography

Do not worry if you are not familiar with these methods; the point is that there are instruments and techniques available for rapid, direct analysis. For many of these methods, however, extensive calibration and development is necessary to determine if interferences affect the accuracy of the results. After validating a method, analyses of similar types of samples can be made with confidence. For example, calibrating an X-ray fluorescence instrument using an NIST Portland cement SRM will be suitable for analyzing unknown cement samples. However, this same standard and method might not be reliable for measurements of contaminants in used motor oil.

In this chapter, I concentrate on understanding the chemical fundamentals of the steps in analytical procedures. This knowledge and practice will give you the skills to solve difficult problems when you encounter them.[2] Actual practice in industry and research will make extensive use of direct approaches whenever possible to increase efficiency.

2.2.3 Sample Preparation (How Analysts Really Spend Their Time)

A sample that requires analysis is often a mixture of many components in a complex matrix. For samples containing unidentified compounds, the components must be separated from each other so that each individual component can be identified and measured by appropriate analytical methods. The components in a mixture can be separated by using the differences in physical or chemical properties of the individual components. Dumping spaghetti and water in a colander separates the two components because the liquid water can run through the colander and the solid spaghetti cannot (assuming that it is not

---

[2]As said by Adams, A. "In retrospect I am always grateful for the benefits of prior intensive practice." *Examples: The Making of 40 Photographs*, Little-Brown: NY, 1983.

grossly overcooked as prepared in some dining halls). Some water will stick to the spaghetti and some spaghetti may go down the drain because the colander is not 100% efficient. An analogous example is the filtering of a solid precipitate to separate it from a solution. This step is also not 100% efficient, but precipitating reagents are chosen to produce precipitates that are easily filtered.

The spaghetti example is a separation based on the states of matter of the two components. Other physical properties that are useful for separations are size (sieving), density (centrifugation), and vapor pressure (distillation). Analytical sample preparation procedures often take advantage of the different chemical properties of each sample component. After dissolving a solid, it is common to filter the solution to remove insoluble components. Common cases are starch binders in pharmaceutical tablets or silicate minerals in soil samples. When we precipitate one particular metal analyte from a solution, we are using the difference in solubility between the analyte and interferences to separate them. Table 2.5 lists some common methods of sample preparation and purification with the primary physical or chemical property for the process. These methods isolate analyte(s) from other sample components or convert a sample to a form that is easier to handle, that is, a solution. These methods range in complexity from simply dissolving soluble substances and filtering insoluble species to instruments that apply a high voltage across a polymeric medium to electrophoretically separate different ions. The common aspect of these methods is that they take a complex mixture and make it simpler or more suitable for measurements.

We use these procedures to achieve an efficiency that is reproducible and close to 100%. However, perfection is uncommon in most endeavors, necessitating the control samples discussed in Section 2.1. The field blanks that we analyze with the sample test portions tell us if a sample procedure step introduces contaminants. The spiked field blanks provide a measure of the amount of analyte that is lost as test portions are processed through a procedure.

TABLE 2.5 Sample Preparation Methods

Processing Method	Mechanism
Filtration	Phase and physical size
Dissolution	Solubility
Digestion	Aggressive dissolution via heat, acid treatment, or oxidation state change
Precipitation	Solubility
Crystallization	Solubility
Centrifugation	Density
Distillation	Vapor pressure
Extraction	Partitioning
Chromatography	Partitioning
Electrophoretic mobility	Charge and molecular size (biomolecules)

Many of the methods in Table 2.5 depend on selecting appropriate reagents so that a given step is nearly 100% efficient, that is, it goes to completion. In isolating an analyte by precipitating it from solution, a precipitating agent that leaves a very low concentration of the analyte ion remaining in solution is selected. We will see in Chapter 8 how to calculate precipitate solubility from K'_{sp} equilibrium constants. If the analytical measurement requires a complexing agent, an agent with a large equilibrium constant, K'_n, for the analyte and small K'_n values for interferences will provide complete and specific complexation. These types of issues are why equilibrium factors can impact analytical measurements and the reason that we will discuss them in depth in Part II of the text.

A combination of methods is necessary in many applications to achieve complete analyte isolation. Common examples are precipitation followed by filtration and rinsing and extraction followed by chromatography. Figure 2.3 shows an example of a sample procedure for the analysis of contaminants in fish, a rather complex sample, even when dead. After sampling the material to be analyzed, either in the field or in the laboratory, the two sample fractions are processed by two different multistep pathways. The different pathways are necessary because the different target analytes, organic versus elemental analytes, have very different chemical properties. The two types of analytes require different sample preparation methods and are measured by different instrumental methods.

The brief outlines in Figure 2.3 omit many of the details of the original sources from which they are adapted: AOAC Official Methods 972.23 (Lead in fish) and 983.21 (Organochlorine pesticide and polychlorinated residues in fish).[3] In the Pb analysis for example, a cloudy solution after dissolving in acid requires addition of an EDTA-containing buffer solution to achieve complete dissolution.[4] Analysis of multiple analytes in a complex sample can require a large number of different protocols. Even similar analytes can require different procedures. Analysis of Hg is difficult because of the loss by volatilization and the ashing procedure used for Pb is not suitable for Hg. Likewise, extractions of different classes of organic compounds can require different solvents.

As a final note, you will see basic sample manipulations such as dissolution, precipitation, filtration, and extraction in many analytical procedures. The purpose of these steps will not always be the same. Figure 2.4 shows highly specialized equipment that is common in well-equipped chemistry laboratories. This particular apparatus simultaneously extracts and filters to obtain an aqueous solution of caffeine and other organic compounds. It is designed to be rapid, so it does not provide a quantitative extraction of caffeine. A quantitative extraction would require boiling the ground sample for some amount of time, followed by filtering with several rinses to collect all soluble analyte. It is not unusual for a laboratory protocol to specify repeating a step

---

[3]Horwitz, W., Ed. *Official Methods of Analysis of AOAC International*, 17th ed.; AOAC International: Gaithersburg, MD, 2000. This source is a two-volume set of validated procedures for a wide range of analytes and sample matrices.

[4]EDTA is a complexing agent that helps solubilize metal ions.

until some condition is met, for example, a filtrate is colorless or a supernatant is clear. For the extraction apparatus in Figure 2.4, a quantitative extraction is not necessary and would no doubt degrade the filtrate for its intended purpose.

The remainder of this chapter provides a detailed description of solutions and the chemistry of manipulating and isolating analytes. The next two sections discuss solvents, solutions, and solubility. Solubility is a general mechanism for isolating analytes or removing impurities by dissolution, precipitation, recrystallization, and extraction. Section 2.5 provides a quantitative description of partitioning, which is the basis for extractions and chromatographic separations. The remaining sections describe the practical application of these principles in sample preparation procedures.

You-Try-It 2.A
The **sample-prep** worksheet in **you-try-it-02.xlsx** contains two sample preparation procedures for the analysis of lipid-soluble vitamins in vitamin tablets. On the basis of the expected amounts in the starting sample, determine the analyte concentrations in the solution to be analyzed.

2.3 SOLVENTS AND SOLUTIONS

To begin thinking about how to use the various methods discussed above for sample preparation and measurements, we must develop some intuition in recognizing the nature of sample matrices, solutes, and solutions. Terminology can be confusing, and even the term *sample* can be vague. The following terms are often used loosely, and the context might be necessary to know the exact

Figure 2.4   A combination extraction/filtration apparatus.

meaning. As an example, some crystalline materials exist as "solid solutions" in which the different atoms are intermixed. They are still solids, and we use the context to know if we are considering a solid or a liquid solution. We will use the descriptions in Table 2.6 as working definitions.

Besides the gaseous, solid, or liquid classification, we can classify sample matrices and analytes as ionic or nonionic. Ionic solids are materials that consist of ions.[5] They will dissolve in water, but solubility varies widely from very high concentrations for soluble salts to negligible amounts for insoluble salts. Ionic solids in general will not dissolve in nonpolar solvents. As the example in Figure 2.3 illustrated, a single sample can have different types of analytes that require very different processing methods.

2.3.1 Solvent Polarity

The solubility of a chemical species will depend on how closely its characteristics match the solvent. Water, pure or adjusted to a specific pH with acid or base, is the solvent of choice for ionic compounds. The phrase *like dissolves like* is a reliable guide as you develop your chemical intuition.

For nonionic compounds, there is a wide range in the solubility of solutes in aqueous and nonaqueous, or organic, solvents. We can differentiate solutes and solvents in terms of their *polarity*. Polarity depends on the differences in

TABLE 2.6 Definitions for mixtures and solutions

Term	Definition
Mixture	*n.* A gas, solid, liquid, slurry, or gel that contains multiple components
Multiphase	*adj.* A mixture in which components have separated into more than one phase, for example, separated organic and aqueous solvents or a solid precipitate in contact with a liquid
Solvent	*n.* A liquid that may consist of one or more miscible components
Miscible	*adj.* Liquids that will coexist as one liquid phase
Immiscible	*adj.* Liquids that separate into separate phases
Solute	*n.* A chemical species that dissolves in a solvent
Solution	*n.* A liquid that contains solutes
Saturated	*adj.* Refers to a solution that is in equilibrium with another phase, solid, liquid, or gas
Solubility	*n.* The concentration at which a solute is saturated in a given solvent
Soluble	*adj.* An imprecise term indicating that a given species will dissolve in a given solvent
Slightly soluble	*adj.* An imprecise term indicating that a given species will dissolve to a small extent in a given solvent
Insoluble	*adj.* An imprecise term indicating that a given species will not dissolve to any appreciable extent in a given solvent

[5]Ionic liquids are ionic solids that melt below room temperature. They usually contain at least one organic ion.

electronegativity of the constituent atoms and how the atoms are arranged in the molecular structure. The polarity of an organic molecule can be predicted by the functional groups that it contains. A *functional group* is a group of atoms in a molecule that contributes distinct properties. The simplest hydrocarbon is a linear alkane, which consists of methylene ($-CH_2-$) groups with a methyl ($-CH_3$) on each end of the chain. Most organic compounds consist of an alkyl chain backbone with other functional groups attached. Table 2.7 lists the names of some common organic functional groups. For these functional groups, I am showing the simplest case of the carbons being fully bonded with hydrogen. It is possible for a functional group to contain another functional group. The carboxyl group consists of a hydroxyl functionality attached to a carbonyl group. The ester example is shown with an R. This is a generic symbol for any functional group and is referred to as an "R group."

The polarity of a molecule, and melting temperature, boiling temperature, and other properties, will depend on the size of the molecule and the functional groups that it contains. Table 2.8 lists a relative measure of solute polarity from nonpolar to polar. As you add functional groups to an alkane, they become more polar.

The longest sequence of carbon–carbon bonds in a hydrocarbon is called the *main chain*. Alkyl groups attached to the main chain are called *side chains* and the structure is said to be branched rather than linear. In addition to the size and functional groups, chain branching will also affect the solubility of a solute in a given solvent. For example, 1-butanol and 2-butanol differ only in the position of hydroxyl group, but their solubilities in water are 7.4% and 12.5%, respectively.

Comparing relative polarities on the basis of functional groups should be done for molecules of similar size and branching. Methanol and octanol both contain a polar hydroxyl group, but the 8-carbon chain of octanol makes it

TABLE 2.7   Names of Organic Functional Groups

Functional Group	Composition
Alkyl	$-C_n H_{2n+1}$
Methylene	$-CH_2-$
Alkenyl or ethylene	$>C=C<$
Alkynyl	$-C\equiv C-$
Aryl	$-C_6H_5$
Ether	$-CH_2-O-CH_2-$
Hydroxyl	$-OH$
Carbonyl	$>C=O$
Amine	$-\ddot{N}H_2$
Carboxyl	$-COOH$
Ester	$-COOR$
Chloro	$-Cl$

TABLE 2.8   Relative Polarity of Different Classes of Organic Compounds

Relative Polarity		Components
Very nonpolar	Alkanes (aliphatic hydrocarbons)	$-CH_2-, -CH_3$
Nonpolar	Alkenes (olefins)	$>C = C<$
Nonpolar	Aromatic hydrocarbons	$-C_6H_5$
Moderate polarity	Ethers	$-CH_2-O-CH_2-$
Moderate polarity	Esters, ketones, aldehydes	$>C = O$
Polar	Alcohols, amines	$-OH, -\ddot{N}H_2$
Polar	Carboxylic acids	$-COOH$
Very polar	Water	H_2O

immiscible with water. A qualitative "polarity index" from 0 (nonpolar) to 9 (water) is often used to classify solvents, and Table 2.9 lists some examples. Tables 2.8 and 2.9 provide trends and guidelines for understanding mixed liquids. Solvents with a polarity index greater than approximately 5 will be miscible with water (see Example 2.1). Less polar solvents will be immiscible with water and separate as a second phase.

Two immiscible phases in contact will have some limited solubility in each other. Table 2.10 lists solubilities of nonpolar organic solutes in water.[6] As expected, the alkanes have much lower solubility than the organic compounds that contain polar functional groups. Likewise, the organic phase will have a low but finite concentration of H_2O dissolved in the organic phase. The exact amount will depend on the polarity of the organic phase.

TABLE 2.9   Polarity Index of Selected Solvents

Compound	Formula	Polarity Index
Pentane, hexane, etc.	C_nH_{2n+2}	0.0
Cyclohexane	C_6H_{12}	0.2
Toluene	$C_6H_5CH_3$	2.4
Benzene	C_6H_6	2.7
n-Butanol	C_4H_9OH	4.0
Acetone	CH_3COCH_3	5.1
Methanol	CH_3OH	5.1
Ethanol	CH_3CH_2OH	5.2
Acetonitrile	CH_3CN	5.8
Dimethylformamide (DMF)	$CHON(CH_3)_2$	6.4
Dimethyl sulfoxide (DMSO)	CH_3SOCH_3	7.2
Water	H_2O	9.0

[6]Values from *IUPAC-NIST Solubility Database*, NIST Standard Reference Database 106, Version 1.0, http://srdata.nist.gov/solubility; copyright US Department of Commerce; accessed August 2022.

TABLE 2.10   Solubility of Selected Organic Compounds in Water at 298 K

Compound	Solubility as g Solute per 100 g Solution
Phenol	≈ 7
1-Hexanol	≈ 0.6
Benzene	0.177 ± 0.004
Toluene	0.053 ± 0.002
Cyclohexane	$(5.8 \pm 0.4) \times 10^{-3}$
Hexane	$(1.1 \pm 0.1) \times 10^{-3}$
Octane	$(1.77 \pm 0.09) \times 10^{-4}$
Decane	$(1.5 \pm 0.5) \times 10^{-6}$

Example 2.1 Solvents.   Predict which of the following solvents will be miscible: benzene, cyclohexane, and methanol.

Looking at the polarity index, benzene and cyclohexane are relatively nonpolar and methanol is relatively polar:

```
            polarity

cyclohexane  0.2
benzene      2.7
methanol     5.1
```

We expect the two nonpolar solvents, cyclohexane and benzene, to be miscible with each other. Likewise, we expect the two solvents with very different polarities, cyclohexane and methanol, to be immiscible. Predicting the miscibility of benzene and methanol is more difficult, but with a difference in polarity index of 2.4, we can guess that they will be miscible.

2.4   INTRODUCTION TO SOLUBILITY

Analogous to the preceding discussion on immiscible solvents, there is a limit to the amount of any solute that you can dissolve in a solvent at a given temperature. If you add too much sugar to your iced tea, you will see some undissolved sugar at the bottom of your glass. The same limit occurs for many ionic solutes. Ionic solids that have limited solubility in water are called *insoluble salts*. It is easier to build up your chemical intuition by recognizing the ions that are common in soluble salts rather than the many insoluble salts. Table 2.11 lists common ions that occur mostly as soluble salts. These solubility rules serve as guidelines, and what we label as a soluble compound can range from being only slightly soluble to very soluble. If you encounter unfamiliar salts in procedures, it is best to look up the solubility in a reference source.

There is no fixed cutoff in concentration for what we call soluble versus insoluble salts. Table 2.12 lists examples of the solubilities of some soluble and insoluble salts in water.[7] These concentrations are the amount of salt with which the solution is said to be saturated, that is, the concentration at which no more salt can be added without forming a precipitate. The silver salt examples illustrate the range of possible solubilities, where silver nitrate is very soluble, silver acetate is slightly soluble, and silver chloride is insoluble. As another example, we know calcium carbonate is insoluble from the abundance of limestone, shells, etc., observed in nature. A comparison of calcium carbonate to the other calcium salts shows the order of magnitude difference between soluble and insoluble compounds. Note that solubility for many ionic compounds will be sensitive to temperature. For calcium carbonate, the solubility will also be dependent on the amount of dissolved CO_2 and the pH. As we will

TABLE 2.11   Solubility Rules

Ionic Solids Containing	Soluble	Insoluble Exceptions
Group 1 cations	All	
NH_4^+	All	
NO_3^-, ClO_3^-, ClO_4^-, CH_3COO^-	All	
Cl^-, Br^-, I^-	Almost all	Ag^+, Pb^{2+}, Hg_2^{2+}
F^-, SO_4^{2-}	Almost all	Group 2 cations, Ag^+, Pb^{2+}, Hg_2^{2+}

TABLE 2.12   Aqueous Solubility of Salts (g/100 g H_2O)

Solid	0°C	20°C	80°C	100°C
NaCl	35.7	35.9	38.0	39.2
NaBr	80.2	90.8	120	121
NaI	159	178	295	302
Na_2CO_3	7.0	21.5	43.9	
KCl	28.0	34.2	51.3	56.3
KBr	53.6	65.3	94.9	104
KI	128	144	192	206
$CaCl_2 \cdot 6H_2O$	59.5	74.5	147	159
$CaBr_2 \cdot 6H_2O$	125	143	295	
CaI_2	64	67.6	78	81
$CaCO_3$		≈0.001		
$AgNO_3$	122	216	585	733
Ag(acetate)	0.73	1.05	2.59	
AgCl	0.00007 (5°C)	0.0002 (25°C)		

[7]Adapted with permission from Speight, J. G., Ed. *Lange's Handbook of Chemistry*, 16th ed.; McGraw-Hill: New York, 2005, except $CaCO_3$ data calculated from K_{sp}.

see in Chapter 8, the total concentration of an ion in solution is also very sensitive to the presence of acidic, basic, and complexing species.

2.4.1 Strong Electrolytes

Aqueous *electrolytes* are soluble chemical compounds that make a solution electrically conductive when they are dissolved in water. Electrolytes can be either organic or inorganic compounds. In either case, they form ions in solution by dissociating or reacting with water.[8] A *strong electrolyte* is an ionic compound that dissociates completely or nearly completely when added to water. A *weak electrolyte* is a compound that remains mostly neutral in solution, but some fraction will exist in ionic forms. The actual conductivity of a solution will depend on both the nature and the concentration of added electrolytes.

As Table 2.12 showed, ionic salts can have a wide range of solubilities in water. You might have noticed that several ions are common in salts of high solubility, for example, Na^+, K^+, Cl^-, and NO_3^-. These ions are common in strong electrolytes because they tend to remain fully dissociated from other ions when dissolved in water. An equivalent description is that they do not react with water to any significant extent. Table 2.13 lists the most common ions with this characteristic. On evaporating solutions, these ions form solid salt compounds with each other and also with ions that are not strong electrolytes. With the exception of the insoluble examples listed in Table 2.11, you can expect a salt containing at least one of these ions to be soluble in water.

It is useful to memorize the ions in Table 2.12 because they identify the most common strong acids, strong bases, and neutral salts.[9] That is, one of the listed cations compounded with hydroxide ion will be a strong base and one of the listed anions coupled with hydrogen ion will be a strong acid. Strong acids, HCl, HNO_3, etc., and strong bases, NaOH, KOH, etc., dissociate completely in water and are strong electrolytes (see Example 2.2). Ca^{2+} (calcium ion) is included in Table 2.13, but it is a borderline case. It forms a small amount of $Ca(OH)^+(aq)$ and also precipitates as $Ca(OH)_2(s)$ at high Ca^{2+} concentration. SO_4^{2-} (sulfate ion) is also a bit of a borderline case as HSO_4^- is a relatively strong weak acid.

The ions of strong electrolytes are often called *spectator ions*, which refers to ions that do not participate directly in a reaction. The ions of strong electrolytes are spectator ions in acid–base equilibria but not in many other types of reactions. As an example, Cl^- does not react with water, so it is a spectator ion in acid–base reactions. However, it forms a precipitate in the presence of Ag^+ and it can undergo redox chemistry to form Cl_2. In these reactions, it is a

---

[8]Nonaqueous electrolytes also exist, but we do not consider them here.

[9]It is not too difficult to remember these ions if you group them in the periodic table. The cations occupy the first two columns. The halide ions are grouped together excluding F^-, and the other three anions are common oxyanions.

TABLE 2.13   Counterions of Strong Bases and Strong Acids

Cations		Anions	
Li^+	Lithium ion	Cl^-	Chloride ion
Na^+	Sodium ion	Br^-	Bromide ion
K^+	Potassium ion	I^-	Iodide ion
Ca^{2+}	Calcium ion	NO_3^-	Nitrate ion
Sr^{2+}	Strontium ion	ClO_4^-	Perchlorate ion
Ba^{2+}	Barium ion	SO_4^{2-}	Sulfate ion

reactant and definitely not a spectator ion. Do not let a preconceived notion about a certain species keep you from recognizing other types of reactions that could occur.

Example 2.2 Solubility.   Predict which of the following solids will dissolve in water: ammonium nitrate, barium sulfate, potassium chloride, sodium hydroxide, and paraffin wax.

Looking at the solubility rules for the salts, all of the salts listed are soluble except for barium sulfate. Paraffin wax is not a salt but a long-chain alkane that is solid at room temperature. It is insoluble in water because of the large difference in polarity between water and an alkane. To summarize,

```
NH4NO3          Soluble
BaSO4           Insoluble
KCl             Soluble
NaOH            Soluble
paraffin wax    Insoluble
```

2.5   EXTRACTION AND PARTITIONING THEORY

Extraction is the transfer of a target analyte from one phase to another. The target analyte might be one specific compound or a class of similar compounds. The starting phase might be the original sample matrix or a solution in which the sample has been dissolved or digested. Table 2.14 lists common purification steps for specific types of sample matrices and analytes. These steps can serve several purposes: to transfer the target analyte to a phase that is compatible with the analytical method, to isolate the analyte from interferences, and to concentrate the analyte. Concentrating the analyte has a direct effect on the sensitivity of the subsequent measurement. If a suitable solvent is not available to extract all of the analyte, the test portion must be digested using thermal and chemical means. The most common digestion methods are for the elemental analysis of metals in solid samples, where the sample matrix can be destroyed without destroying the elemental analytes.

TABLE 2.14 Methods of Analyte Isolation

Sample Matrix	Target Analyte	Method
Air	Particulate matter	Filtration, impaction, or impingement
	Soluble compounds	Trap in solution bubbler
	Volatile compounds	Adsorption on a sorbent
Liquid, aqueous	Metal ion	Precipitation and filtration
	Organic compounds	Liquid–liquid or solid-phase extraction
Liquid, organic	Organic compounds	Liquid–liquid or solid-phase extraction
Solid, inorganic	Metal ion	Dissolution or digestion
Solid, organic	Metal ion	Ashing, acid digestion, or solvent extraction
	Organic compounds	Solvent extraction

The methods to perform extractions vary depending on the sample matrix and nature of the analyte. Air samples are forced through a solution or over a sorbent, such as activated carbon. Gas sorbents are often contained in or coated on the walls of a tube, which is called a denuder. Analytes are recovered from a sorbent by heating or using a solvent. Filtration by impaction and impingement are the trapping onto a solid or liquid surface, respectively, using mechanical force. In remediation contexts, these processes are called *scrubbing* or *stripping*. Air sampling is a common task to monitor hazardous vapors in industrial workplaces and in environmental studies.

For solid samples, it is often possible to extract analytes from a test portion without completely dissolving the material. Doing so is especially important for organic analytes to avoid decomposition of molecules by overly harsh treatment. Common extraction solvents are hot water (as in the extraction of ground coffee in Figure 2.4), organic solvents, and supercritical fluid (SCF) CO_2. After filtering to remove sample matrix, these procedures isolate analytes in a solution for further cleanup and analysis. As an example, boiling tea leaves in water extracts the tannins, theobromine, and caffeine (the good stuff) out of the leaves and into the water. These organic analytes are "cleaned-up" by transferring them to a nonpolar organic solvent to eliminate salts that interfere in the subsequent measurement.

Processing liquid samples is our easiest case to handle, and the earlier description of direct analysis listed several examples where measurements are made with little to no sample preparation. Even with a liquid sample matrix, it might be necessary to perform sample preparation steps. Filtering obviously removes particulate matter for separate analysis or to protect instruments. Extraction procedures can cleanup samples to remove interferences, concentrate the analyte(s), and transfer analytes to a different solvent that is compatible with the measurement method. For all types of matrices, the fraction of analyte recovered from the sample matrix after sample preparation must be determined in validation experiments.

Collection, digestion, and extraction procedures are sample and analyte specific, and there are a huge number of published procedures. Rather than describe many examples, we will focus here on liquid–liquid and solid-phase extractions. This focus will provide the basis for further discussion of chromatographic separations. In general, differences in solubility allow target analytes to be separated from a sample matrix or from other classes of compounds. In both liquid–liquid and solid–liquid extractions, we say that the solute (analyte) *partitions* between the two immiscible phases. *Partitioning* is an equilibrium condition in which the solute is distributed or partitioned between two phases depending on its solubility in each phase. As a thermodynamic process, the relative amounts that partition between the two phases can be described with an equilibrium constant. We illustrate the quantitative description of partitioning using liquid–liquid extractions in the following subsection. The term *partitioning* is also used to refer to specific forms of chromatography (normal-phase and reversed-phase). Use the context of a description to determine if the term is being used for the general distribution of a solute between two phases or a more specific process.

2.5.1 Liquid–Liquid Extractions

Liquid–liquid extraction uses two immiscible solvents to transfer one class of solutes from one solvent to the other. The solvent to which we are transferring solutes is called the extracting solvent or the *extractant*. In some cases, another species, such as a complexing agent that binds a metal ion, is the actual extractant and the solvent is called a *diluent*.

The purpose of laboratory extractions can be to transfer the analyte to the extractant solvent, or to transfer unwanted species to the extractant solvent. For example, after steam distilling plant matter to obtain organic natural products, the aqueous distillate is extracted with a nonpolar organic solvent to transfer the organic compounds to the organic solvent. This process can concentrate the organic compounds in a new solvent and separate them from inorganic cations and anions. In an example of synthesizing a new organic compound, the organic reaction mixture will be extracted with water or brine to remove salts and unwanted compounds. The desired reaction product remains in the nonpolar organic solvent for further work-up steps. Since ionic species will strongly transfer to the aqueous phase, the pH of the extractant water or brine can be adjusted to enhance transfer of undesired species that are weak acids or weak bases.

For approximately equal volumes of the two immiscible liquids, a given solute will partition to a greater extent into the solvent in which it is more soluble. Most organic compounds will partition into an organic solvent and ionic and very polar solutes will partition primarily into the aqueous phase. Figure 2.5 shows a schematic of a liquid–liquid extraction being performed in a separatory funnel. The separatory funnel is shaken gently to mix the immiscible solvents and speed up the equilibration of solutes between the two phases. The flask is placed in the stand as shown in the figure to allow the two solvents to separate fully. The two phases are separated by opening the stopcock to drain the aqueous phase into a new container. A second container is placed under the separatory funnel and the stopcock is reopened to collect the organic phase. (Note that some organic solvents, such as dichloromethane, are more dense than water and the two solvents will be reversed from what is shown in Figure 2.5.)

If a given solute is extracted very efficiently, the amount of extractant phase can be made small to concentrate the analyte in the new phase. If an extraction process is not efficient for a given solute and extracting solvent, the process can be repeated multiple times, which is referred to as *sequential extractions*. The number of sequential extractions that are needed for a given solute depends on the efficiency and the desired recovery of solute. A mathematical expression to calculate fractional distribution of a solute in two solvents is given below.

2.5.2 Conventions for Writing Equilibria and Equilibrium Constants

Before discussing partitioning, it is useful to refresh our memories of the general concept of equilibrium. The coexistence of water and ice at 0°C is an example of a physical equilibrium. We briefly discussed solutions that were saturated, in which a salt has been added beyond its solubility so that some solid exists in contact with the solution. The concentration of the ions in a saturated solution is an equilibrium condition. Chemical equilibrium is a thermodynamically

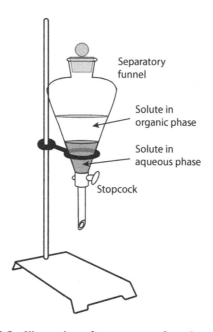

Figure 2.5 Illustration of a separatory funnel extraction.

stable condition in which two or more related species or phases coexist. All chemical species exist in equilibrium with other forms, although in many cases the equilibrium is so extreme in one direction or the other that we neglect the minor species. Strong electrolytes are a case in point; NaCl(aq) might exist in aqueous solution, but the concentration is so small to be immeasurable and we consider only Na^+ and Cl^- being in solution. We refer to these reactions as having *gone to completion*, and we use them to advantage in sample preparation and the classical methods that are discussed in Chapter 3.

The distribution of a solute between two phases, i.e., partitioning, is also an equilibrium condition. As above for extreme cases, the distribution of Na^+ and Cl^- between water and a nonpolar organic solvent is so extreme that we assume these strong electrolyte ions totally remain in the aqueous phase. For solutes, S, that partition to measurable amounts, we write the partitioning or distribution of S between two immiscible solvents using the following chemical equilibrium:

$$S(\text{phase1}) \rightleftharpoons S(\text{phase2})$$

For the typical case in which phase 1 is an aqueous solution (aq) and phase 2 is an organic solvent (org), we write

$$S(\text{aq}) \rightleftharpoons S(\text{org})$$

The quantitative measure of the extent to which a solute partitions between two phases is given by an equilibrium constant K or K'. K is the thermodynamic equilibrium constant and K' is a formal equilibrium constant, which we use when working with molar concentrations. Chapter 5 explains the difference in detail, but for these introductory examples, we will use K' and concentration units. There are several general rules for writing K expressions for any type of reaction. Example 2.3 illustrates the rules.

Example 2.3 K' Expressions. Rules for writing K' expressions:

1. The formal equilibrium constant is always written with the concentrations of the products over the reactants:

$$\text{Reactants} \rightleftharpoons \text{Products}$$

$$K' = \frac{[\text{products}]}{[\text{reactants}]}$$

2. The stoichiometric coefficients of the reaction are exponents in the K' expression, for example,

$$2\,\text{Reactants} \rightleftharpoons \text{Products}$$

$$K' = \frac{[\text{products}]}{[\text{reactants}]^2}$$

3. We do not include solvent or pure solid or liquid substances in a K' expression. For the partitioning equilibrium,

$$S(aq) \rightleftharpoons S(org)$$

The equilibrium constant does not contain either water or the organic solvent and is simply

$$K'_D = \frac{[S(org)]}{[S(aq)]} \tag{2.1}$$

where I have added subscript D to indicate that this constant is for a particular type of equilibrium.

2.5.3 Partition Theory

The equilibrium constant expression for partitioning, K'_D, is called the *distribution constant* or *partition ratio*:[10]

$$K'_D = \frac{[S(org)]}{[S(aq)]} \tag{2.2}$$

Other names for this quantity are quite common in manufacturer information and the scientific literature, including partition coefficient and distribution coefficient. The usefulness of this equilibrium constant is that it allows us to predict the relative distribution of a solute between two phases.

K'_D for a liquid–liquid extraction contains [S(org)] and [S(aq)], the equilibrium molar concentrations of solute S in the organic and aqueous phases, respectively. In addition to the partitioning of a solute between two immiscible liquid phases, this terminology is used for other types of partitioning such as the distribution of a metal ion between the aqueous phase and an immobilized state when adsorbed on a solid surface. Such chemical systems can be quite variable because the partitioning will be dependent on pH, metal complexing agents, and the chemical nature of soil particles.[11] In all cases, the K'_D values are valid only for systems that are at equilibrium. For liquid–liquid extractions in the laboratory, this means allowing sufficient time to reach the equilibrium condition as determined by validation experiments.

---

[10]I have added the prime to be consistent with all other equilibrium constants given in this book. The approved IUPAC nomenclature uses only K_D.
[11]Understanding Variation in Partition Coefficient, K_d, Values, EPA 402-R-99-004A, https://www.epa.gov/radiation/understanding-variation-partition-coefficient-kd-values; accessed August 2022.

The values of K'_D can be estimated from the solubility of a solute in the two immiscible solvents. Let us use caffeine as an example; 2.2 g will dissolve in 100 ml of cold water and 14 g will dissolve in 100 ml of dichloromethane. The K'_D is approximately

$$K'_D = \frac{14\,\text{g}/100\,\text{ml}\ \text{CH}_2\text{Cl}_2}{2.2\,\text{g}/100\,\text{ml}\ \text{H}_2\text{O}} = 6.4$$

Actual measurement of distribution constants will vary from these estimates because the solubilities can be very sensitive to temperature. Partitioning can also depend on concentrations of salts and other substances that might complex with the solute to be extracted. For species that have acidic or basic groups, the pH of the aqueous phase should be buffered so that the solute is neutral to be extracted into the organic phase.

The fractional amount of one form of a chemical species in equilibrium with all other possible forms is called an *alpha fraction*, α. We will use these ratios extensively in later chapters. The fraction of solute S remaining in phase 1, $\alpha_{S(aq)}$, is

$$\alpha_{S(aq)} = \frac{\text{moles S(aq)}}{\text{moles S(total)}} = \frac{\text{moles S(aq)}}{\text{moles S(aq)} + \text{moles S(org)}} \qquad (2.3)$$

On substituting concentration × volume for moles we get

$$\alpha_{S(aq)} = \frac{[S(aq)]V_{aq}}{[S(aq)]V_{aq} + [S(org)]V_{org}} \qquad (2.4)$$

This expression can be rearranged to obtain

$$\alpha_{S(aq)} = \frac{1}{1 + \dfrac{[S(org)]V_{org}}{[S(aq)]V_{aq}}} \qquad (2.5)$$

Finally, substituting K'_D for the ratio of equilibrium concentrations gives

$$\alpha_{S(aq)} = \frac{1}{1 + K'_D\left(\dfrac{V_{org}}{V_{aq}}\right)} \qquad (2.6)$$

Note that the fraction of S remaining in phase 1 depends on both the partition ratio and the ratio of the volumes of the two phases. Extraction efficiency, meaning the transfer of the maximum solute into as little extracting solvent as possible, is much better by performing several extractions of smaller portions of the extracting phase rather than doing the equivalent total volume just once. Examples 2.4, 2.5, 2.6, and 2.7 illustrate the difference and typical application of this calculation. When doing multiple extractions, Equation (2.6) is modified to the following, where n is the number of sequential extractions.

$$\alpha_{S(aq)} = \left[\frac{1}{1 + K_D' \dfrac{V_{org}}{V_{aq}}} \right]^n \tag{2.7}$$

In this expression, V_{org} and V_{aq} are the volumes for each individual extraction. Some texts write this expression in other forms, which are useful for determining certain quantities such as the total volume of the organic phase used in the extraction. In those cases, V_{org} or other factors in Equation (2.7) might be defined differently. Be careful to determine how terms are defined when comparing your calculations to results in other books or reference sources.

For a specific example, let us look at the partitioning of solutes between water and 1-octanol, $C_8H_{17}OH$, which are immiscible. This equilibrium constant is used as a predictor of the extent to which chemicals will partition from solution to naturally occurring organic matter in the environment and the extent to which drugs will cross lipid bilayers. Table 2.15 lists equilibrium constants for selected solutes.[12] These constants are usually tabulated as log P, but I have also converted them to K_D'. The small alcohols and acids, which are miscible with water, partition somewhat equally between water and octanol. There is a clear difference going down the table at 1-hexanol, where the nonpolar character of the alkyl chain becomes more important than the polar functional group. As you might have predicted, the linear alkanes have much higher values of K_D' due to their very nonpolar character. Oleic acid is a long-chain fatty acid, and as the K_D' value shows, the long alkyl chain causes it to partition very strongly into the organic phase even though it has a polar acid group.

Example 2.4 Extraction Calculation.   Predict the fraction of solute left in the aqueous phase when methanol, benzene, and octane are extracted from 100 ml of water using 30 ml of octanol as the extracting solvent.

Since $n = 1$, this calculation simplifies to

$$\alpha_{S(aq)} = \frac{1}{1 + K_D' \dfrac{V_{org}}{V_{aq}}}$$

[12]Adapted with permission from Sangster, J. "Octanol-Water Partition Coefficients of Simple Organic Compounds," *J. Phys. Chem. Ref. Data* **1989**, *18*, 1111–1227. Copyright 1989 American Institute of Physics.

TABLE 2.15   Octanol–Water Partition Coefficients

Compound	$\log P$	K'_D
Methanol	−0.74	0.18
Ethanol	−0.30	0.50
Acetone	−0.24	0.58
Acetic acid	−0.10	0.68
1-Propanol	0.25	1.8
1-Hexanol	2.03	1.1×10^2
Benzene	2.13	1.3×10^2
1-Octanol	3.07	1.2×10^3
Cyclohexane	3.44	2.8×10^3
Hexane	4.00	1.0×10^4
Octane	5.15	1.4×10^5
Oleic acid, $C_{17}H_{33}COOH$	7.64	4.4×10^7

Entering the solvent volumes and the value of K'_D from Table 2.15 for methanol (meOH) gives

$$\alpha_{meOH(aq)} = \frac{1}{1 + 0.18\left(\dfrac{30\,ml}{100\,ml}\right)} = 0.95$$

The other solutes are calculated in the same manner, with the results being

$$\alpha_{meOH(aq)} = 0.95 \text{ or } 95\%$$

$$\alpha_{benzene(aq)} = 0.025 \text{ or } 2.5\%$$

$$\alpha_{octane(aq)} = 2.4 \times 10^{-5} \text{ or } 0.002\%$$

Keep in mind that this calculation gives a fractional result of the solute remaining in water. Obviously, the concentration of octane in water is quite low and the amount in the extractant will likewise be a small amount. The calculation is useful to determine if a certain extracting solvent or certain solvent volumes will achieve a quantitative extraction. From these results, octanol provides a quantitative extraction of octane from water. For benzene, one extraction achieves 97.5% recovery. Extracting methanol from water is a challenge and is generally not attempted. Methanol is isolated from aqueous solution by distillation for large amounts or by adsorption onto a sorbent if present at low concentrations.

You-Try-It 2.B
The single-extraction worksheet in you-try-it-02.xlsx contains a table with octanol–water log P values for various solutes. Determine the fraction of solute remaining in the aqueous phase and the fraction transferred to the organic phase for each of the solutes.

Example 2.5 Multiple Extractions.   Predict the fraction of solute left in water for the extraction of methanol, benzene, and octane from 100 ml of water when using three sequential 10-ml portions of octanol.
    Again using Equation (2.7),

$$\alpha_{S(aq)} = \left[\frac{1}{1 + K'_D \dfrac{V_{org}}{V_{aq}}} \right]^n$$

Entering $n = 3$, the solvent volumes, and the value of K'_D from Table 2.15 for methanol gives

$$\alpha_{meOH(aq)} = \left[\frac{1}{1 + 0.18\left(\dfrac{10\,ml}{100\,ml}\right)} \right]^3 = 0.95$$

The other solutes are calculated in the same manner, with the results being

$$\alpha_{meOH(aq)} = 0.95 \text{ or } 95\%$$

$$\alpha_{benzene(aq)} = 3.6 \times 10^{-4} \text{ or } 0.04\%$$

$$\alpha_{octane(aq)} = 3.6 \times 10^{-13} \text{ or } 4 \times 10^{-11}\%$$

The extraction of methanol into octanol is so poor that changing the procedure makes no real difference. Compared to a single extraction, the sequential extraction provides a significant improvement for benzene. Either procedure provides a quantitative extraction for octane.

Example 2.6 Extractant Volume.   If we require 99.9% of the benzene to be extracted from 100 ml of water, what amount of octanol is necessary if we do only a single extraction?

From Table 2.15, K'_D for benzene partitioning between octanol and water is listed as 130. Since $n = 1$ and we require $\alpha_{benzene(aq)} = 0.001$, this calculation simplifies to

$$0.001 = \frac{1}{1 + 130\left(\dfrac{V_{org}}{100 \text{ ml}}\right)}$$

Rearranging and solving gives

$$V_{org} = 770 \text{ ml}$$

This extraction requires a large amount of the organic phase. A better approach will be to pursue sequential extractions to reduce the total volume of the octanol.

Example 2.7 Extractant Volume.   Repeat the previous extraction using 10-ml portions of octanol. How many sequential extractions will be necessary to achieve removal of 99.9% of the benzene from 100 ml of water?

In this case, n is unknown and our volumes are $V_{org} = 10$ ml and $V_{org} = 100$ ml. Setting up the expression with $\alpha_{benzene(aq)} = 0.001$ gives us

$$0.001 = \left[\frac{1}{1 + 130\left(\dfrac{10 \text{ ml}}{100 \text{ ml}}\right)} \right]^n$$

The easiest way to solve for n is to find alpha for $n = 1, 2, 3, \ldots$ which is easily done in a spreadsheet and copying the formula into multiple rows.

n	alpha	% Extracted
1	0.071	93
2	0.0051	99.5
3	0.00036	99.96
4	0.000026	99.997

The % extracted value is the fraction removed from water and is found from $(1 - \alpha) \times 100\%$. The results show that three sequential 10-ml extractions with octanol will remove more than 99.9% of the benzene from the water. The total amount of extractant in this procedure is 30 ml, much smaller than the amount required when using only one extraction step.

You-Try-It 2.C
The multiple-extractions worksheet in you-try-it-02.xlsx contains a copy of Table 2 from worksheet 2.B. The exercise is to determine the number of sequential extractions that are necessary to extract $> 95\%$ and $> 99\%$ of each solute from the aqueous phase to the organic phase.

You-Try-It 2.D
The extraction-volume worksheet in you-try-it-02.xlsx contains an example of the table of results for different conditions in extracting acetic acid from water with diethyl ether. Determine optimal conditions for this extraction using 1-octanol in place of the ether.

2.5.4 Ion Pairs

In general, ionic species do not partition from an aqueous solution into a non-polar solvent. For weak acids and weak bases, we can adjust pH of the aqueous phase to make these species neutral and gain some selectivity in what can transfer to a nonpolar solvent. Some solution species have a permanent charge independent of pH. Common examples are quaternary ammonium ions, NR_4^+, where the R groups could be identical or all different. Salts of these compounds serve as disinfecting cleaners and antimicrobial compounds and also form important biomolecules such as choline. An ionic species can cross the interphase boundary into an organic solvent if it associates strongly with a counterion to form an *ion pair*. The ion pair then has an overall neutral charge. If the ions in an ion pair have sufficient nonpolar character, the ion pair can have a high solubility in an organic solvent. The ions of strong electrolytes, such as NaCl, will not associate in an ion pair because the Na^+ and Cl^- remain fully hydrated in aqueous solution. For quaternary ammonium ions, suitable ion pair counterions might be the ionized forms of benzoic, citric, or oleic acid, which have significant nonpolar character. Using an appropriate counterion to form an ion pair provides a method to convert a charged molecule or biomolecule to a neutral form and apply efficient separation methods in organic solvents.

An ionic species can also cross the interphase boundary to an ion pair in the organic phase if another ion exchanges into the aqueous phase. To illustrate this extraction concept, consider the problem of removing the radioactive Cs^+ that is present in aqueous solution in waste tanks at legacy facilities that processed nuclear materials. For long-term disposal, the Cs^+ must be concentrated and then solidified in a stable matrix. Simply evaporating the waste solution would be incredibly energy intensive and leave the Cs^+ in a complex solid or slurry containing many other metals and waste products. The scheme in Figure 2.6 shows a Cs^+–crown ether complex that is soluble in an organic solvent. The

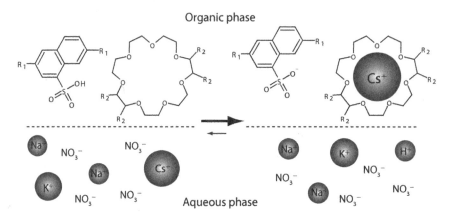

Figure 2.6   A crown ether to extract Cs^+ from aqueous solution.

dashed line represents the boundary between the aqueous and organic phases. The left side of the figure shows the Cs^+ in the aqueous phase and the right side of the figure shows the Cs^+ complexed with the crown ether in the organic phase. Using this crown-ether complexing agent can strip Cs^+ out of the aqueous solution and into a smaller volume of an organic solvent. The Cs^+-containing organic solution can be concentrated further with much less energy and it contains fewer components that could interfere in solidifying the Cs^+.

Note that the sulfonic acid, RSO_3H, that is present in the organic phase donates a proton into the aqueous phase. This proton exchange provides charge balance for the extraction process and produces a counter-anion to form the ion pair with the Cs^+–crown ether complex. The R_1 group is $C_{12}H_{25}$, and the R_2 group is a fused phenyl ring with other substituents. These large organic side chains help to make the ion pair very soluble in the organic solvent. A crown ether was developed for this extraction because the size of the central cavity is tailored to bind Cs^+ with very high selectivity. The relative size of the other cations in Figure 2.6 differs from Cs^+, and they are not extracted. The large forward arrow that indicates the equilibrium lies to the right is for Cs^+ only. Leaving the nonradioactive cations in the aqueous phase increases the overall cost efficiency of this process.[13]

2.6   INTRODUCTION TO STATIONARY PHASES

Quantitative liquid–liquid extractions can be difficult to achieve when working with small volumes of test portions, which is often the case in analytical applications. The analyte can be lost in the glassware and it is difficult to separate the two liquid phases without some loss. A more efficient and more convenient separation method uses an extracting phase that is immobilized on an inert

[13]Chemical Separations Group, Oak Ridge National Laboratory; https://www.ornl.gov.

solid particle. These particles are contained in a plastic or glass column with a stopcock at the bottom. Opening and closing the stopcock allows or pauses the flow of a liquid over the particles. The solid material is called the *stationary phase* and the liquid phase is called the *mobile phase*. The stationary phase remains the same in a given column and different materials are needed for different classes of analytes. The mobile phase is variable and might be pure solvent for rinsing, the sample solution to transfer analyte to the stationary phase, or a different solvent or solvent composition to elute analyte from the column. *Elution* is the term for the step or process in which the solute in the mobile phase exits the column for collection. Analogously, the mobile phase is often called the *eluent*.

Figure 2.7 shows the general arrangement for using a stationary phase. The columns are glass or plastic and the stationary phase is held in place by a porous frit or plug of glass wool. The scale of the column and the amount of stationary phase depend on the application. A lab-scale purification of gram-size quantities might use a column that is 10 cm to 1 m in length. A large purification column will be reused for multiple separations. A cleanup-step purification for an ultratrace analysis might use a column that holds only a few ml of solvent and is discarded after use to prevent contamination of subsequent samples.

There are two practical implementations for using an immobilized stationary phase, solid-phase extraction (SPE) and column chromatography. SPE and column chromatography often use the same stationary-phase materials, and the chemical interactions with these materials are the same in both formats. They differ in that SPE is used in a stepwise manner whereas chromatography is a continuously flowing process. The stationary and mobile phases in SPE are chosen to completely retain one class of analytes on the stationary phase and allow all other solutes to wash through the column. This one class of retained

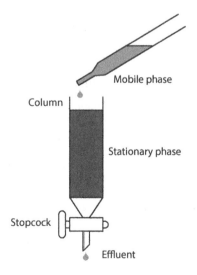

Figure 2.7   Schematic of a solid stationary-phase in a column.

solutes is then collected by changing the mobile phase to elute the analytes in a small amount of eluting solvent. The process "cleans up" and concentrates the analytes in a new solvent. In column chromatography, the conditions are adjusted so that solutes are retained on the stationary phase to varying degrees. The optimal case is that each solute will have a different distribution between the stationary and mobile phases, and different solutes will travel through the column at different rates. Each individual solute can then be collected in a separate *fraction* as they elute from the column at different times. Here fraction refers to some amount of the eluent as it exits the column. Robotic fraction collectors can automate this collection process.

Note that in these two processes, the term *retention* has slightly different meanings. In both cases, it refers to the solutes interacting with and being held on the stationary phase. In SPE, retention is complete for the retained class of solutes. This allows nondesirable species to be washed away. The analytes are then eluted by changing the mobile phase. In column chromatography, the retention varies for each solute. We call the time for a given solute to elute from a column its retention time. Retention time has no meaning in SPE.

Before discussing the details of SPE and column chromatography, the following discussion gives an overview of the variety of stationary phases that are available. Table 2.16 lists the most common classes of stationary phases and the basis of their interaction with solutes. The wide range of interactions can provide a stationary phase for purifying almost any type of analyte from a sample matrix. Most of these stationary phases can be used in both SPE and column chromatography applications. The two exceptions are size-exclusion and affinity chromatography. Size-exclusion, also called gel-filtration or gel-permeation, chromatography is almost always used in a continuous-flowing column format. Although called affinity chromatography, the *lock-and-key* mechanism of isolating one specific biomolecule is really an SPE process. These two types of separations are discussed more fully in later chapters.

These different types of stationary phases are discussed in more detail in the following two sections and in the last chapter on analytical separations.

TABLE 2.16 Types of Stationary Phases for SPE and Chromatography

Name (Interaction)	Stationary Phase	Analytes
Adsorption (surface attraction)	Solids (usually alumina or silica)	Organic molecules, isomers
Ion-exchange (electrostatic)	Charged group on a polymer	Inorganic and organic ions
Normal-phase (relative polarity)	Polar group on an inert support	Organic molecules
Reversed-phase (relative polarity)	Nonpolar group on an inert support	Organic molecules
Size-exclusion (size)	Inert porous polymer	Polymers and biomolecules
Affinity (lock and key)	Biomolecules on an inert support	Biomolecules

Column chromatography in this chapter refers to gravity or flash columns that are used in sample preparation. The last chapter of the text discusses thin-layer chromatography (TLC) and high-performance liquid chromatography (HPLC), which use the same types of stationary phases. These methods are used to separate, identify, and quantitate analytes. HPLC instruments use a pump to force the mobile phase through more efficient columns. By adding a detector at the column exit, HPLC achieves separation and measurement in one step. The solid materials used for adsorption-based separation, e.g., high-purity silica, SiO_2, or alumina, Al_2O_3, particles, serve as both the active surface and the column packing. The other stationary phases consist of the stationary phase bonded to inert support particles. In either case, the size of the packing material will be different for gravity versus pressurized columns.

Ion-exchange stationary phases are polymers that have a charged functional group.[14] They retain atomic or molecular ions in the mobile phase based on the relative electrostatic interaction with the charged groups on the polymeric stationary phase. The stationary phase polymer will have the opposite charge compared to the analytes to be retained. A cation-exchange polymer immobilizes a negative charge to attract cations from the mobile phase. An anion-exchange polymer contains a positive charge to attract anions from the mobile phase. The polymer backbone is stable in organic solvent for cases when a water/organic mobile phase provides a more efficient separation.

Table 2.17 categorizes adsorption and partition stationary phases in two general classes, reversed-phase and normal-phase.[15] Some manufacturer's literature combine adsorption with normal stationary phases, since they both use nonpolar organic mobile phases. I've included silica and alumina adsorption phases in the normal-phase list. Normal-phase partition chromatography uses a polar stationary phase and a nonpolar organic solvent as the mobile phase. Reversed-phase partition chromatography uses a relatively nonpolar stationary phase and a polar mobile phase. These terms are historical because the normal-phase using organic solvents was developed first.

TABLE 2.17 Adsorption and Partition Stationary Phases

	Reversed Phase	Normal Phase
Mobile phase	Aqueous-based solvent	Nonpolar organic solvent
Analyte characteristics	Nonpolar to moderate polarity	Moderate to polar polarity
Stationary-phase	Phenyl	Silica
functional groups	Cyano Octyl (C8) Octadecyl (C18)	Alumina Diol Amine Cyano

[14]Manufacturer literature often refers to these polymers as ion-exchange resins.
[15]Adapted from Sigma-Aldrich, SPE Tubes–Phase Selection Guide.

Commercial reversed-phase stationary phases are more technically advanced than other choices and they generally provide the best separations. However, they cannot separate solutes that are not soluble in aqueous-based solvents. Solutes that are only soluble in nonpolar organic solvents require normal-phase adsorption or partitioning stationary phases for separation. To clarify a nonintuitive point for normal-phase separations, the analyte characteristics in Table 2.17 refers to the polarity of functional groups in the different nonpolar solute molecules. Some stationary phases are usable in both reversed-phase and normal-phase separations, with the difference being the most suitable mobile phase for a given class of solutes.

2.7 SOLID-PHASE EXTRACTION (SPE)

Like other sample preparation steps, SPE is a tool to isolate one or more analytes from a sample matrix in order to quantitate those analytes. An SPE extraction can serve several purposes, including one or a combination of the following:

- separate analytes from interferences (sample cleanup)
- concentrate analytes in a smaller volume of solvent
- transfer analytes to a new solvent

Interferences might be sample matrix components that introduce a measurement bias, are not compatible with subsequent steps in a method, or degrade instrument performance over time. Concentrating analytes during sample preparation, i.e., before measurement, is typically called *preconcentrating*. Analytes are preconcentrated in SPE by passing a large amount of test solution through an SPE cartridge. All retained analytes are then collected in a small volume of the eluting mobile phase. Preconcentration factors can range from 2 to over 1000, although somewhere between 5 and 50 is more common. A solvent change occurs in SPE because the eluting solvent is always different than the test solution solvent. Solvent exchange is sometimes the main purpose of performing an SPE step. A common example is to exchange analytes into a solvent that is compatible with a gas-chromatographic or HPLC instrumental method.

The most common SPE extractions separate ions from neutral solutes or polar from nonpolar compounds. The wide variety of interactions listed in Table 2.17 provides numerous options for sample cleanup. In general, we can consider the process simply as sample components *sticking* or *not sticking* to the stationary phase. Cleanup of a test portion is usually performed with a stationary phase that retains the analytes, and this approach is described in detail below. It is also possible to use a stationary phase that retains interferences for removal from the test solution. Analytes simply pass through the stationary phase, no preconcentration or solvent exchange occurs, and the cartridge is discarded.

A complete commercial SPE column, or *cartridge*, consists of a plastic body, the stationary phase, porous frits to retain the stationary phase, and a stopcock (refer to Figure 2.7). The stationary phase will generally occupy about ≈20% of the cartridge to leave the tube mostly open for loading sample solution. The solvent passes through the column by gravity or, to increase the flow rate, by suction. Figure 2.8 shows a schematic of an SPE cartridge on a port of a vacuum manifold. The suction provided by the vacuum increases the speed of the solvent flow. The multiple ports allow an analyst to simultaneously process multiple test solutions through separate cartridges. Containers to collect the wash solvent and the eluent are placed in the vacuum manifold below each cartridge (not shown in Figure 2.8).

Stationary phase is also available supported on disks that are placed in a filter apparatus for faster processing or coated on the surface of well plates for small sample volumes. Solid-phase microextraction (SPME) uses stationary phase coated on a fiber for preconcentrating trace and ultratrace amounts of analytes. The fiber is housed in a syringe needle for protection. After exposure to a test solution for some amount of time, the SPME fiber can be inserted directly into an instrument to desorb analyte for measurement.

Cartridges, and other formats, come in a variety of sizes. The amount of stationary phase needed for a separation depends on the amount of solute to be retained. For the case of retaining the analyte of interest, overloading a cartridge that contains insufficient stationary phase results in loss of the analyte. The amount of a test portion at which the analyte is lost is called the *breakthrough volume*, V_B. This breakthrough volume will depend on the analyte concentration and other experimental conditions. Stationary phases typically have a maximum loading on the order of 1% of sorbent weight, and analyte weight should be kept well below the maximum. For example, if you anticipate as much as 500 µg of analytes in a test solution, you would use a cartridge containing 100 mg of the stationary phase. SPE breakthrough volume and analyte recovery are two important parameters to determine in validation experiments.

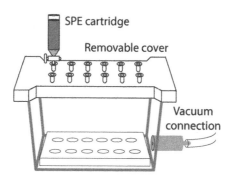

Figure 2.8 SPE cartridge on a multiport vacuum manifold.

There are five steps in a typical SPE cleanup:[16]

1. Condition
2. Equilibrate
3. Load analytes
4. Rinse
5. Elute analytes

A fresh cartridge is placed on a port of the vacuum manifold (refer to Figure 2.8), a collection vessel is positioned below it, and vacuum is applied. The stationary phase is first rinsed with a solvent to clean and condition it. Methanol is a common conditioning solvent to penetrate and wet the stationary phase. This wetting allows subsequent solutions to also penetrate for the solutes to interact with the functional group on the stationary phase. The equilibrating solvent is the same or similar as the test solution. Passing it through the column removes the conditioning solvent and equilibrates the stationary phase to the same conditions for loading analyte. For ion-exchange stationary phases, the equilibrating solvent will also have an appropriate pH for retaining the analytes on loading.

Test solution is loaded into the top of the cartridge and the stopcock is opened to pass the solution through the stationary phase. Loading can continue multiple times as long as breakthrough does not occur. If all goes according to plan, 100% of the target analytes are retained on the stationary phase. The column is then rinsed with the same or a similar solvent as the test solution to remove solutes that do not stick to the stationary phase. At this point it is common to stop adding mobile phase and allow air to pass through to dry residual solvent.

The stopcock is closed and the SPE cartridge is moved over a clean and empty collection tube. A measured amount of a different mobile phase, the eluting solvent, is added to the cartridge. The stopcock is opened to elute the analytes into the collection vessel. For partition-based separations, the mobile-phase composition is changed to a different polarity so that the analytes are more soluble in the eluent than in the stationary phase. Ions are eluted from ion-exchange columns by using a solvent with a pH that changes the charge on the analytes or the stationary phase to neutral so there is no longer an electrostatic attraction. The eluent containing the analytes are saved for further processing or analysis and the SPE cartridge is usually discarded to avoid cross-contamination in subsequent samples.

Reversed-phase SPE with C8 or C18 stationary phases finds widespread use, but I'll use those stationary phases as examples in the last chapter. Ion-exchange SPE has numerous variations, so I will discuss it in depth. As a

---

[16]Specific procedures for some commercial SPE cartridges might omit the equilibration or add additional washes.

reminder, *weak acids* and *weak bases* can be neutral or charged depending on the solution pH. Weak acids are neutral at lower pH and anions when deprotonated at higher pH. Weak bases include NH_4^+ and organic amines that are protonated and charged at low pH and neutral at higher pH. The transition between the protonated and unprotonated forms of weak acids and weak bases is given by the species pK_a. At a solution pH that equals pK_a for a given species, 50% will be protonated and 50% will be deprotonated. Chapter 5 discusses this topic in depth. For SPE applications, a species can be made charged or neutral by adjusting the solution pH.

Figure 2.9 shows a microscopic depiction of different solutes, denoted by the different shapes, after loading onto a cation-exchange SPE column. Mobile-phase molecules are not shown. The one solute that has a positive charge, represented by triangles, is attracted and retained on the stationary phase. The other neutral or negative solutes pass through the column at the flow rate of the mobile phase. After washing out the other solutes, the positive analyte ion is eluted from the column by adding a strong acid to protonate the carboxylate functionality on the stationary phase. With the stationary phase now at neutral charge, there is no attraction for the positive analyte and it is carried out of the cartridge to be collected. The counteranion of the strong acid in the mobile phase balances the charge when the analyte cation is released from the stationary phase.

The example in Figure 2.9 shows a stationary phase with a deprotonated carboxylic acid functional group, −COOH, which is a weak acid. This stationary phase is used for strong cation analytes that are not affected by pH. However, there are four types of ionic species, strong and weak cations and strong and weak anions. A different type of stationary phase is appropriate for each of these four cases. Table 2.18 lists examples of commercially available ion-exchange stationary phases. The strong and weak descriptors for these stationary phases refer to the functional group on the polymer. Recall

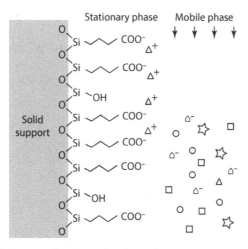

Figure 2.9 Microscopic schematic of a cation-exchange stationary phase.

TABLE 2.18   Ion exchange Stationary Phases

Stationary Phase	Functional Group	Functional Group Charge	Analytes
Strong cation exchange (SCX)	Sulfonate	(−)	Weak bases
Weak cation exchange (WCX)	Carboxylate	(−) or 0	Strong cations
Strong anion exchange (SAX)	Quaternary ammonium	(+)	Weak acids
Weak anion exchange (WAX)	Tertiary ammonium	(+) or 0	Strong anions

that a "strong" ion does not associate with a counterion in solution. (Common inorganic examples of ions that form strong electrolytes were listed in Table 2.13.) The strong cation exchange group, sulfonate, remains negative independent of pH of the mobile phase. Likewise, the strong anion exchange group, quaternary ammonium, remains positive independent of pH of the mobile phase. The strong ion exchange polymers are used for weak acid and weak base analytes. The analytes are eluted by changing the mobile phase pH to make the analytes neutral. The weak ion exchange polymers are used for analytes that remain charged independent of solution pH. These analytes are eluted by changing the mobile phase pH to make the stationary phase neutral. In Table 2.18, the zero for functional group charge indicates making the stationary phases neutral to elute the analytes. The availability of strong and weak ion exchangers provides flexibility to separate a wide range of analytes in different pH ranges.

To illustrate a validated solid-phase extraction procedure, the following excerpt is taken from the cleanup procedure for the analysis of morphine, codeine, and other opioids in biological fluids.[17] The stationary phase is a cation-exchange polymer, meaning that cations stick to the stationary phase. The opioids of interest have a nitrogen atom that is protonated in acidic pH. Retention occurs at low pH to give the analytes a positive charge. After washing, they are eluted at high pH to make them neutral.

1. Place labeled SPE cartridges in the extraction manifold.
2. Add 2 ml methanol and aspirate.
3. Add 2 ml pH 6.0 phosphate buffer and aspirate. Important! Do not permit SPE sorbent bed to dry. If necessary, add additional buffer to rewet.
4. Pour specimens into appropriate SPE columns. Aspirate slowly so that the sample takes at least 2 min to pass through the column.

[17]Opioid Quantitation and Confirmation by GCMS, Virginia Department of Forensic Science, *220-D100 Toxicology Procedures Manual*, uncontrolled copy, 2019.

5. Add 1 ml of 1 M acetic acid to each column and aspirate. Dry the columns under full vacuum/pressure for at least 5 min.
6. Add 6 ml of methanol to each column and aspirate. Dry the columns under full vacuum/pressure for at least 2 min.
7. Wipe the SPE column tips with wipes. Place labeled 10 ml conical test tubes in the manifold test tube rack. Be sure SPE column tips are in the designated conical tube.
8. Add 2 ml of 2% ammonium hydroxide in ethyl acetate to each column. Collect eluent in conical test tubes by column aspiration or gravity drain.
9. Evaporate eluates to dryness to continue analysis.

Steps 2 and 3 in this procedure clean and condition the column. The test portion solution is loaded onto the column in step 4, and the solvents in steps 5 and 6 wash away unwanted sample components. In step 8, the basic solution deprotonates the analytes to release them from the stationary phase. The collected analytes (*eluates*) are taken to dryness for additional processing and analysis. This description omits many other sample preparation steps and the use of internal standards and control samples to ensure accuracy in the method.

As another example, the drug Ritalin is sold in the form of methylphenidate hydrochloride, meaning that it is protonated and has a chloride counteranion. Figure 2.10 shows both forms.

Ritalin has a pK_a of approximately 8.8, meaning that at a pH near or less than 8.8, a significant fraction of the molecule is in solution as the protonated molecule. The pH-dependent equilibrium between the free base and the protonated forms will affect the ability to extract this molecule into an organic phase. This same pH-dependent effect allows us to use SPE efficiently for many organic molecules. The neutral form will stick to a nonpolar stationary phase, allowing salts and impurities to be washed away. Adjusting the pH to protonate or deprotonate a given molecule will then rinse it from the column to be collected.

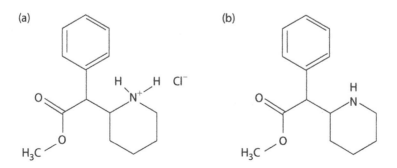

Figure 2.10   (a) Methylphenidate HCl and (b) the free-base form.

As a final example, affinity-based methods can have very high selectivity for cleanup or isolation of biological molecules. They operate by the lock-and-key concept where multiple interactions can make the binding very specific. Two common examples are antigens and antibodies and complementary strands of nucleic acids. Select proteins can be tagged by genetic engineering to incorporate an amino acid sequence that is retained on a stationary phase. One protocol is to tag proteins with polyhistidine, which can make multiple bonds with a metal ion. A metal such as Ni^{2+} is immobilized on a stationary phase to retain the tagged protein and untagged proteins are washed from the column. The tagged protein is then eluted by adding mobile phase containing imidazole, which competes for the Ni^{2+} binding sites to release the protein. We will revisit this type of affinity interaction in immunoassays after discussing complexation in Chapter 7.

You-Try-It 2.E
The percent-recovery worksheet in you-try-it-02.xlsx contains a sample preparation procedure for norepinephrine and epinephrine (stress hormones) in blood plasma. An internal standard is added during an SPE step to monitor the recovery in this purification step and to calibrate the analytical measurement. The exercise is to calculate the percentage recovery of the internal standard and the concentrations of the two analytes.

2.8 COLUMN CHROMATOGRAPHY

Analogous to solid-phase extraction, chromatography is a technique to separate solutes that are dissolved in a solution. SPE is a stepwise process to isolate one class of solutes in a mixture from very different sample components. Chromatography is based on the same principle of solutes partition between a stationary and a mobile phase. The difference in chromatography is that no solutes are retained 100% on the stationary phase. If the stationary and mobile phases are chosen well, different solutes partition to varying extents in the stationary phase. The different solutes then travel at different rates and elute from the column at different times. Chromatography can not only separate different classes of solutes as in SPE, but it can be refined to separate each individual solute. For example, a certain stationary phase might separate a series of nonpolar solutes from each other based on small differences in their polarity. It is not uncommon to use both SPE and chromatography in sample cleanup procedures for samples that contain a wide variety of analytes.

The most common use of column chromatography in chemistry is to isolate a reaction product from reactants and side products. The number of solutes is not large and the purified product is collected as it exits the column. Similar uses are found in biochemistry to isolate one particular protein or other biomolecule for further study. SPE is a more common sample cleanup method in analytical applications with two exceptions. Size-exclusion separations are performed in the continuous mode of chromatography and cases where the amount of analyte overloads an SPE cartridge requires a column separation.

Commercial vendors provide a variety of adsorption, partition, ion exchange, and size-exclusion stationary phases for different applications. As in SPE, stationary phases are selected to match the analytes and to be compatible with the solvent. For chemical separations, adsorption chromatography using silica or alumina is the most common stationary phase. The cost of these materials is much less than for the functionalized reversed-phase stationary phases. In biochemical purifications, size-exclusion (gel-filtration) or ion-exchange are common approaches.

Column packings are often categorized by mesh size. The mesh designation is the number of wires per inch of the sieve that is used to sort powders by size. A larger mesh number indicates smaller particles. A common mesh size for a gravity column is 60–120 mesh, which has particles of 250–125 μm diameter. Smaller diameter column packing provides more efficient separation by providing more uniform paths between particles. Smaller column packings also make the slow slower. *Flash chromatography* uses air pressure (on the order of 100 psi) to increase the flow of the mobile phase through the column. Typical packings for these columns are 230–400 mesh, or particles of 40–60 μm. Flash chromatography is also used for preparative-scale separations of milligram to gram quantities, but with increased efficiency. Manufacturer literature provides guidance on stationary phase selection, packing diameters, and column size for different applications.

Simple column chromatography consists of a glass or plastic tube with a glass or plastic frit on the bottom to hold the stationary phase in place. A stopcock at the exit provides manual control of the solvent flow. The test portion is loaded onto the top of the column and a continuous flow of mobile phase sweeps all solutes through the column. Sample components that interact weakly with the stationary phase stay predominantly in the mobile phase and travel through the column quickly. Other components that partition more strongly to the stationary phase take a longer time to pass through the column. As the compounds elute, they can be collected in different fractions from the column effluent as a function of time. Figure 2.11 shows snapshots in time of a mixture of several components passing through a liquid chromatography column. In this figure, the test portion is loaded at the top of the column at time = 0. Fresh solvent is added continuously and the components in the mixture begin passing through the column. At time = 2 min, the sample components have separated into three distinct bands. These continue through the column at different rates and are collected separately in fresh collection containers. At time = 8 min the darkest band is still near the top of the column. This component could be eluted from the column faster now by changing the mobile phase solvent that is added to the top of the column. Besides cleanup procedures in sample preparation, simple column LC is also used in preparative scale work to purify and isolate some components of a mixture.

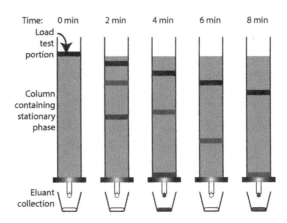

Figure 2.11   Schematic of a column chromatographic separation.

    The schematic in Figure 2.12 illustrates the separation process at the molecular level. The figure represents a snapshot in time as two components, A and B, are loaded onto a column and interact with a reversed-phase C18 stationary phase. In this snapshot, at the top of the column, the two components are not yet separated. The horizontal zigzags represent the alkyl chain of the stationary phase. The stationary phase is bonded to an inert particle, represented by the vertical line to the left. Mobile-phase solvent molecules are not shown, but they are flowing downward in the void space on the right and also penetrate between the alkyl chains. The A and B molecules that are shown embedded in the stationary phase represent molecules that are immobilized. As the A and B molecules remaining in the mobile phase are swept down the column, the immobilized solutes will be in contact with pure solvent and return to the mobile phase. The degree to which a mixture component partitions into the stationary phase determines the relative

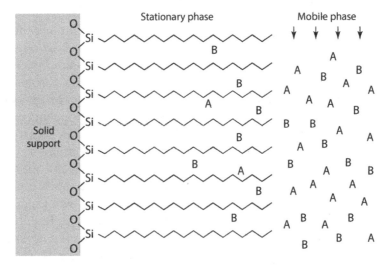

Figure 2.12   Microscopic schematic of two solutes interacting with a stationary phase.

rate of its progress through the column. Owing to the different degrees of partitioning between mobile and stationary phases, as indicated by the number of A and B molecules shown in the stationary phase, A and B travel through the column at different rates. Since component A remains mostly in the mobile phase, it will elute from the column first. The stronger interaction of component B with the stationary phase causes it to be retained on the column longer. Given this result, predict the relative polarity of components A and B.

The analyst's job is to choose appropriate mobile and stationary phases for the different solutes to progress through the column at slightly different rates in the shortest total time. The different solutes are separated by collecting fractions of the eluent as a function of time. For sample preparation purposes, many of the fractions will be discarded, keeping only the ones that contain analytes. The advantage of chromatography is that the result is equivalent to doing a large number of extractions, that is, achieving high separation efficiencies in short times. It also separates similar solutes from each other for subsequent measurement.

To summarize this section, SPE and column chromatography provide powerful methods to separate analytes from interferences, which we often call cleanup procedures. Selection of a stationary phase from the wide variety of available choices is usually done on the basis of performance for the specific types of analytes and interferences that are present in a test portion. This chapter has described the basic principles but has not discussed the large number of specialized commercial products, such as columns for environmental, pharmaceutical, and clinical analyses. Luckily, manufacturers provide application notes with validated methods for most types of analytes. The application notes are especially useful for mixed-mode columns. These columns contain stationary phases with more than one type of functional group, for example, both reversed phase and ion exchange. They allow a single SPE cartridge or chromatography column to be able to separate a range of solute types, including polar, nonpolar, and ionic. After eluting neutral molecules, ions can be eluted by changing the pH of the mobile phase as in ion-exchange methods.

Chapter 2. What Was the Point?   The main points of this chapter were to provide an overview of common sample processing methods and to discuss the concept of partitioning, which underlies extractions and separations. Many of the principles of these methods underlie other similar instrumental methods of analysis.

PRACTICE EXERCISES

1. Identify the following chemicals as being solid, liquid, or gaseous at 25°C and 1 atm pressure. Make a two-column table with ionic compounds in one column and nonionic compounds in the other. For the ionic compounds, predict if they will be soluble or insoluble salts in water. For the

organic compounds, classify them as polar, moderately polar, or nonpolar.
(a) methane
(b) octane
(c) water
(d) beeswax
(e) acetic acid
(f) sodium acetate
(g) K$_2$CrO$_4$
(h) methyl *tert*-butyl ether (MTBE)
(i) ethanol
(j) perchloric acid
(k) NaOH
(l) KCl
(m) CaCO$_3$

2. You have the following laboratory equipment:
   - distilled water, strong acid, strong base, and organic solvents
   - an analytical balance and a hot plate
   - beakers, flasks, and volumetric glassware
   - a wire mesh screen, funnels, and coffee filters
   - SPE cartridges with a C18 stationary phase

   Suggest sample preparation and analytical procedures to make the following measurements:
   (a) the amount of insoluble starch in aspirin tablets
   (b) the weight fraction of sand in a sand and gravel mix
   (c) the concentration of NaCl in seawater
   (d) the amount of caffeine in a can of cola

3. For each of the following solutes, predict if it will partition predominantly into water or octanol when using equal amounts of the two solvents.
   (a) acetic acid
   (b) hexane
   (c) 1-hexanol
   (d) methanol
   (e) oleic acid

4. Estimate the solubility of 1-octanol in water using the data in Table 2.15 and Equation (2.7).

5. The extraction of an organic acid from 100 ml of aqueous solution using 30 ml of organic solvent leaves 10% of the acid in the aqueous phase.
   (a) What is K'_D for this acid under the extraction conditions?
   (b) What percentage of the acid do you predict will remain in the aqueous phase if you do four separate extractions using 7.5 ml of the organic solvent each time?

(c) How many sequential extractions must you do to achieve 99.9% removal of the acid from the aqueous phase if you use 10.0 ml portions of the organic solvent each time?

6. You are approaching a deadline and you have time to do only two sequential extractions for a sample. What is the minimum total volume of organic solvent that is needed to transfer 99.9% of the solute from 100 ml of an aqueous solution if K'_D is 66?

7. You have an aqueous solution at pH = 11 containing the sodium salts of benzoic acid (pK_a = 4.204), acetic acid (pK_a = 4.756), and phenol (pK_a = 9.98). At pH = 11, these three compounds are present as the benzoate, acetate, and phenolate ions, respectively. To what pH could you acidify this solution to extract the phenol and not the two acids into an organic solvent?

8. The forensics analysis of opioids in biological fluids involves a solid-phase extraction (SPE) followed by measurement with an instrumental method (gas chromatography mass spectrometry, GC-MS). List the control samples that should be prepared and measured to obtain accurate results in such a procedure. Include in your answer the purpose of each control sample.

9. Section 2.6 described how to adjust mobile phase pH to elute analytes from strong and weak cation-exchange SPE cartridges. How should the mobile phase pH be adjusted to first retain and then elute analytes when using strong and weak anion-exchange SPE cartridges? What other anions besides OH^- might be useful to elute retained singly charged anions from an anion-exchange column?

10. Suggest appropriate aqueous buffer mobile phases for the following SPE cleanup procedures. In each case, the analytes should be retained for subsequent elution. Amino acids are neutral in pH between approximately 2.2 and 9.5. At pH below 2.2, the carboxylic acid is protonated and the amino acid is a cation. At pH greater than 9.5, the amino group is deprotonated to create an anion.
    (a) Separating free amino acids from proteins and salts using a C18 reversed-phase SPE cartridge.
    (b) Concentrating amino acids in water using a strong cation exchange (SCX) cartridge.
    (c) Concentrating amino acids in water using a strong anion exchange (SAX) column.
    (d) Br^-, BrO_3^-, Cl^-, ClO_3^-, NO_3^-, and SO_4^{2-}. Given that all of these solutes are anions of strong acids, explain why the stationary phase should be a weak anion exchanger.

11. Suggest suitable solvents for the mobile phase in the liquid chromatographic separation of the following solutes. Predict the order of elution of the alcohols for the two different columns.

    (a) A series of alcohols, methanol through butanol, using a cyano normal-phase column

    (b) A series of alcohols, methanol through butanol, using a C8 reversed-phase column

12. Reversed-phase stationary phases provide excellent separations for non-polar compounds. Many biological samples contain a mixture of neutral and charged molecules at a given pH. Separating neutral and charged molecules on a reversed-phase column can often be accomplished by adding a salt that converts the charged molecules to an ion pair. Two common types of ion-pair reagents are

- quaternary amines, for example, tetrabutylammonium dihydrogen phosphate
- alkylsulfonates, for example, sodium heptane-1-sulfonate

    Use these examples to answer the following questions.

    (a) Draw the structures of the two ion-pair reagent examples.

    (b) Explain how addition of an ion-pair reagent affects the time that an analyte is retained on a reversed-phase column.

    (c) Choose the type of ion-pair reagent that is appropriate for dopamine, tyramine, and tryptamine in pH = 7 buffer.

    Choose the type of ion-pair reagent that is appropriate for a mixture of ribonucleosides such as 5'-AMP, 3',5'-c-AMP, 3'-AMP, 5'-ADP, and 5'-ATP, in pH = 7 buffer.

13. A calibration standard for dried seafood (NIST Standard Reference Material 1566b) contains

    (a) Cd: 2.48 ppm

    (b) Cu: 71.6 ppm

    (c) Fe: 205.8 ppm

    (d) Pb: 0.038 ppm

    (e) Zn: 1424 ppm

    This SRM and other samples are to be analyzed using the following procedure.[18]

- Place 0.30 g dry powder, 5 ml of concentrated HNO_3, and 2 ml of 30% H_2O_2 in a microwave digestion vessel.
- Microwave for 30 min as per manufacturer guidelines and allow to cool.
- Transfer to 25.0-ml volumetric flask and dilute to mark with deionized water.
- Dilute with 3 M HNO_3 if further dilution is necessary.
- Measure Pb and Cd by graphite furnace atomic absorption spectroscopy (GFAAS).

---

[18]Adapted from AOAC Official Method 999.10 Lead, Cadmium, Zinc, Copper, and Iron in Foods.

- Measure Zn, Cu, and Fe by flame atomic absorption spectroscopy (FAAS).

For each analyte in the SRM, calculate if further dilution is necessary if the spectrometers have the following working ranges for these metal ions in solution:

- GFAAS (Cd, Pb): 0.001–0.05 ppm FAAS (Cu): 0.002–0.2 ppm FAAS (Fe, Zn): 0.05–5 ppm.

14. What kind of bear will dissolve in water but not in oil?

15. Use this procedure for psychoactive psilocin to answer the following questions.[19] The enzymatic step converts psilocybin, a phosphate ester form, to psilocin.

   - Grind dried mushroom samples as finely as possible, and weigh a 200-mg test portion into a 10-ml test tube.
   - Add 500 μg cocaine as internal standard.
   - Add 3 ml phosphate buffer (pH 6) and 100 μl beta-glucuronidase solution. Incubate the mixture for 90 min at 47°C.
   - Centrifuge or filter and collect the clear supernatant.
   - Condition a C18 SPE cartridge with two 3-ml portions of methanol followed by 3 ml of pH = 6 phosphate buffer.
   - Load clear supernatant from enzymatic hydrolysis onto the SPE column.
   - Wash with 2 ml pH = 6 phosphate buffer followed by 2 ml 0.005 M HCl.
   - Elute with 2 ml dichloromethane–methanol (65/35, v/v) + 5 ml of concentrated ammonium hydroxide (freshly prepared).
   - Evaporate the eluent at 60°C under a gentle stream of nitrogen and reconstitute residue in 500 μl ethyl acetate.

   (a) What is the charge of the psilocin at pH = 6?
   (b) Heating the sample to too high a temperature can degrade the psilocin but does not degrade the internal standard. Will the measurement be erroneously high or low if the heat is too high? What approach can serve as a check for analyte loss?
   (c) Measurement results, in relative units, are psilocin = 4130 and cocaine = 82,110. The measurement technique responds to psilocin and cocaine with equal sensitivity. What is the concentration in parts per million of psilocin in the dried mushroom sample?
   (d) Discuss why the internal standard is used in this procedure. Include the possible bias that it prevents and potential errors that it will not correct.

16. Organophosphates, for example, tri-isobutylphosphate and tris(2-chloroethyl) phosphate, are plasticizers and chlorinated flame retardants that

[19]Adapted from Marit Pufahl, "Solid-Phase Extraction of Psilocin from Hallucinogenic Mushrooms," Varian Application Note.

occur as very low concentration pollutants in natural waters. Moderate polarity SPE cartridges with a poly(styrene-*co*-divinylbenzene) stationary phase provide a method to concentrate organophosphates for analysis. A typical procedure follows; note that the analyte is concentrated from greater than 1 l of water to 1 ml of organic solvent.[20]

- Place new SPE cartridges on vacuum manifold, one for each water sample.
- Condition cartridges with 1 ml methanol, followed by 1 ml 50:50 methanol/acetonitrile.
- Pass 1.5–2.5 l water samples through cartridges.
- Dry the cartridge using nitrogen.
- Elute analytes with three portions of 0.33 ml 50:50 methanol/ acetonitrile.

For each of the following solutes, predict if it will be retained strongly, retained weakly, or not retained on the cartridge. You may assume that the pH of the natural water is near 6. For each solute, indicate if there will be any difference in retention if the water sample is acidified to pH of less than 2.

(a) Hexane
(b) Sodium acetate
(c) Sodium benzoate
(d) Phosphoric acid
(e) Acetone

17. How does the limit of detection (LOD) compare to the lower limit on a control chart?

18. You notice an increase in the number of ants in your apartment. You suspect that something is attracting them into your living space so you devise a simple test to detect ant attractants—you lick your finger, swipe your finger on a surface to sample, and test for sweetness by taste. (A questionable method at home and definitely NOT appropriate in a laboratory). With your measurement method in hand, what is your sampling plan? Give your rationale. You notice a spilled bottle of soda on a closet floor. How will you modify your sampling plan?

19. Your supervisor tasks you with monitoring the purity of a metal ore, which has been crushed and partly mixed by the supplier. The purity determines the refining conditions, and ultimately the cost of producing metal product. The ore is delivered to the processing plant in five rail cars each week. Your budget and facilities allow you to measure 10 samples before the ore is transferred into the plant. Describe a sampling plan and justify your reasoning.

[20]Adapted from Regnery, J.; Korte, E. "Determination of Organophosphates in Lake Water," Varian Application Note SI-02094.

CHAPTER 3

CLASSICAL METHODS

Learning Outcomes

- Identify common types of reactions in aqueous solution.
- Balance chemical reactions to determine reaction stoichiometry.
- Calculate analyte concentrations from gravimetric, titration, or coulometric data.

3.1 INTRODUCTION

Although I title this chapter "Classical Methods," many of the analytical methods described in the following sections are commercially available as automated instruments.[1] Researchers and equipment manufacturers continue to develop and improve these classical or "wet-chemical" methods for current analytical problems.

Classical methods determine the amount of an analyte in a sample by relating it to a known amount of a pure standard. In the simplest case, an unknown amount of a pure material is determined by weighing against a standard weight. The measured mass in this *gravimetric* procedure is the basis for the analytical result. Although this procedure is simple in principle, separating an analyte from other

---

[1]For example, try an Internet search for titrators or autotitrators. Some common commercial brands are Brinkman, Mettler-Toledo, and Metrohm.

Basics of Analytical Chemistry and Chemical Equilibria: A Quantitative Approach, Second Edition. Brian M. Tissue.
© 2023 John Wiley & Sons, Inc. Published 2023 by John Wiley & Sons, Inc.
Companion Website: www.wiley.com/go/tissue/analyticalchemistry2e

sample components and converting it to a pure form depends on appropriate chemical transformations and careful laboratory techniques. Similarly, maintaining a standard weight requires careful handling and environmental control so that the mass of the standard does not change over time. At the laboratory level, analysts purchase standard weight sets or contract a technical service to calibrate analytical balances. Standard weights are known with high accuracy by being traceable through secondary standards to the SI definition of the kilogram.[2]

In other types of classical methods, a standard solution is added to a dissolved test portion until the analyte and the standard react completely. The analyte concentration in this *titration* procedure is determined by a volume measurement, knowledge of the standard concentration, and the reaction stoichiometry. This type of analysis is very useful because volumetric glassware can deliver liquid volumes with high accuracy (Section 1.3). Accurate mass measurements are still necessary to prepare the standard solution for this approach.

Analytical laboratories are dominated by instrumentation, but wet-chemical methods are needed in several cases:

- Many measurements are made for analytes at the percentage level.
- They are used to certify analytical standards.
- They can be portable, easy to use, and rapid for screening or detection.

For samples where a component is present on the percentage level, instrumental methods are overly sensitive. Compared to weighing an agricultural product before and after drying, extracting water to determine moisture content by an instrumental method introduces additional sources of errors. Another characteristic of classical methods is that they are directly traceable to certified weights or concentration standards. This direct relationship to a verified quantity makes them useful for certifying analytical standards. These secondary standards are then used to calibrate instrumentation and validate analytical methods. Finally, many field measurements are based on wet-chemical methods because the simple equipment is easy to transport and requires no electrical power. Measurements made by wet-chemical methods can also be more robust and less susceptible to environmental fluctuations than instrumental methods. These factors might be significant when making measurements in real time in a factory or food processing plant.

This chapter discusses the following classical methods:

- *Gravimetry*: determining equivalence to a known *mass*
- *Titration*: determining equivalence to a known *volume* of a standard solution
- *Coulometry*: determining equivalence to a known amount of *electrical charge*

---

[2]For the SI definition of the kilogram, see https://www.nist.gov/si-redefinition/kilogram/kilogram-future; accessed August 2022.

These methods are implemented differently, but they share the requirement that the underlying reactions are quantitative, that is, they go to completion. Given a complete reaction, a fixed stoichiometric relationship between analyte and standard can be used to convert measurement of a mass, volume, or electric current to the amount of analyte. The accuracy of a measurement is then traceable to the known purity of chemical standards and reagents and to the known accuracy of laboratory equipment such as analytical balances and volumetric glassware (refer to Chapter 1).

Before discussing the use of chemical reactions for analytical purposes, Section 3.2 provides a refresher on different types of reactions. If you are well-versed with this material, skip through the section quickly. If your memory of these reactions has dimmed, read this section carefully because we use these reactions throughout the text. I know from experience that forgetting general chemistry concepts impedes learning new analytical topics. Not recognizing the nature of a given compound makes it impossible to think about the chemistry that might occur. Use a general chemistry text or an online resource to determine the properties of unfamiliar chemicals or reactions.

3.2 REVIEW OF CHEMICAL REACTIONS

3.2.1 What Is a Chemical Reaction?

A chemical reaction occurs when substances, the *reactants*, collide with enough energy to rearrange the atoms to form different compounds, the *products*. Figure 3.1 illustrates this process in a simple cartoon, including the release of energy. The products, C and D, persist because they are more stable, that is, they have a lower energy, than reactants A and B. The change in energy that occurs when a reaction takes place is called the *thermodynamics* of the reaction. The rate or speed at which a reaction occurs is described by the *kinetics* of the reaction. We will look at these topics in more depth when we discuss redox reactions in Chapter 9. In analytical measurements, we prefer to use rapid reactions for practical reasons. If reactions are not rapid, analytical protocols will specify the times or reaction conditions that are necessary to allow the reaction to go to completion.

Figure 3.1   Reactants A and B changing to products C and D.

When referring to chemical species in a reaction, we use letters in parentheses to designate the form of the reactants and products:

- *Solids*: $H_2O(s)$, $AgCl(s)$, $CaCO_3(s)$
- *Liquids*: $H_2O(l)$, $Br_2(l)$
- *Gases*: $H_2O(g)$, $H_2(g)$, $Cl_2(g)$
- *Aqueous solution*: $NH_4^+(aq)$, $Cl^-(aq)$, $Br^-(aq)$, $Br_2(aq)$, $CH_3COOH(aq)$

We often think of substances in their room-temperature form, but I have included water in three phases as a reminder that the physical state will depend on reaction conditions. I do not show water with the aqueous solution examples, that is, $H_2O(aq)$, because we usually omit the designator when a substance in a reaction is also the solvent. Both $Br_2(l)$ and $Br_2(aq)$ are included as examples. The liquid phase is designated when the bromine concentration exceeds its solubility so that pure $Br_2(l)$ is in contact with an aqueous solution. In addition, note that both ions and neutral molecules are common in aqueous solution.

3.2.2 Gas-Phase Reactions

We start with some simple reactions to further illustrate the "language of chemistry." Gases can react with other gases, liquids, or solids to form new products. Besides the combustion methods discussed here, many of the instrumental methods discussed in Part III of the text require the analyte to exist as gas-phase atoms or molecules for measurement. These instruments—atomic spectrometers, mass spectrometers, and gas chromatographs—usually use thermal energy to vaporize and sometimes decompose samples.

Combustion of methane:

$$CH_4(g) + 2O_2(g) \rightarrow CO_2(g) + 2H_2O(l)$$

Heating calcium carbonate (limestone) to make calcium oxide (lime):

$$CaCO_3(s) + heat \rightarrow CaO(s) + CO_2(g)$$

Hydrogen fuel cell (oxidation, reduction, and overall reactions):

$$2H_2(g) \rightarrow 4H^+(aq) + 4e^-(aq)$$
$$\underline{O_2(g) + 4H^+(aq) + 4e^-(aq) \rightarrow 2H_2O(l)}$$
$$2H_2(g) + O_2(g) \rightarrow 2H_2O(l)$$

These examples show the most stable products, but the actual reaction products are dependent on the reaction conditions. Ideally the combustion of methane produces only water and carbon dioxide. This ideal case occurs if the combustion takes place in a 1:2 stoichiometric ratio of methane to oxygen or if there is an excess of oxygen. Combustion in oxygen-deficient conditions can produce other products such as carbon monoxide, $CO(g)$, and carbon soot particles. The third example includes intermediate species that do not appear in the overall reaction. Showing these intermediates provides some information about the mechanism of the reaction, which is not apparent from the overall reaction.

Producing multiple products, including desired products and undesired "side products," is very common in gas-phase reactions and in organic synthesis. The synthetic chemist will do her best to optimize product yield, and she will get some amount of pure product. The analytical chemist who converts some fraction of analyte to a form that is not measurable by a specific technique has essentially failed. Analytical reagents and reaction conditions are chosen to achieve quantitative, i.e., 100%, conversion of an analyte to the measurable form. Adding an excess of reagents is a common approach to drive a reaction to completion. Validated analytical procedures will list the conditions and the time requirements for each step in a procedure to go to completion. Of course, when things do not work as expected for control samples, reaction conditions are a prime suspect for fixing problems.

It is not unusual to need multiple reaction steps to achieve complete conversion of an analyte to a pure form. Gravimetric reagents can be specific to quantitatively precipitate one analyte from aqueous solution. However, the resulting solid might be a heterogeneous precipitate. A subsequent annealing at high temperature, known as ignition, is needed to convert the product to one pure form for weighing.

In the examples above, I placed stoichiometric coefficients in front of the reactants and products to "balance" the reaction. We balance reactions so often in chemistry that it is good to have a general procedure for doing so. Atoms combine in integral ratios to form stable molecules. The number of each atom must be equal on each side of a reaction. This requirement is called *mass balance*. You will sometimes see imprecise formulas for reactants or intermediate products, such as NO_x, which refers to a mixture of NO and NO_2.[3] The other property that must be equal on each side of a reaction is the total charge. This requirement is called *charge balance* or electrical neutrality. If reactions could occur to produce an excess of positive or negative charges, we would all get quite a charge out of life! Example 3.1 gives a stepwise illustration on how to balance a reaction.

---

[3]Similarly, some solids exist with a mixture of metal valence states resulting in a nonstoichiometric ratio, for example, WO_{3-x} or $YBa_2Cu_3O_{7-x}$, where x is the degree of oxygen deficiency.

Example 3.1 Combustion of Octane.   The products of the reaction of an alkane and O_2, assuming complete combustion, are $CO_2(g)$ and $H_2O(l)$. Balance the reaction for the combustion of octane.

The unbalanced reaction is

$$C_8H_{18}(l) + O_2(g) \rightarrow CO_2(g) + H_2O(l)$$

Start balancing this reaction with carbon. There are eight carbons on the left, so we need eight carbons on the right:

$$C_8H_{18}(l) + O_2(g) \rightarrow 8CO_2(g) + H_2O(l)$$

Next balance the number of hydrogen atoms. There are 18 hydrogen atoms on the left, so we need 18 hydrogen atoms on the right:

$$C_8H_{18}(l) + O_2(g) \rightarrow 8CO_2(g) + 9H_2O(l)$$

Finally balance the oxygen atoms. There are 25 oxygen atoms on the right, so we need 25 oxygen atoms on the left:

$$C_8H_{18}(l) + 12\frac{1}{2}O_2(g) \rightarrow 8CO_2(g) + 9H_2O(l)$$

The reaction is now mass balanced. The charge on each side of the reaction is 0, so the reaction is charge balanced and we are done. You will also see this balanced reaction written so that there are no fractional stoichiometric factors:

$$2C_8H_{18}(l) + 25O_2(g) \rightarrow 16CO_2(g) + 18H_2O(l)$$

Both reactions are balanced and usage of one or the other is simply a matter of personal preference. Note that the phase designators are for standard temperature and pressure, even though the combustion occurs in the gas phase.

We will not discuss many gas-phase reactions in this text, but they are quite important for elemental analysis where an organic compound is combusted to determine the ratio of C, N, O, and other elements. The Examples 3.2 and 3.3 illustrate the types of information that can be obtained from combustion methods.

Example 3.2 CHNO Analysis.   Determination of C, H, N, and other non-metals is accomplished by combusting a test portion completely to produce gases that are trapped and measured by instrumental methods. What is the empirical formula of 34.9 mg of a purified substance that produces the following results for each element? Predict the identity of this substance.

Element	Measured Mass, mg
C	17.27
H	1.81
N	10.07
O	5.75

Divide each of the weights by the formula weight of the element to convert to millimoles. Convert the results to integers by dividing by the lowest number of millimoles, in this case 0.3595 mmol of oxygen.

Element	Formula wt (g/mol)	Moles, mmol	Integral Value
C	12.011	1.438	4
H	1.0079	1.797	5
N	15.007	0.7189	2
O	15.999	0.3595	1

The empirical formula is $C_4H_5N_2O_1$.

On searching a reference source, you do not find a substance with this formula. In this case, multiply each subscript by 2, then 3, and so on until you find a match. The first two possibilities are $C_8H_{10}N_4O_2$ and $C_{12}H_{15}N_6O_3$. You find no match for the 12-carbon formula, but the 8-carbon formula matches caffeine and enprofylline. Identifying this substance will require further qualitative analysis, such as determining the melting point or recording some type of spectrum.

Example 3.3 Nitrogen/Protein Analysis. Determining C, H, and N is accomplished by combustion at 950°C or higher in pure oxygen followed by measurement of the resulting gases (nitrogen oxides are reduced to N_2). What is the percentage of N and protein in a test portion of animal feed given the following results? Multiply the percent nitrogen by 6.25 to obtain percent protein.[4]

Sample Weight, g	Measured Nitrogen, g
0.2185	0.0734
0.2543	0.0847
0.2802	0.0938

Since the analyzer returns results as weights, obtaining a percentage is found by dividing the nitrogen measurement by total sample weight and multiplying by 100%, for example,

[4]AOAC Official Method 990.03, *Protein (Crude) in Animal Feed: Combustion Method.*

$$\% \, N = \frac{0.00734\,g}{0.2185\,g} \times 100\% = 3.36\%$$

```
Sample        Measured      Percent
Weight, g    Nitrogen, g   Nitrogen, %
  0.2185       0.0744         3.36
  0.2543       0.0847         3.33
  0.2802       0.0938         3.35
```

The average and standard deviation of these three measurements are

$$3.35 \pm 0.01\% \, N$$

Multiplying both the average and standard deviation by the 6.25 factor to convert to percent protein gives

$$20.9 \pm 0.1\% \, protein$$

Although the calculation for this protein analysis is simple, the actual measurement is more involved. The instructions in an application note from the instrument manufacturer include

- measurement of blanks until a steady reading is given,
- measurement of three to five standards,
- measurement of the sample test portions,
- measurement of the standard again.

The blank and standard measurements are necessary to prevent drift from affecting the accuracy of measurement.

3.2.3 Limiting Reagent

A *limiting reagent* is a reactant that is completely consumed in a reaction. The excess reactants that are not limiting reagents will remain in some residual amount. In the examples above, and in most of the wet-chemical methods, we want our reagents in excess to achieve complete conversion of the analyte to a desired form (see Example 3.4). If we are adding a complexing agent to a solution containing a metal ion to form a colored complex, we want an excess of the complexing agent so that all of the metal ion, the limiting reagent, is converted to the form that we measure. There are other cases where we use a limiting reagent to form a mixture. To make an acid–base buffer solution, we can add a specific amount of strong base (the limiting reagent) to a weak acid to produce a mixture of weak acid and conjugate base to obtain a certain pH.

An important question to ask yourself in evaluating analytical procedures: what can happen when reactants are mixed? Many of the equilibrium mixtures that we use, such as acid–base buffer solutions, are made by a limiting-reagent reaction. Once the limiting reagent is consumed in the initial reaction, we can then consider what equilibria can occur to determine equilibrium concentrations of reactants and products. Misidentifying the reactants, products, and their starting concentrations will make it impossible to calculate equilibrium concentrations correctly.

Example 3.4 A Limiting-Reagent Calculation.   One step in the production of copper metal from copper ore is the conversion of sulfide to oxide:

$$Cu_2S(s) + O_2(g) \rightarrow Cu_2O(s) + SO_2(g)$$

What is the amount of O_2 necessary for complete reaction of 1.0 kg Cu_2S? First, look at the reaction. There are equal numbers of Cu and S atoms on each side of the reaction, but not of O. Try to balance the O by placing a 2 in front of Cu_2O. Now the Cu is unbalanced so also place a 2 in front of Cu_2S

$$2Cu_2S(s) + O_2(g) \rightarrow 2Cu_2O(s) + SO_2(g)$$

This reaction is not quite balanced, but it is getting closer. We need another S atom on the right, so place a 2 in front of SO_2. Now we can finish balancing this reaction by placing a 3 in front of the O_2

$$2Cu_2S(s) + 3O_2(g) \rightarrow 2Cu_2O(s) + 2SO_2(g)$$

Convert 1.0 kg of Cu_2S to moles by dividing by the formula weight:

$$(1.0\,kg\ Cu_2S)\frac{1000\,g}{1\,kg}\frac{1\,mol\,Cu_2S}{191.22\,g\,Cu_2S} = 5.23\,mol\ Cu_2S$$

Since the balanced reaction requires 3 mol of O_2 for every 2 mol of Cu_2S,

$$(5.23\,mol\ Cu_2S)\frac{3\,mol\ O_2}{2\,mol\ Cu_2S} = 7.8\,mol\ O_2$$

To prevent O_2 from being a limiting reagent in this conversion, at least 7.8 mol of O_2 must be provided per kilogram of Cu_2S.

3.3 REACTIONS IN AQUEOUS SOLUTION

This text focuses on reactions and chemical equilibria in solution, but the general principles also apply to gas-phase and solid-state reactions. Many types of reactions are used to convert an analyte to a measurable form. The first step in deciphering an analytical procedure is to identify the nature of the reactants and the reaction type. Table 3.1 categorizes the most common reactions that we will encounter in aqueous solution.[5] Supporting reagents that use one of these reactions will also be added to a test solution to control some aspect of an equilibrium. If a procedure instructs you to add 1 M NaOH, you probably know that you are making a solution more basic. Other reagents might control the redox state of an analyte, prevent an analyte from precipitating at high pH, or remove interferences.

We will treat each of these reaction types in detail, both as reactions that go to completion and as equilibria. A reaction that goes to completion is written with a right arrow ($\rightarrow$). The direction of the reaction indicates that the reactants on the left are transforming into the products on the right. The amounts of products generated from a reaction will depend on any limiting reagents as discussed above. Once a reaction is complete, then we can consider how the products and any remaining reactants will come to equilibrium. A general review of equilibrium concepts is presented next, followed by subsections with the conventions for describing equilibria of each of these reaction types.

3.3.1 Equilibrium Terminology

Equilibrium is the thermodynamic condition at which a chemical system is not changing. The chemical potential, or driving force, for the reactants to go to products is equal to the chemical potential for the products to return to reactants. State variables such as temperature and pressure are constant, the number of phases is not changing, and the concentrations of the chemical

TABLE 3.1 Reaction Types in Aqueous Solution

Reaction Type	Reactants	Process	Equilibrium Constants
Neutralization	Acids and bases	Transfer of protons	K_w', K_a', and K_b'
Complexation	Metal ions and ligands	Transfer of ligands	K_n' and β_n'
Precipitation	Cations and anions	Phase change to solid	K_{sp}'
Redox	Everything	Transfer of electrons	K'

[5]The metal–ligand example listed for complexation is only one of many types of complexes, but it is a common case in analytical applications.

species in the system are not changing. Reactions are still occurring on the microscopic scale, but the macroscopic concentrations are not changing. To indicate this thermodynamic condition, we replace the arrow in the reaction with right/left harpoons $\rightleftharpoons$. Other symbols are also common, for example, $\leftrightarrow, \Leftrightarrow, \updownarrow$, or $\Updownarrow$; they all indicate that the reaction is an equilibrium.

When we say that a system is at equilibrium, we always mean that it is in a thermodynamic equilibrium. A chemical system can be metastable, that is, stable kinetically, but not at equilibrium. A balloon containing a mixture of $H_2(g)$ and $O_2(g)$ at room temperature is not at equilibrium. However, it does not explode without a spark. Similarly, equilibrium and a "steady state" are not necessarily the same thing. An open system in which reactants are entering a reaction vessel and products are leaving the reactor can maintain steady-state concentrations of reactant and product, but this system is not at equilibrium.

A chemical species will always exist in equilibrium with other forms of itself. The other forms may exist in undetectable amounts, but they are always present. These other forms arise due to the natural disorder of nature, which we call entropy (it is impossible to be perfect). As an example, pure water consists of the molecular compound, H_2O, in equilibrium with some amount of the dissociated hydroxide ions, OH^-, and protons, H^+.

$$H_2O(l) \rightleftharpoons OH^-(aq) + H^+(aq)$$

$$2H_2O(l) \rightleftharpoons OH^-(aq) + H_3O^+(aq)$$

These two equilibrium reactions are equivalent. Using the hydronium ion, H_3O^+, serves to indicate that the proton is hydrated, or surrounded tightly, by water molecules. In this example, the ionized forms of water are easily detected with a pH meter. The pH of pure, degassed water at room temperature is 7.0 due to an $H_3O^+(aq)$ equilibrium concentration of 1×10^{-7} M. Since the stoichiometry of water dissociation is 1:1, pure water also has $[OH^-] = 1 \times 10^{-7}$ M.

Our quantitative measure of the extent to which reactants and products coexist is the equilibrium constant K or K'. K is the *thermodynamic equilibrium constant*, which we can look up in tables. K' is a *formal equilibrium constant*, which we use when working with molar concentrations. Later chapters explain the difference in detail and for now we will use K' in these introductory examples. Recall the three general rules for writing K' expressions for any type of reaction

- The formal equilibrium constant is always written with the concentrations of the products over the concentrations of the reactants.
- The stoichiometric coefficients of the reaction are exponents in the K' expression.
- We do not include solvent and pure substances, solids and liquids, in a K' expression.

Right away you see that a large K' indicates that the equilibrium lies to the right. Knowing what this means about the concentrations depends on how we write the equilibria. Each different type of reaction has a convention for writing the direction of the reaction. A related quantity is the *reaction quotient, Q*. It has the same mathematical form as the equilibrium constant, but we calculate it for nonequilibrium concentrations of reactants and products. It provides a quantitative measure of how far a system is from equilibrium at a given snapshot in time as a reaction is approaching equilibrium concentrations.

3.3.2   Neutralization

The Brønsted-Lowry definition describes acids as proton donors and bases as proton acceptors. Mixing any acid with any base results in neutralization because the base will accept the proton that the acid can provide. For the case of the strong electrolytes HNO_3 and $NaOH$ in aqueous solution, the reaction is

$$H^+(aq) + NO_3^-(aq) + Na^+(aq) + OH^-(aq) \rightarrow NO_3^-(aq) + Na^+(aq) + H_2O$$

The nitrate and sodium ions are spectator ions in this reaction. They may be left out to show the *net reaction* as

$$\underset{\text{proton}}{H^+(aq)} \ + \ \underset{\text{base}}{OH^-(aq)} \ \rightarrow \ \underset{\text{protonated base}}{H_2O}$$

For this neutralization reaction of a strong acid and a strong base, the pH in the resulting solution will depend on whether HNO_3 or $NaOH$ was the limiting reagent (see Example 3.5).

Example 3.5 Neutralization Reaction with a Limiting Reagent.   What are the amounts of all reactants and products for the following reaction if we mix 0.050 mol of $NaOH$ with 0.03 mol of H_2SO_4 in water?

NaOH and H_2SO_4 are strong electrolytes, so the neutralization reaction will go to completion. The overall reaction is

$$Na^+(aq) + OH^-(aq) + H^+(aq) + SO_4^{2-}(aq) \rightarrow Na^+(aq) + SO_4^{2-}(aq) + H_2O$$

We can simplify this reaction by removing the spectator ions. The net reaction for this neutralization is

$$OH^-(aq) + H^+(aq) \rightarrow H_2O$$

Since there are two protons for each sulfuric acid, there are 0.06 mol of H^+. The first line below the net reaction shows the starting amounts that are mixed. The next line shows the amounts after the reaction is complete.

$$OH^-(aq) \quad + \quad H^+(aq) \quad \rightarrow \quad H_2O$$

0.050 mol	0.060 mol	
≈ 0 mol	0.010 mol	0.050 mol

The amount of each solute in solution after the reaction is

≈ 0 mol OH^-	0.050 mol Na^+
0.010 mol H^+	0.030 mol SO_4^{2-}

OH^- was a limiting reagent in this case. I use ≈ 0 mol for OH^- because there is a small equilibrium amount always present in water. Note that the total amount of positive charge (0.01 M H^+ + 0.05 M Na^+) equals the total amount of negative charge in solution (2×0.03 M SO_4^{2-}). This condition must always be true and confirming it is a good way to catch calculational errors.

3.3.3 Acid–Base Equilibria

Equilibria of weak acids and weak bases are written as the acid or base reacting with water to produce hydronium or hydroxide ions, respectively. The equilibrium constants are labeled as K'_a and K'_b, where the subscript "a" indicates an equilibrium for an acid hydrolysis reaction and the subscript "b" indicates a base hydrolysis reaction. We follow the general rules to write the equilibrium constant expressions, products over reactants and the solvent, water, omitted. Generic formulas that show the convention with equilibrium expressions are

$$HA(aq) + H_2O \rightleftharpoons A^-(aq) + H_3O^+(aq)$$

$$K'_a = \frac{[A^-][H_3O^+]}{[HA]}$$

$$B^-(aq) + H_2O \rightleftharpoons HB(aq) + OH^-(aq)$$

$$K'_b = \frac{[HB][OH^-]}{[B^-]}$$

The following two examples show equilibria for acetic acid and the weak base, ammonia. The generic weak base shown above had a negative charge. Ammonia is an example of a neutral base that becomes charged when protonated.

$$CH_3COOH(aq) + H_2O \rightleftharpoons CH_3COO^-(aq) + H_3O^+(aq)$$

$$K'_a = \frac{[CH_3COO^-][H_3O^+]}{[CH_3COOH]}$$

$$NH_3(aq) + H_2O \rightleftharpoons NH_4^+(aq) + OH^-(aq)$$

$$K'_b = \frac{[NH_4^+][OH^-]}{[NH_3]}$$

3.3.4 Complexation

A complex is a solution-phase association between an electron-pair acceptor and an electron-pair donor. Protonation of a base by H^+ is actually a complexation reaction, but we treat acids and bases separately because of their wide use in chemistry. We use the following terms to describe complexes:

Complex: a solution-phase association between two or more chemical species, often a central metal ion and surrounding ligands.
Ligand: a neutral or anionic species with a pair of unbonded electrons that can bond to the empty orbital of a metal ion.
Coordination number: the number of bonds made to a central metal ion, usually 2, 4, or 6.
Chelate: a specific type of complex in which one or more ligands contain multiple unpaired electrons and make multiple bonds with the central metal ion.

The most common type of complex that we will encounter will be a metal ion surrounded by one or more ligands. The following examples show formation of 1:1 complexes that are typical for chelation in the titration of calcium with EDTA,[6] and biological complexes such as the iron–heme complex in hemoglobin:

$$Ca^{2+}(aq) + EDTA^{4-}(aq) \rightarrow Ca(edta)^{2-}(aq)$$
$$Fe^{2+}(aq) + HEME(aq) \rightarrow Fe(heme)^{2+}(aq)$$

---

[6]EDTA is ethylenediaminetetraacetic acid. When used in a chemical formula, ligand abbreviations are written in lowercase letters.

3.3.5 Complex Equilibria

Complex equilibria are always written in the direction of the complex forming. This convention then specifies the form of the equilibrium constant, which is called a formation or stability constant. Stability constants are tabulated in two different forms, stepwise and cumulative. Symbols for these constants vary in different sources, but I will use the IUPAC recommendation of K_n' and β_n' for stepwise and cumulative constants, respectively. The n subscript indicates the number of ligands in the product on the right. As an example, consider the equilibrium:

$$Co^{3+}(aq) + 6NH_3(aq) \rightleftharpoons Co(NH_3)_6^{3+}(aq)$$

Following the rules for writing equilibrium constants gives

$$\beta_6' = \frac{\left[Co(NH_3)_6^{3+}\right]}{\left[Co^{3+}\right]\left[NH_3\right]^6}$$

where the β_6' symbol indicates that this is a cumulative stability constant for six ligands. This example shows six ammonia ligands complexing a metal ion. Since the ammonia is neutral, the complex retains the same charge as the metal ion. For ligands that have negative charges, the overall charge of the complex is the sum of the metal and all ligands, for example,

$$Ag^+(aq) + CN^-(aq) \rightleftharpoons AgCN(aq)$$

$$AgCN(aq) + CN^-(aq) \rightleftharpoons Ag(CN)_2^-(aq)$$

In this second case, the resulting complex has an overall charge of -1, the same as the sum of charges on the left side of the reaction. The equilibrium constant expressions for these two stepwise complexation reactions are written with the K' symbols:

$$K_1' = \frac{[AgCN]}{\left[Ag^+\right]\left[CN^-\right]}$$

$$K_2' = \frac{\left[Ag(CN)_2^-\right]}{[AgCN]\left[CN^-\right]}$$

For the first step in an equilibrium, K_1' and β_1' are identical. For the cumulative reaction between Ag^+ and two CN^-, the reaction and equilibrium constant expressions are

$$Ag^+(aq) + 2CN^-(aq) \rightleftharpoons Ag(CN)_2^-(aq)$$

$$\beta_2' = \frac{[Ag(CN)_2^-]}{[Ag^+][CN^-]^2}$$

3.3.6 Precipitation

Many ions react to form insoluble precipitates (refer to Section 2.4). When the concentrations of the ions in solution rise above the solubility limit, the ions combine to form solid particles that precipitate from solution. Recognizing these insoluble precipitates requires some intuition based on solubility rules or by looking up equilibrium constants. Mixing two solutions that contain soluble salts can result in precipitation. For the case of mixing solutions of the strong electrolytes $AgNO_3$ and $NaCl$,

$$Ag^+(aq) + NO_3^-(aq) + Na^+(aq) + Cl^-(aq) \rightarrow AgCl(s) + NO_3^-(aq) + Na^+(aq)$$

After removing the spectator ions, the net reaction is

$$\underset{\text{cation}}{Ag^+(aq)} + \underset{\text{anion}}{Cl^-(aq)} \rightarrow \underset{\text{ionic solid}}{AgCl(s)}$$

The stoichiometry of a precipitation reaction will depend on the precipitate (see Example 3.6), which must be a neutral solid:

$$a\,M^{m+}(aq) + b\,X^{x-}(aq) \rightarrow M_aX_b(s)$$

In general $\{a \times (m+)\} + \{b \times (x-)\} = 0$.

Example 3.6 Precipitation Reaction.   What reaction occurs as an acidic solution of iron(III) chloride, $FeCl_3$, is neutralized with a solution of sodium hydroxide, $NaOH$?

As $NaOH$ is added to neutralize the acid in the solution, a point is reached where the ions present are

$$Fe^{3+}(aq), Cl^-(aq), Na^+(aq), \text{ and } OH^-(aq)$$

Since we know $FeCl_3$ is a soluble salt, $Cl^-(aq)$ is a spectator ion, as is $Na^+(aq)$. The possible reaction is the precipitation of $Fe(OH)_3$. The unbalanced reaction is

$$Fe^{3+}(aq) + OH^-(aq) \rightarrow Fe(OH)_3(s)$$

There is one Fe on each side of the reaction, so we only need to balance the hydroxide. There are three hydroxides on the right, so we need three hydroxides on the left:

$$Fe^{3+}(aq) + 3OH^-(aq) \rightarrow Fe(OH)_3(s)$$

The reaction is now mass balanced. The charge on the left side of the reaction is +3 + 3(−1) = 0, which equals the charge on the right side of the reaction. The reaction is mass and charge balanced, so we are done.

3.3.7 Precipitate Equilibria

Insoluble salts in contact with water will maintain some equilibrium concentration of the constituent ions in the aqueous solution. The concentrations of the ions remaining in solution are described by the equilibrium constant K'_{sp}. The "sp" subscript indicates that the equilibrium is a precipitation reaction, and K'_{sp} is called the *solubility product*. We always write the equilibrium reaction of an insoluble salt as the precipitate dissolving

$$M_aX_b(s) \rightleftharpoons aM^{m+}(aq) + bX^{x-}(aq)$$

where M^{m+} is the cation, usually a metal, X^{x-} is the anion, and a and b are their stoichiometric coefficients, respectively. Now, writing the equilibrium constant following the usual rules,

$$K'_{sp} = [M^{m+}]^a[X^{x-}]^b$$

Per the general rules for writing K' expressions, the concentration of the solid compound, M_aX_b, does not appear in the expression.

There is no precipitate equilibrium until ion concentrations exceed the value of K'_{sp}. Using our previous example,

$$Ag^+(aq) + Cl^-(aq) \rightarrow AgCl(s)$$

When a sufficient amount of chloride ion is added to a solution that contains silver ion, solid silver chloride will precipitate from solution. The concentrations of Ag^+ and Cl^- that remain in solution are controlled by the equilibrium reaction and the value of K'_{sp}.

$$AgCl(s) \rightleftharpoons Ag^+(aq) + Cl^-(aq)$$
$$K'_{sp} = [Ag^+][Cl^-]$$

3.3.8 Redox Reactions

Essentially all chemical species and materials can undergo reduction and oxidation (redox) reactions. Organic materials do not combust instantly at room temperature, but can do so in the presence of oxygen if ignited. Iron metal does not instantly convert to rust, Fe_2O_3, but will do so over time. Look around and think about the oxidation state of what you see. Much of what we see in our environment is fully oxidized—calcium is present as Ca^{2+} not Ca metal, silicon is present as SiO_2 not Si. Material that is not fully oxidized is metastable in air at moderate temperatures. These materials have usually formed from biological or geological processes with the input of energy.

In the reactions that we consider, you can think of the reactants as competing with other reactants to form products. Redox chemistry is the same because different chemical species have different affinities for electrons. Redox reactions involve the transfer of electrons from one reactant to another. One reactant will be oxidized and another reactant will be reduced. Mixing solutions that contain different electroactive elements will allow one species to lose one or more electrons to a species that has the greatest affinity for electrons. We therefore break redox reactions into two half reactions. For the reaction of zinc metal with Cu^{2+} in solution,

$$\begin{array}{cccc} Zn(s) & +\ Cu^{2+}(aq) & \rightarrow\ Zn^{2+}(aq) & +\ Cu(s) \\ \text{electron} & \text{electron} & \text{oxidized} & \text{reduced} \\ \text{provider} & \text{acceptor} & \text{product} & \text{product} \end{array}$$

Two electrons are transferred from each zinc atom to each copper ion. The zinc metal is oxidized to zinc ions, and the copper ions are reduced to copper metal. We can write the following half reactions:

$$Zn(s) \rightarrow Zn^{2+}(aq) + 2e^-$$

$$Cu^{2+}(aq) + 2e^- \rightarrow Cu(s)$$

Combining the half reactions to eliminate the electrons gives the overall reaction as shown above.

3.3.9 Redox Equilibria

Writing redox equilibria follows the same rules as for any other type of equilibrium; for example, for the reaction

$$Zn(s) + Cu^{2+}(aq) \rightleftharpoons Zn^{2+}(aq) + Cu(s)$$

the equilibrium constant expression is

$$K' = \frac{[Zn^{2+}]}{[Cu^{2+}]}$$

Again, note that the pure solids, Cu(s) and Zn(s), do not appear in the expression.

In the other types of equilibria, the appropriate equilibrium constant is our measure to calculate equilibrium concentrations. In redox chemistry, it is more useful to relate relative concentrations of reactants and products as voltages. We will leave the details of this topic to Chapter 9. For the redox examples in this chapter, we can assume that the reactants are chosen because they go to completion.

You-Try-It 3.A
The limiting-reagent worksheet in you-try-it-03.xlsx contains different types of reactions that are useful in analytical procedures. Determine the minimum amount of reagent that is necessary for each reaction to go to completion.

With this review of reaction types and equilibrium conventions, we are ready to use chemical reactions for analytical measurements. The next several sections highlight the use of chemical reactions to obtain a quantitative result. When you see an analytical protocol with reagents added in different steps, think about what type of reaction is occurring and why such a reagent might be necessary.

3.4 GRAVIMETRY

Gravimetry, or gravimetric analysis, is a quantitative method to determine an analyte concentration by weighing a pure, solid form of the analyte. The solid form is obtained by adding a precipitating reagent to a solution that contains the analyte. The precipitating agent should be selective for the target analyte. Recall from Section 1.6 that one requirement for a primary standard was a high formula weight. Gravimetric reagents often produce high formula masses for the same reason, to obtain greater precision when weighing. Since gravimetric analysis is made using a certified weight, it is a principal method for preparing and certifying primary standards.

Table 3.2 outlines a basic gravimetric procedure to determine the concentration of an analyte. The first step is not necessary in some applications, such as determining particulate matter in air or total suspended solids in water. In these cases, some amount of air or water is simply filtered to collect the solid matter. An analyte in solution is removed from solution by precipitation with a suitable reagent. For an analyte that is present as a component of a solid, the solid must be dissolved to perform the precipitation.

The precipitate is collected by filtering, rinsed to remove impurities, and dried. If necessary, a high-temperature ignition converts the precipitate to a pure form. *Ignition* involves heating to a high temperature with a flame or a muffle furnace (a high-temperature oven). All steps in a gravimetric procedure have the goal of achieving quantitative collection and conversion to a pure substance. When this requirement is met, the mass of the weighed substance can be related back to the number of moles of analyte in the original sample (see Example 3.7). As the following example illustrates, a real gravimetric laboratory procedure will have many more steps than are listed in the outline in Table 3.2.

TABLE 3.2 Simplified Precipitation Procedure

Step	Purpose
Dissolve and precipitate	Separate analyte from other sample components
Filter	Collect and wash to remove impurities
Dry or ignite	Convert precipitate to pure form
Weigh on analytical balance	Obtain experimental measurement
Calculate concentration	Determine unknown using reaction stoichiometry

Gravimetric Determination of Iron
1. Clean several crucibles, dry in oven, and bring each to constant weight.
2. Accurately weigh several portions of sample into separate beakers.
3. Add dilute HCl to each test portion to dissolve completely.
4. If any solid particulate is visible, filter to remove.
5. Add hydrogen peroxide to test solution to oxidize all iron in solution to +3 oxidation state.
6. Heat solution to remove excess peroxide.
7. Add ammonium hydroxide to the hot solution to precipitate iron hydroxide.
8. Add slurry of filter aid to the test solution.
9. Decant solvent through ashless filter paper.
10. Wash precipitates with hot water three to five times and pour through filter paper.
11. Filter and collect all solid with a quantitative transfer.
12. Allow precipitate and filter paper to dry and then transfer to constant-weight crucible.
13. Char filter paper under heat lamp or with low heat burner and then ignite sample to convert iron hydroxide to iron oxide.
14. Cool crucible in desiccator and then weigh.
15. Repeat to determine constant weight of the crucible with iron oxide.
16. Calculate percentage of iron in each test portion.

To explain some of the details in the example procedure, *constant weight* is the repetitive drying and weighing of a crucible until the weight does not change. This process is necessary for the clean crucibles and when they contain the precipitate so that the difference in weights is due only to the precipitate. The hydrogen peroxide is added to oxidize all iron to Fe^{3+} because $Fe(OH)_3(s)$ has a much lower K'_{sp} than $Fe(OH)_2(s)$. The filter aid in step 8 is added because iron hydroxide tends to form small colloidal particles that clog the filter paper. The filter aid and filter paper are both removed by combustion in step 13. The thorough washing in step 10 is necessary to remove soluble interferences.

Handbooks provide detailed procedures for gravimetric analysis, including precipitating agents and common interferences. A few examples are listed in Table 3.3.[7] Note that the weighed form is often different from the initial precipitate; for example, calcium oxalate and Al and Fe hydroxides are converted to their oxide. The conversion provided by the ignition step creates a more stable form that is free of waters of hydration and other impurities. Handbooks and procedures will list "gravimetric factors" for specific gravimetric methods. These factors include the formula weights of the precipitate and analyte so that multiplying a precipitate weight by the gravimetric factor gives the analyte weight.

TABLE 3.3   Precipitating Agents for Selected Elements

Analyte	Precipitating Agent	Weighed Form, After Ignition If Applicable
Mg	$NH_4H_2PO_4$	$Mg_2P_2O_7$
Ca	Oxalic acid, $(COOH)_2$	CaO
Sr, Ba, Ca, Pb	H_2SO_4	$SrSO_4$, $BaSO_4$, $CdSO_4$, $PbSO_4$
Al, Fe	NH_4OH	Al_2O_3, Fe_2O_3
Ag	HCl	AgCl
Cl^-	$AgNO_3$	AgCl
SO_4^{2-}	$BaCl_2$	$BaSO_4$
PO_4^{3-}	$MgCl_2$	$Mg_2P_2O_7$

Example 3.7 Gravimetric Calculation.   To determine the composition of aluminum scrap metal, which is alloyed with copper, you dissolve 0.2755 g of metal chips in nitric acid and precipitate the aluminum with ammonium hydroxide. The copper remains in solution as an ammine complex. On drying and igniting the precipitate, you find 0.4111 g of Al_2O_3. What is the composition of the aluminum scrap assuming that it contains only Al and Cu?

From the weight of Al_2O_3 we can determine the amount of Al in moles that was present in solution

---

[7]Adapted with permission from Dean, J. A. *Analytical Chemistry Handbook*, McGraw-Hill: New York, 1995. See this reference for an extensive listing.

$$0.4111\text{g Al}_2\text{O}_3\left(\frac{1\,\text{mol Al}_2\text{O}_3}{101.96\,\text{g Al}_2\text{O}_3}\right)\left(\frac{2\,\text{mol Al}}{1\,\text{mol Al}_2\text{O}_3}\right)=8.06395\times10^{-3}\,\text{mol Al}$$

Converting to weight of Al

$$8.06395\times10^{-3}\,\text{mol Al}\left(\frac{26.981\text{g Al}}{1\,\text{mol Al}}\right)=0.21757\text{g Al}$$

The fraction of Al in the alloy is this weight of Al divided by the total sample weight:

$$\frac{0.21757\text{g Al}}{0.2755\text{g sample}}=0.7897$$

which we express as weight percent by multiplying by 100%:

$$0.7897\times100\%=78.97\%$$

Since there are only two components in the metal, the fraction of Cu in the alloy is

$$100\%-78.97\%=21.03\%$$

We discuss solubility in detail in Chapter 8, but Figure 3.2 shows the reality for any insoluble precipitate. Equilibrium concentrations of the ions of the insoluble precipitate will remain in solution. These equilibrium concentrations are characteristic for each insoluble precipitate, as described by the value of K'_{sp}. The precipitates that are useful for gravimetric analysis are those that have very low solubility, that is, they have very small values of K'_{sp}. As an example, AgCl has a K'_{sp} of approximately 1.8×10^{-10}. In the absence of competing equilibria, AgCl will have equilibrium $[Ag^+]$ and $[Cl^-]$ of 0.000013 M

$$K'_{sp}=1.8\times10^{-10}=[Ag^+][Cl^-]$$
$$[Ag^+]=[Cl^-]=1.3\times10^{-5}$$

Using a total of 500 ml of water for washing, and assuming that equilibrium is established during washing,

$$(1.3\times10^{-5}\,\text{M Ag}^+)(0.500\,\text{l})\frac{1\,\text{mol AgCl}}{1\,\text{mol Ag}^+}(143.321\,\text{g/mol})$$
$$=9.3\times10^{-4}\,\text{g AgCl}$$

or an approximate loss of 1 mg of AgCl. If the precipitate mass is on the order of 0.5 g, the error due to the equilibrium concentrations that remain in solution is ≈0.2%. Using more sample to obtain a larger precipitate reduces the

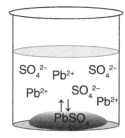

Figure 3.2 Schematic of solution equilibrium for an insoluble salt.

error. Samples with a low analyte concentration that produce small amounts of precipitate will have correspondingly greater uncertainty.

More problematic than losses due to the equilibrium concentrations are impurities that coprecipitate with the analyte of interest. Avoiding this problem usually requires some knowledge of the sample matrix. Looking at Table 3.3, alkaline earth metals will interfere in the precipitation of $PbSO_4$. Similarly, using Ag^+ for Cl^- precipitation to AgCl is problematic if other halides, and a few other anions, are present in the solution.

Validating analytical methods for different types of samples can provide information on the impact of interferences so that procedures can be modified to reduce systematic errors. If a certain sample component interferes in a precipitation, it can often be removed by using a precipitation or complexometric agent that is selective for the interfering substance. In the same manner, a weighed precipitate can be redissolved to determine the contribution of the interfering component by another analytical procedure. The amount of the interference is then subtracted from the initial precipitate weight for a more accurate determination of the analyte. For high accuracy, subsequent analysis of the precipitate is done using instrumental methods to identify and quantitate interfering components.

You-Try-It 3.B
The gravimetry worksheet in you-try-it-03.xlsx contains examples of gravimetric procedures. Predict the precipitate weight for each of the listed samples.

3.5 TITRATION

Titration is a quantitative method to determine the concentration of an analyte in solution by reacting it with a *titrant*, a standardized reagent solution. The exact concentration of the titrant is known by preparing it from a primary standard or by standardizing it versus a primary standard. The wide use of titration arises because we can measure liquid volumes very accurately. As discussed in Section 1.6, volumetric glassware is one of our primary tools for maintaining accuracy in our analytical methods.

The experimental measurement in a titration is to determine the *equivalence point*, the point in the reaction at which all analyte is consumed by reaction with the titrant. The number of moles of analyte that was present in the test portion is calculated from the volume of titrant needed to reach the equivalence point, the titrant concentration, and the reaction stoichiometry. The unknown concentration is then found by dividing by the volume of the solution before beginning the titration. If the test portion was diluted or dissolved, these factors are included to get analyte concentration in the original sample.

Titration methods have been developed using neutralization, complexation, precipitation, redox, and other types of reactions. The only requirement is that the titration reaction goes to completion. For the general reaction where A is the analyte and B is the titrant,

$$a\mathrm{A} + b\mathrm{B} \rightarrow \text{products}$$

The calculation for the amount of analyte is

$$\text{moles}_\mathrm{A} = \text{moles}_\mathrm{B}\left(\frac{a}{b}\right) = V_\mathrm{B}c_\mathrm{B}\left(\frac{a}{b}\right) \tag{3.1}$$

where V_B is the volume of titrant to reach the equivalence point, c_B is the concentration of the titrant, and (a/b) is the stoichiometric ratio between the analyte and the titrant. The calculation is often written as $M_1V_1 = M_2V_2$, where M is molar concentration and V is volume. This expression is correct for the special case of $a = b$. Do not forget to include the stoichiometric ratio in calculations when the titration reaction does not have a 1:1 stoichiometry.

As in gravimetry, "titrimetric factors" that include stoichiometric ratios can be found tabulated in reference sources for specific titration reactions. We will not use these tabulated factors; our main concern is to understand the chemical processes that can occur before, at, and after the equivalence point.

3.5.1 Titration Error

In practice, we must use some experimental method to detect the equivalence point. The detection method might respond at a slightly different titrant volume from the true equivalence point. We call our experimental determination the *end point*. We call the difference between the titrant volume required to reach the end point and the volume to reach the true equivalence point the *titration error*:

$$\left| V_\text{end point} - V_\text{equivalence point} \right| \tag{3.2}$$

Validated titration methods will specify reagents or conditions that minimize this error. We must understand this issue if we are developing a new analytical procedure or adapting an existing method for a different analyte. The end point of a titration is often determined by a visual indicator that changes color or other appearance. Using a neutralization titration as an example, adding the titrant near the equivalence point causes the solution pH to change rapidly. This pH change is detectable with indicators that change color as a function of pH. Acid–base indicators are weak acids that change color when they gain or lose their acidic proton(s). Table 3.4 lists a few common indicators with the color of their acidic and basic forms, that is, the solution appearance before, at, and after the end point.[8] The last column gives the pH range over which the color change occurs.

Multiple indicators are available to be able to choose one that matches the equivalence point pH for different acids and bases. The pH on neutralizing a strong acid with a strong base will be 7.0. The pH at the equivalence point for weak acids and weak bases will be different, depending on the K_a' and K_b' values. Neutralizing acetic acid with strong base has an equivalence point pH near 8. Using methyl orange indicator for this titration will introduce a larger titration error than using bromothymol blue or phenolphthalein. We revisit this point graphically when we look at titration curves in Section 3.6. Table 3.5

TABLE 3.4 Acid–Base Indicators and Their Characteristic Colors

	Acidic Form	End Point	Basic Form	pH Range
Methyl orange	Red	Orange	Yellow	3.1–4.4
Bromocresol green	Yellow	Green	Blue	4.0–5.6
Methyl red	Red	Yellow	Yellow	4.4–6.2
Bromothymol blue	Yellow	Green	Blue	6.2–7.6
Phenolphthalein	Colorless	Light pink	Red	8.0–10

TABLE 3.5 Indicators for Other Types of Titrations

Analyte	Titrant	Titration Type	End Point Detection
Metal ions	EDTA	Complexometric	Metal indicator (color change)
Cl^-	Ag^+	Precipitation	CrO_4^{2-} (red precipitate)
Ba^{2+}	SO_4^{2-}	Precipitation	Arsenazo III indicator
Fe^{2+}	MnO_4^-	Redox	Persistence of permanganate color
$S_2O_3^{2-}$	IO_3^-	Redox	Starch indicator (blue color)
I_2	$S_2O_3^{2-}$	Redox	Starch indicator (blue color)

[8]Adapted with permission from Dean, J. A. *Analytical Chemistry Handbook*, McGraw-Hill: New York, 1995. See this reference for an extensive listing.

lists titration indicators that are available for other types of reactions and Example 3.8 gives a sample calculation.

Example 3.8 Titration Based on a Precipitation Reaction. Hg_2^{2+} can be determined by titrating with NaCl solution, which precipitates Hg_2^{2+} as $Hg_2Cl_2(s)$. The net titration reaction is

$$Hg_2^{2+}(aq) + 2Cl^-(aq) \rightarrow Hg_2Cl_2(s)$$

15.23 ml of 0.2205 M NaCl is required to reach the end point in titrating a 100.0-ml test portion. What is the Hg_2^{2+} concentration in this 100.0-ml solution?

The measured volume of the titrant tells how much Cl^- was necessary to reach the end point:

$$\text{moles}_{Cl^-} = (0.01523\ 1)(0.2205\,M\ Cl^-) = 3.3582 \times 10^{-3}\,\text{mol}\ Cl^-$$

From the stoichiometry of the titration reaction, we can convert moles of titrant to moles of analyte:

$$\text{moles}_{Hg_2^{2+}} = (3.3582 \times 10^{-3}\,\text{mol}\ Cl^-)\left(\frac{1\,\text{mol}\,Hg_2^{2+}}{2\,\text{mol}\,Cl^-}\right)$$

$$= 1.6791 \times 10^{-3}\,\text{mol}\ Hg_2^{2+}$$

Finally, we can use the moles of analyte determined by the titration to calculate the Hg_2^{2+} concentration in the test portion:

$$c_{Hg_2^{2+}} = \frac{1.6791 \times 10^{-3}\,\text{mol}\ Hg_2^{2+}}{0.1000\ 1} = 0.01679\,M\ Hg_2^{2+}$$

You-Try-It 3.C
The titration worksheet in you-try-it-03.xlsx contains titration results for a number of analytical measurements. Calculate the unknown concentration for each case.

Manual titration is done with a burette, which is a long, graduated tube to deliver a precise volume of titrant. The amount of titrant used in the titration is found by subtracting the volume of titrant in the burette at the start of the

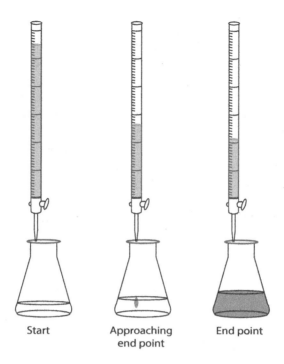

Start   Approaching   End point
end point

Figure 3.3 Titration with a visual indicator.

titration from the volume at the end point. This difference in the readings is the volume of titrant delivered to the titration solution to reach the end point. An important factor to get accurate results is to align your eye with the meniscus of the liquid to read the burette volume reproducibly. Figure 3.3 illustrates this sequence of steps to perform a titration. The addition of titrant can be done rapidly until nearing the end point, and then more slowly to determine the final volume on reaching the end point. The solution is stirred or swirled by hand for rapid mixing of the titrant with the analyte.

Instrumental methods are also available to detect the end point for many types of titration reactions. The advantage of using an instrument is that it can record a signal as a function of titrant volume. Plotting this data can provide a more accurate determination of the end point than recording a single point based on a visual indicator. This difference in accuracy is especially relevant if the end point is gradual or indistinct.

The most familiar instrument is the pH meter for neutralization titrations. The pH meter uses an ion-selective electrode for H^+. Complexometric or precipitation titrations can often use ion-selective electrodes that are specific for one of the ions involved in the titration reaction. We describe the principles of these potentiometric methods in Chapter 9. Redox or ORP electrodes, which use a platinum or gold electrode, can detect the large change in redox potential of a solution on reaching the end point of a redox titration. Using an indicator that changes color allows the use of absorption of light for detection.

Another advantage of using an instrumental method for end point detection is that it can be integrated into automated instruments. These instruments, called autotitrators or simply titrators, are rapid and especially useful for making repetitive titrations. An autotitrator contains a microprocessor that controls delivery of the titrant, stops at the end point, and calculates and displays the concentration of the analyte. The microprocessor control can even implement GLP-compliant procedures for user access, method validation, etc. The end point is usually detected by some type of electrochemical measurement, but some specialized titrators use optical methods.

Some examples of titrations for which autotitrators are available include

- acid or base determination with end point detection by conductivity or potentiometry (pH);
- complexometric titration of metal ions with EDTA monitoring a color indicator photometrically;
- determination of water by Karl Fischer reagents by a coulometric titration (more details below).

3.5.2 EDTA Titration

Metal ions are easily titrated using ethylenediaminetetraacetic acid, EDTA, to form the 1:1 metal–EDTA complex. EDTA is the common abbreviation, but the actual titrant will be a solution of a soluble salt such as Na_4EDTA. The form of EDTA depends on pH and exists as the −4 anion at high pH. Table 3.6 lists the formation constants for EDTA with different metal ions.[9] The high values indicate a strong interaction so that the complexation goes to completion (in the absence of interferences). At the equivalence point of an EDTA titration, essentially all metal is bound in the metal–EDTA complex. Figure 3.4 shows the structure of EDTA with a complexed metal ion, M (metal charge not shown). The ethylene backbone is on the left in this figure. The lone pair electrons on each of the two nitrogen atoms make coordinate covalent bonds to the metal ion. The structure is such that the acetic acid groups can "wrap around" to also form bonds with the metal ion. The total of six bonds between the EDTA and the metal ion gives rise to the strong interaction.

In a metal–EDTA titration, the concentration of free metal ion will decrease rapidly at the equivalence point. This rapid change is detectable with an indicator electrode that responds to metal ion concentration or with an indicator that changes color when it binds or loses a metal ion. Calmagite indicator has two –OH groups with acidic protons. The color of calmagite changes depending on whether these protons are present or displaced by a metal ion. Figure 3.5 shows the structure of calmagite with a calcium ion

---

[9]Adapted with permission from Speight, J. G., Ed. *Lange's Handbook of Chemistry*, 16th ed.; McGraw-Hill: New York, 2005.

replacing the protons. Although I use the equilibrium symbol, the reaction goes almost completely to the EDTA–metal ion complex at the equivalence point. Calmagite is also useful for magnesium ions and is a common indicator for titrating water hardness. The calmagite–metal ion complex has a red color. In the absence of a metal ion at pH = 10, one proton is present on calmagite and the color of the indicator is blue. This indicator is useful for titrating Ca^{2+} and Mg^{2+} with EDTA titrant because the EDTA binds Ca^{2+} and Mg^{2+} more strongly than does the indicator. On reaching the equivalence point, all metal ion is complexed by EDTA, and the indicator changes from red to blue.

TABLE 3.6 β_1 Values for Metal–EDTA Complexes

Metal	Formula	β_1
Aluminum	Al(edta)$^-$	1.3×10^{16}
Calcium	Ca(edta)$^{2-}$	1×10^{11}
Chromium(II)	Cr(edta)$^{2-}$	4×10^{13}
Chromium(III)	Cr(edta)$^-$	1×10^{23}
Cobalt(II)	Co(edta)$^{2-}$	2.0×10^{16}
Cobalt(III)	Co(edta)$^-$	1×10^{36}
Copper(II)	Cu(edta)$^{2-}$	5×10^{18}
Iron(II)	Fe(edta)$^{2-}$	2.1×10^{14}
Iron(III)	Fe(edta)$^-$	1.7×10^{24}
Lead	Pb(edta)$^{2-}$	2×10^{18}
Magnesium	Mg(edta)$^{2-}$	4.4×10^{8}
Mercury	Hg(edta)$^{2-}$	6.3×10^{21}
Nickel	Ni(edta)$^{2-}$	3.6×10^{18}
Strontium	Sr(edta)$^{2-}$	6.3×10^{8}
Zinc	Zn(edta)$^{2-}$	3×10^{16}

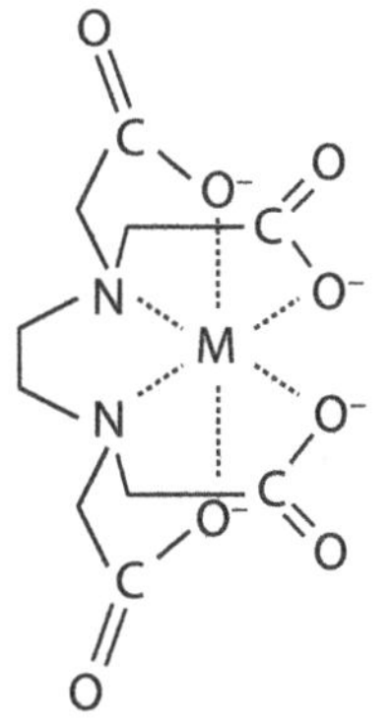

Figure 3.4 Metal–EDTA complex.

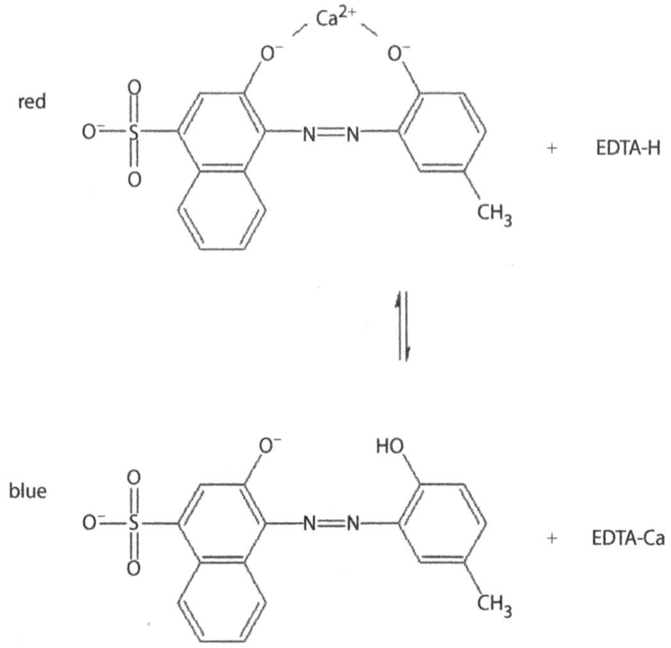

Figure 3.5 Calmagite indicator for EDTA titration.

3.5.3 Back Titration

For some analytes, a direct or forward titration is either not feasible or the end point detection is not clear. In such cases, the titration is done in an indirect manner by adding excess reagent. This reagent reacts completely with the analyte, and a titration is then used to determine the amount of the reagent remaining in solution. This type of titration is called a *back titration*.

An example is titrating an insoluble solid such as $CaCO_3$, $Mg(OH)_2$, or MgO. In these cases, the neutralization of the solid base with an acid titrant requires a wait after each addition of acid, making the determination of the end point a slow process. These types of samples can be titrated much more quickly by reacting the solid with a known amount of acid. The excess acid after neutralizing the analyte is then titrated with a standard base. Many metal–EDTA and other types of titrations will also be performed by back titration for the same reasons, that is, to eliminate a slow process in the titration. The process is best illustrated by the calculation in Example 3.9.

Example 3.9 Antacid Back Titration.   0.250 g of an antacid tablet is mixed with 30.0 ml of standardized 0.1500 M HCl. Titrating the excess acid required 19.05 ml of 0.1022 M NaOH titrant. What is the neutralizing power of this antacid tablet expressed as moles of acid per gram of antacid?

The procedure for this analysis is to add an excess of acid to neutralize all the base in the antacid test portion. The total amount of acid is

$$(0.0300 \ 1)(0.1500 \, M) = 0.00450 \, moles$$

The number of moles of base in the test portion is the total number of moles of standard acid minus the amount left unreacted as determined by the titration

$$0.00450 \, moles - (0.01905 \ 1)(0.1022 \, M) = 0.00255 \, moles$$

The neutralizing power for this 0.2500 g test portion is 0.00255 moles of acid. For 1.00 g of the antacid, the neutralizing power will be four times this value:

$$0.00255 \, moles \times 4 = 0.0102 \, moles \ acid/g \ antacid$$

3.6   TITRATION CURVES

Using an instrumental method to record a signal as we add the titrant allows us to plot a full titration curve. Figure 3.6 shows a titration curve for a strong acid with 0.100 M NaOH, as recorded with a pH meter. The equivalence point is determined from a large change in signal for a small addition of titrant. This titration required 0.0350 l of titrant, or 0.00350 mol of OH^-. It occurred at a pH of 7, as expected for titration of a strong acid with a strong base. Since the stoichiometry of this neutralization reaction is 1:1, the titration shows that there were 0.00350 mol of H^+ in the original solution.

We can identify three different regions in this titration experiment. Before the equivalence point, the pH is determined by the concentration of unneutralized strong acid. At the equivalence point, the pH = 7 and changes rapidly for small additions of titrant. Past the equivalence point, the pH is determined by the concentration of the excess strong base that we are adding.

The sharpness of the equivalence point will depend on the concentration of the analyte. The shaded rectangle in Figure 3.6 shows the pH range over which the phenolphthalein indicator changes color. It is slightly different from the equivalence point, but since pH is changing so rapidly near the equivalence point, the titration error is small. That is, the volume to reach the experimental end point is very nearly equal to the true equivalence point. Figure 3.7 shows the titration curves of two strong acid solutions at different concentrations. The curve for the acid at a low concentration does not have as sharp a transition as at the equivalence point. In this case, the titration error is significant, and a better indicator should be found.

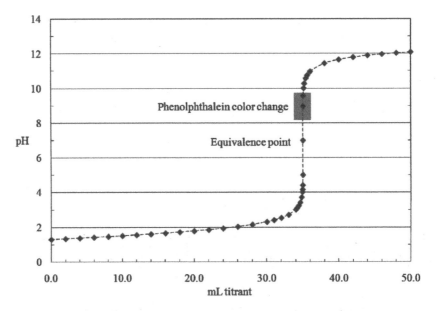

Figure 3.6 Titration curve of a strong acid with a strong base.

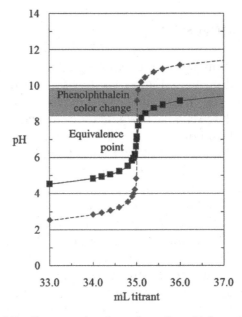

Figure 3.7 Concentration dependence for acid–base titration.

One method to improve our ability to pinpoint the position of the equivalence point is to use a Gran plot. The following equation is derived from the K_a' equilibrium constant expression for titrating an acid with a base:

$$V_{base} \times 10^{-pH} = K_a'(V_e - V_{base})$$

Plotting $V_{base} \times 10^{-pH}$ versus V_{base} gives a line with slope of K_a and an x-intercept of V_e. Figure 3.8 shows an example, where the x-scale has been expanded near the equivalence point. When curvature makes the equivalence difficult to determine, the data in the Gran plot is extrapolated as a line to find the x-intercept.

You-Try-It 3.D
The titration worksheet in you-try-it-03.xlsx contains a set of titration data. Plot the data as signal versus added titrant and as a Gran plot to determine the equivalence point.

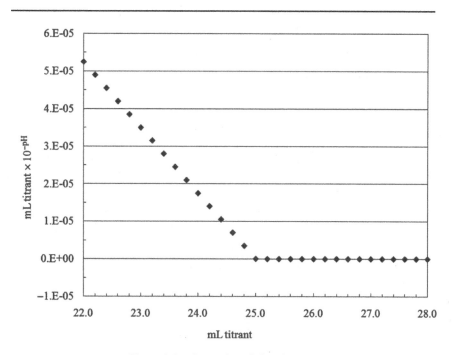

Figure 3.8 Gran plot of titration curve.

3.7 COULOMETRY

We finish this chapter with a specialized titration method that is common for some types of analytes. Coulometry is an analytical method for measuring the concentration of an analyte in solution by completely converting the analyte from one oxidation state to another. In this sense, it is similar to a redox titration. However, it uses an electrical source to supply electrons, which is very precise and accurate. Coulometry is an absolute measurement similar to gravimetry; it requires no chemical standards or calibration. It finds use in absolute concentration determinations to prepare standards and in Karl Fischer titrations of water in samples. There are two approaches to coulometry:

Potentiostatic: uses a constant potential at the working electrode to quantitatively oxidize or reduce the analyte.

Amperostatic: uses a constant current to quantitatively oxidize or reduce the analyte.

Both of these methods require a quantitative oxidation or reduction of the analyte, that is, there can be no interferences present. They do not require a direct oxidation or reduction of the analyte at an electrode. It is possible, and common, to electrochemically convert a precursor reagent to the titrant, which then reacts with the analyte quantitatively.

3.7.1 Coulometric Titration

Owing to the depletion of the analyte at an electrode surface, it is very difficult to completely oxidize or reduce all of an analyte by electrochemistry. Because of this condition, called *concentration polarization*, coulometry is usually implemented by generating an intermediate reagent that undergoes redox reaction with the analyte. The intermediate reagent is electrochemically generated from an excess of a precursor, preventing concentration polarization.

An example is the electrochemical oxidation of I^- (the precursor) to I_2 (the intermediate reagent). The I_2 can chemically oxidize analytes such as organic molecules. Figure 3.9 shows a schematic of a simple setup for a coulometric titration. Note that there is no burette or other titrant solution. The titrant is generated in situ by the generator electrode. A typical titration will measure the time required to reach the end point when using a constant current source. The amount of charge transferred to reach the end point is then,

$$\text{charge} = (\text{current})(\text{time})$$

where charge is expressed in coulombs, C, the charge carried in 1 s at a constant current of 1 amp. This current is related to analyte concentrations by the

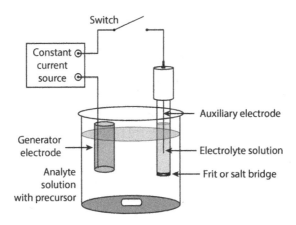

Figure 3.9 Schematic of a coulometric analysis.

Faraday constant and the redox stoichiometry of the reaction. The Faraday constant, $F = 96485.3$ C/mol, is the amount of charge of one mole of electrons.

The point at which all of the analyte has been converted to the new oxidation state is the end point. It is determined by some type of indicator or instrument that is also present in the solution. For the coulometric titration of ascorbic acid, vitamin C, with I_2, soluble starch is used as a visual indicator:

$$C_6H_8O_6 \quad + \quad I_2 \quad \rightarrow \quad C_6H_8O_6 \quad + \quad 2I^- + 2H^+$$
$$\text{Ascorbic acid} \qquad\qquad \text{dehydroascorbic acid}$$

At the end point, all ascorbic acid is consumed so that excess I_2 will be present in solution. The I_2 forms a blue or violet complex with the starch. Example 3.10 gives a sample calculation.

Example 3.10 Coulometric Titration. A coulometric titration of ascorbic acid required 0.3974 C of charge. How many moles of ascorbic acid were present in the sample?

The relevant reactions are

$$2I^-(aq) + 2e^- \rightarrow I_2(aq)$$

$$C_6H_8O_6(aq) + I_2(aq) \rightarrow C_6H_6O_6(aq) + 2I^-(aq) + 2H^+(aq)$$

The number of electrons transferred in the titration is determined using the Faraday constant

$$\frac{0.3974\,C}{96485\,C/\text{mol}} = 4.119\,\mu\text{mol}\,e^-$$

The number of moles of ascorbic acid is determined using the titration result and the stoichiometric factors:

$$(4.119\,\mu\text{mol}\,e^-)\left(\frac{1\,\mu\text{mol}\,I_2}{2\,\mu\text{mol}\,e^-}\right)\left(\frac{1\,\mu\text{mol}\,C_6H_8O_6}{1\,\mu\text{mol}\,I_2}\right) = 2.059\,\mu\text{mol}\,C_6H_8O_6$$

3.7.2   Karl Fischer Coulometric Titration

The Karl Fischer titration is a very common and useful coulometric method for analyzing water from 1 ppm to 5%. This concentration range covers the amount of water in organic solvents, oils, agricultural material, consumer products, and other types of samples. The Karl Fischer titration uses the following scheme to quantitate water. The first step of the reaction is the formation of an alkylsulfite intermediate, where RN is a base such as pyridine or imidazole:

$$CH_3OH + RN + SO_2 \rightarrow [RNH]SO_3CH_3$$

The intermediate reagent, $[RNH]SO_3CH_3$, is an alkylsulfite salt that reacts with water, the analyte to be quantitated, and iodine. The final products are soluble salts:

$$[RNH]SO_3CH_3 + H_2O + I_2 + 2RN \rightarrow [RNH]I + [RNH]SO_4CH_3$$

The I_2 is generated electrochemically and the reaction occurs until all water is consumed. The end point can be detected when I_2 remains in solution, and the equivalence factor for the reaction is 10.72 C per 1 mg H_2O.

Chapter 3. What Was the Point?   This chapter reviewed the main types of chemical reactions that occur in aqueous solution. These reactions are used in analytical procedures for sample preparation and to quantitate analytes. The two most common procedures are gravimetry and titration. Coulometry is a type of titration that has widespread use for some analytes, for example, water. A key concept underlying these wet-chemical methods is to keep track of reaction stoichiometry in calculations to convert experimental measurements to analyte concentrations.

PRACTICE EXERCISES

1. For the following pair of equilibria, what is the relationship between K_1 and K_2?

$$A \rightleftharpoons B+C \qquad 2A \rightleftharpoons 2B+2C$$
$$\quad K_1 \qquad\qquad\qquad K_2$$

2. As a refresher, write the equilibria and K expressions for your favorite weak acid, weak base, metal–ligand complex, and precipitate. If you do not have favorites, look at the ingredients on bottles of shampoo and cleaning products and choose some. What similarities do you notice? What details are different?

3. Specify the type of reaction that occurs when the following pairs of reactants are mixed. Balance the reaction and write an equilibrium constant expression for the resulting chemical equilibrium.
   (a) Acidic solutions of $KMnO_4$ and $FeCl_2$.
   (b) 10.0 ml of 0.05 M HCl + 10.0 ml of 0.05 M NaOH.

(c) 10.0 ml of 0.05 M HCl + 10.0 ml of 0.05 M NH_3.
(d) 10.0 ml of 0.05 M HCl + 10.0 ml of 0.05 M $AgNO_3$.
(e) 10.0 ml of 0.05 M $CaCl_2$ + 10.0 ml of 0.05 M $(NH_4)_2C_2O_4$.

4. 3.459 g of plant food was dissolved in water and mixed with a solution of $MgSO_4$ followed by slow addition of ammonia. The resulting precipitate was rinsed with isopropyl alcohol and allowed to dry. Weighing showed that 2.947 g of $MgNH_4PO_4 \cdot 6H_2O$ had been collected. What is the weight percentage of phosphorus, P, in the plant food?

5. Three portions of a sample of iron ore are analyzed using the procedure given in Section 3.4. The results are as follows:

Ore Weight, g	Fe_2O_3 Mass, g
0.4793	0.1225
0.5032	0.1271
0.5318	0.1352

What is the weight percentage of iron in this ore? Give the mean and the standard deviation.

6. 100.0 ml of a silver nitrate solution is added to 98.75 g of a water sample taken from the Chesapeake Bay near the Potomac River. (The density of this water sample is 1.01 g/ml.) After filtering, washing, and drying the precipitate, you have 1.9482 g of AgCl. You may assume that all chloride was present as NaCl. What is the salinity of this water sample expressed as
(a) Weight percent (wt%)
(b) Parts per thousand (‰)
(c) Molarity (M)

7. 0.0200 g of a vitamin C tablet was dissolved in 100 ml of 0.1 M KCl that also contained a small amount of soluble starch. Electrolysis of this solution at a constant current of 48.25 mA produced a blue end point at 413.6 s. What is the weight percentage of ascorbic acid in the vitamin tablet? (Hint: write the iodine/iodide half reaction to get the electrochemical stoichiometric factor.)

8. Al and Fe can be determined in plant matter through a combination of gravimetry and titration. 50.02 g of dried and ground plant matter is digested in acid and filtered to remove insoluble silicate. Excess $(NH_4)_3PO_4$ is added to the filtrate. The resulting precipitate is collected, washed, and ignited at 550°C. The result is 66.3 mg of $FePO_4$ and $AlPO_4$. This precipitate is redissolved in H_2SO_4, and Zn is added to reduce all iron to Fe^{2+}. Titrating with 0.002045 M $KMnO_4$ requires 34.91 ml to reach the endpoint.
(a) What is the balanced titration reaction?
(b) What amount of Fe^{2+}, in moles, was present in the solution that was titrated?

(c) What amount of Fe, in grams, was present in the precipitate?

(d) What amount of Al, in grams, was present in the precipitate?

(e) What was the concentration of Fe and Al in the plant matter as weight percentage?

9. The titration curve of a $Ca(OH)_2$ solution has an endpoint at 35.0 ml of titrant. The volume of the original solution was 50.00 ml and the titrant was 0.1000 M HCl.

   (a) What was the concentration of OH^- in the unknown solution?

   (b) Which one of the following indicators will produce the smallest titration error? (The pH range over which the indicators change color are listed in parentheses.) Explain your answer.

       i. Bromocresol blue (3.8–5.4)

       ii. Methyl red (4.2–6.2)

       iii. Phenol red (6.8–8.3)

       iv. Phenolphthalein (8.0–9.6)

   (c) List the predominant ionic species (concentration $> 10^{-7}$ M) in the titration solution after adding 30.0 ml of titrant?

   (d) List the predominant ionic species (concentration $> 10^{-7}$ M) in the titration solution after adding 40.0 ml of titrant?

   (e) After performing this titration several times you find that your results are somewhat erratic. You suspect that you might have lost some $Ca(OH)_2$ precipitate somewhere during the sample preparation. Describe another way that might be more reliable to perform this measurement.

10. A Volhard titration uses an excess of Ag^+ to precipitate all of the Cl^- in a solution. The remaining Ag^+ can be back titrated with SCN^-, which reacts to form AgSCN(s). A small amount of Fe^{3+} serves as an indicator. At the end point, SCN^- reacts with Fe^{3+} to form the red iron–thiocyanate complex. Adding 35.55 ml of 0.1015 M $AgNO_3$ appeared to precipitate all of the chloride from a 50.00-ml test portion. Back titrating this solution required 11.04 ml of 0.1177 M NaSCN to detect the red end point. What was the Cl^- concentration in the 50.00-ml solution?

11. A 1.000 g solid test portion containing $CaCO_3$ is dissolved in dilute acid containing a few drops of calmagite indicator and diluted to 50.00 ml with a pH = 10 buffer. 25.00 ml of 0.1347 M EDTA was then added to the 50.00-ml solution. Back titration of the excess EDTA required 18.05 ml of 0.09550 M Mg^{2+} solution. What was the weight percentage of $CaCO_3$ in the solid sample?

12. 1.062 g of brass filings were dissolved in concentrated HNO_3. After cooling, the solution was diluted in a 250.0 ml volumetric flask. Excess of EDTA was added to a 25.00-ml test portion. The excess EDTA was titrated with standard 0.0200 M $Pb(NO_3)_2$ solution and then 2,2'-bipyridyl was added to displace the EDTA from the Cu^{2+}. (The advantages of

this procedure are that the EDTA solution does not need to be standardized and the 2,2′-bipyridyl releases the Cu^{2+} but not other metals that can interfere in the titration.) Titration of the released EDTA was done with the standard 0.0200 M $Pb(NO_3)_2$ solution using xylenol orange indicator. Three trials of the experiment yielded titrant volumes of 55.5, 57.1, and 56.0 ml of the $Pb(NO_3)_2$ standard. What is the weight percent of Cu in the brass sample?

13. Sulfite, SO_3^{2-}, in food can be determined by reacting with HCl to produce SO_2, which is then bubbled through a H_2O_2 solution to produce sulfuric acid. The H_2SO_4 is then titrated with a standard NaOH solution. The reactions are

$$SO_3^{2-}(aq) + 2H^+(aq) \rightarrow H_2SO_3(aq)$$

$$H_2SO_3(aq) \rightleftharpoons SO_2(g) + H_2O$$

$$SO_2(g) + 2H_2O_2(aq) \rightarrow H_2SO_4(aq) + 2H_2O$$

(a) What are the two reaction types used in this analysis?
(b) What is the stoichiometry between the sulfite analyte and the OH^- titrant?
(c) What indicator do you recommend to perform this titration?
(d) A confirmation step suggests adding $BaCl_2$ solution after titrating. What reaction occurs and why is this step useful?

14. Discuss the relative advantages and disadvantages of performing a titration using coulometry.

15. Discuss the relative advantages and disadvantages of performing an analysis by titration versus gravimetry. Why would you choose one approach rather than the other?

16. Which is the one factor that will always limit the ultimate uncertainty that can be reported in a quantitative chemical analysis?

17. Discuss possible errors that can occur in quantitative chemical measurements if the reaction on which the analysis is based does not go to completion. Answer this question for both gravimetry and titration and predict if results will tend to be less than or greater than the actual analyte concentration.

CHAPTER 4

MOLECULAR SPECTROSCOPY

Learning Outcomes

- Describe the regions of the electromagnetic spectrum and their typical interaction with matter.
- Identify the major components of UV/Vis absorption and fluorescence spectrometers.
- Extract qualitative and quantitative information from optical spectra.
- Calculate analyte concentration from absorbance or fluorescence data.

4.1 INTRODUCTION

Absorption and fluorescence spectroscopy are instrumental methods, but I place them early in the text rather than in Part III because they are common and powerful "tools-of-the-trade." If you are using this text in conjunction with laboratory work, it serves the practical purpose of introducing the concepts of spectroscopy before you use it in the laboratory. Many analytes, especially metal ions, require a reagent to produce a colored or fluorescent complex. This reaction must go to completion for the measurement to be accurate, so in this sense molecular spectroscopies do fit in Part I of the text. Likewise, we assume that any necessary sample preparation steps are quantitative and go to completion before we can make a spectroscopic measurement. Thinking about

Basics of Analytical Chemistry and Chemical Equilibria: A Quantitative Approach, Second Edition. Brian M. Tissue.
© 2023 John Wiley & Sons, Inc. Published 2023 by John Wiley & Sons, Inc.
Companion Website: www.wiley.com/go/tissue/analyticalchemistry2e

the final measurement step in depth will also provide us a better perspective when we consider the impact of chemical equilibria on results. We will see that even in "simple" cases, interfering equilibria can be a significant complicating factor that must be controlled to make accurate spectroscopic measurements.

Using electromagnetic (EM) radiation is one of only a few general strategies that we have for detecting and quantitating analytes (see Chapter 1). Having said that, there are many different types of spectroscopic techniques that scientists use to study matter. In this text we discuss only a few examples that are most common in chemistry laboratories.

Spectroscopy is the use of EM radiation or charged particles to study the structure (qualitative analysis) or composition (quantitative analysis) of matter. *Spectrometry* is a synonym, although usually signifying a quantitative measurement. This definition is a bit too broad, but because there are so many spectroscopic techniques throughout the EM spectrum, it is easier to be broad and exclude related scientific tools than to try to formulate a precise but overly long definition. Two areas that this definition includes, which are not spectroscopic methods, are microscopy and diffraction. Microscopy uses EM radiation or electrons to form an image to study the morphology of a sample. Diffraction uses the interference of EM radiation or electrons to determine the atomic structure of molecules or materials. Microscopy and diffraction are not spectroscopic methods, but spectroscopy is often performed in conjunction with these methods. Two examples are the fluorescence microscope and energy-dispersive X-ray spectroscopy, which is performed using an electron microscope to measure elemental composition. We discuss spectroscopic transitions in detail in Section 4.4, but for now keep in mind that spectroscopy is useful for analytical measurements because we can measure the change in EM radiation when analytes absorb, emit, or scatter the radiation.

4.2 PROPERTIES OF EM RADIATION

EM radiation is composed of individual units of energy called *photons*. A photon consists of an electric field component, $\vec{\mathbf{E}}$, and a magnetic field component, $\vec{\mathbf{B}}$, which are perpendicular to each other and to the direction of propagation. The electric and magnetic fields oscillate in time at a fixed *frequency*, ν, the Greek letter "nu." The unit of frequency is oscillations per second, s^{-1}, which is also called hertz, Hz. Figure 4.1 shows a schematic to

Figure 4.1   EM radiation traveling the distance of one wavelength.

visualize how the $\vec{E}$ and $\vec{B}$ fields oscillate as the photon travels some distance, indicated by the dashed line. The vertical arrows show the magnitude of $\vec{E}$ and the horizontal arrows show the magnitude of $\vec{B}$. The total distance along the direction of propagation shown in the figure corresponds to one complete cycle of the oscillation of the $\vec{E}$ and $\vec{B}$ fields. We call this distance one *wavelength* with the symbol λ, the Greek letter "lambda."

Figure 4.1 also illustrates the characteristic of *polarization* for EM radiation. For a light beam of only a single photon, not very realistic but simple to visualize, the electric field oscillates in only one plane, and we say that the light is polarized. A light beam containing any number of photons, all having their electric vectors in the same direction, is polarized. A light beam consisting of photons that have their electric field vectors oriented in all random directions is said to be unpolarized. Light sources, such as the sun and lamps, are unpolarized because they emit photons with a random distribution of orientations for the electric field vectors. Unpolarized light can be polarized by selecting a single polarization with an absorbing or reflective polarizing optic. Some light sources, such as lasers, will produce a polarized light beam due to the polarizing optics in the device. The next time you are outside on a sunny day, find out if your sunglasses are polarized by looking at the blue sky toward the sun (but not directly at the sun) and with your back to the sun. You will see that the light scattered from the atmosphere has a different degree of polarization depending on whether the light is forward or backward scattered.

We can think of a photon as a traveling pulse of energy, with the energy contained in the oscillating electric and magnetic field. Since matter consists of positively and negatively charged particles, nuclei and electrons, respectively, the electric field can exert a force on the electronic system. Since the nuclei are massive compared to the electrons, we make an approximation that an applied electric field affects only the electrons. Spectroscopic transitions will occur when the energy of a photon is equal to the energy of a transition in the matter. The energy, E, of a single photon is proportional to its oscillation frequency

$$E = h\nu \tag{4.1}$$

where the proportionality factor, h, is Planck's constant (6.6261×10^{-34} J/s).

The range of possible frequencies, and energies, of EM radiation is very large. We classify different frequency ranges in different regions of the EM spectrum separately, described fully in Section 4.3. In general, high frequencies correspond to high energy photons and lower frequencies correspond to lower energy radiation. To illustrate this concept with an example, UV light can cause sunburn because the energy of the photons is high enough to cause damage to molecules in the skin. Visible light, such as sunlight passing through a glass window that blocks the UV light, has lower frequencies and will not cause sunburn.

All photons travel through a vacuum at a constant velocity of 2.9979×10^8 m/s, which we call the speed of light, c. When light passes through other media, the velocity of light decreases. Since most of us do not live in a vacuum, we need a measure to describe the velocity of light in any medium, and the established method is to use a simple ratio. Since the energy of any given photon is fixed, the frequency of a photon does not change. Thus, for a given frequency of light, the wavelength must decrease as the velocity decreases.

The decrease in light velocity is described by the refractive index, n, which is the ratio of c to the velocity of light in another medium, v:

$$n = c/v \tag{4.2}$$

Since the velocity of light is lower in other media than in a vacuum, n is always a number greater than 1. Table 4.1 lists the refractive index for a number of examples.[1] Refractive index is an intrinsic physical property of a substance. It is therefore a useful measurement to check the identity or to monitor the purity of a substance. For simple solutions, that is, having only one major component, it can be used to determine the solution density and concentration of a solute. The refractive index of a material is measured with a refractometer, usually against air. For better accuracy, the measurements can be corrected for vacuum. The difference between n_{air} and n_{vacuum} is only significant in the fourth decimal place.

For anisotropic materials, such as quartz or calcite crystals, light of different polarizations will experience different refractive indices. These indices are called the ordinary refractive index, n_o, and the extraordinary refractive index, n_e (see Table 4.1). Optical components made of such anisotropic, or birefringent, materials can be used to isolate or control different polarizations of a light beam.

TABLE 4.1 Refractive Index, n, of Selected Transparent Media

Medium	n	Medium	n
Air	1.0003	Calcium fluoride	1.434
Water	1.333	Crystalline quartz	1.544 (n_o)
20% sucrose in water	1.364		1.553 (n_e)
40% sucrose in water	1.400	Glass (various compositions)	1.5–1.7
60% sucrose in water	1.442	Alumina (sapphire)	1.767
Carbon disulfide	1.632	Diamond	2.42

Crystal data adapted with permission from Shannon, R. D.; Shannon, R. C.; Medenbach, O.; Fischer, R. X. "Refractive index and dispersion of fluorides and oxides," *J. Phys. Chem. Ref. Data* **2002**, *31*, 931. Copyright 2002 American Institute of Physics.

[1]Refractive index at the wavelength of the sodium D line (589.3 nm). The sucrose solutions are listed as wt%.

The speed of light, c, will change in different media, but in any given medium it remains constant. The distance that a photon must travel to complete one wavelength is given by

$$c = \lambda v \tag{4.3}$$

This relationship is easy to remember by checking the units (see Example 4.1):

$$\text{speed(m/s)} = \text{length(m)} \times \text{frequency(s}^{-1}) \tag{4.4}$$

Given this relationship, we can also write the energy of a photon as

$$E = \frac{hc}{\lambda} \tag{4.5}$$

So, we see that the energy of a photon is directly proportional to its frequency and inversely proportional to its wavelength:

$$E \propto \nu \text{ and } E \propto 1/\lambda \tag{4.6}$$

For practical and historical reasons, we often use different conventions for discussing the wavelength, frequency, or energy of photons in different regions of the EM spectrum. Equations (4.1) and (4.3) give us the relationships to interconvert between different units.

Example 4.1 Radio Wave Calculation. FM radio is the portion of the EM spectrum of frequencies spanning 87–108 MHz. Using the above equations, calculate the spectral range in wavelength.

We want the relationship between frequency and wavelength:

$$c = \lambda v$$
$$2.9979 \times 10^8 \, \text{m/s} = \lambda \left(87 \times 10^6 \, \text{s}^{-1} \right)$$
$$\lambda = 3.45 \, \text{m}$$
$$c = \lambda v$$
$$2.9979 \times 10^8 \, \text{m/s} = \lambda \left(108 \times 10^6 \, \text{s}^{-1} \right)$$
$$\lambda = 2.78 \, \text{m}$$

So the wavelengths of radio waves in the FM spectral band range from 2.78 to 3.45 m, which shows why a longer antenna is useful for FM radio reception.

4.3 ELECTROMAGNETIC SPECTRUM

Chemists use most of the EM spectrum for one purpose or another. For convenience, we classify radiation of different frequencies into different spectral regions. Figure 4.2 and Table 4.2 list the different regions of the EM spectrum. Note that the boundaries between some spectral regions are not well defined and different reference sources might use different cutoffs between regions. Table 4.3 lists many of the common prefixes used for the EM spectrum (these prefixes were introduced in Chapter 1).

Of considerable importance to chemists are spin-flip transitions (marked by the * in Table 4.2) that can occur between energy levels that are split by a magnetic field. Atoms and molecules with an unpaired electron will absorb microwaves (electron spin resonance) and nuclei will absorb radio waves

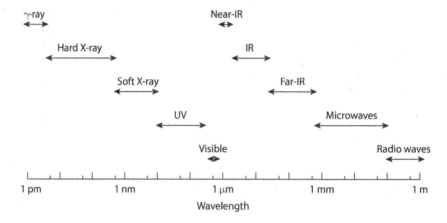

Figure 4.2 Regions of the EM spectrum.

TABLE 4.2 Classification of Different Regions of the Electromagnetic Spectrum

Radiation	Frequency Range (Hz)	Wavelength Range	Type of Transition
γ-rays	$10^{20} - 10^{24}$	<1 pm	Nuclear
X-rays	$10^{17} - 10^{20}$	1–1 pm	Core electrons
UV	$10^{15} - 10^{17}$	400–1 nm	Valence electrons
Visible	$4.3 - 7.5 \times 10^{14}$	700–400 nm	Valence electrons
Near-infrared	$1 - 4 \times 10^{14}$	2.5 µm–700 nm	Valence electrons, overtones
Infrared	$10^{13} - 10^{14}$	25–2.5 µm	Molecular vibrations
Microwaves	$3 \times 10^{11} - 10^{13}$	1 mm–25 µm	Molecular rotations*
Radio waves	$<3 \times 10^{11}$	>1 mm	Free electrons (metals)*

TABLE 4.3 Common Prefixes Encountered with EM Radiation

Prefix	Name	Factor	Prefix	Name	Factor
p	Pico	10^{-12}	k	Kilo	10^{3}
n	Nano	10^{-9}	M	Mega	10^{6}
µ	Micro	10^{-6}	G	Giga	10^{9}
m	Milli	10^{-3}	T	Tera	10^{12}

(nuclear magnetic resonance, NMR). NMR spectroscopy is used extensively to determine the structure of molecules.

The nature of a photon in the different spectral regions does not change—EM radiation of any frequency has the characteristics of oscillating electric and magnetic fields as described earlier. However, photons in different spectral regions have very different energies, and the interaction with matter is very different. The rightmost column in Table 4.2 lists the aspect of matter with which the photons in the different regions interact. UV and X-ray photons have sufficient energy to interact with electrons and break bonds in molecules. Visible light does not have sufficient energy to do so for most molecules, hence the inability to cause sunburn as mentioned previously. IR and microwave radiations interact with the vibrations and rotations in matter, respectively. Either of these spectral regions can be used to pump energy into an aqueous solution. They are quite good for cooking; however, they will not break bonds directly. Given enough time, IR and microwave radiation can raise the solution temperature enough to cause chemical changes.

As another example, the human eye can only detect radiation that is in the visible region of the EM spectrum, hence the name for this type of radiation. Visible light extends from 400 nm to approximately 700 nm. There is no fundamental difference between a photon with a wavelength of 350 nm and one with a wavelength of 400 nm. We make a distinction because our eyes are sensitive to the 400-nm photons directly but not to photons of shorter wavelength. This short wavelength cutoff for vision is due to the absorption of photons with wavelength less than 400 nm by the lens of the eye. The long wavelength cutoff is due to the decrease in sensitivity of the photoreceptors in the retina for wavelengths longer than 700 nm. This cutoff is gradual, and light at near-IR wavelengths beyond 700 nm can be detected by the eye if the light source is very intense. The eye does not detect IR photons because the energy of IR photons is insufficient to induce electronic excitation in the photoreceptors in the retina. Although we cannot see long-wavelength light, we can feel it as heat and detect it with a thermometer or other devices (Figure 4.3). The mechanisms of these other detection schemes are different from the detection process in our eye.

Figure 4.3   Visible or IR light will spin a Crookes radiometer.

4.4 SPECTROSCOPIC TRANSITIONS

4.4.1 Resonance

The interaction of EM radiation with matter—atoms and molecules in gas, liquid, or solid phases—can cause transitions between energy levels of the matter. The result is the absorption, emission, or scattering (redirection) of the radiation. The condition for a strong interaction, called *resonance*, is that the energy of a photon matches the difference between the energy levels of the matter.

> *Absorption* causes a transition from a lower to a higher energy level with transfer of energy from the radiation field to internal energy of matter. The energy absorbed by the matter might be reemitted as light, redistributed to the surroundings as heat, or a combination of both.
>
> *Emission* occurs when a transition from a higher to a lower energy level occurs by the matter emitting a photon. For molecules, this process is usually called fluorescence. Emission of light requires that matter have extra internal energy, that is, it is already in an excited energy level. Processes that excite matter to higher energy levels include absorption of light, thermal excitation, interactions with electrons, and chemical reactions.
>
> *Scattering* is the redirection of photons by matter. Scattering might or might not occur with a transfer of energy, that is, the scattered photons might have the same or a slightly different energy as the incident photons.

Let us not worry about the details of the energy levels at this point. Figure 4.4 shows a simple schematic of the occupancy of electronic energy levels before and after the absorption of a photon. The squiggly arrows represent photons of the same frequency and the horizontal lines represent allowed electronic states. The lower line represents the energy of the ground electronic state and the upper line is one of many excited electronic states. The vertical half arrows represent electrons with the points indicating spin up or spin down. The change after absorption shows an electron in the excited state with the loss of a photon. The excited electron is metastable, meaning that it remains in the excited state for some amount of time, but not forever. It will return to the ground state by reemitting a photon or by converting the energy to heat.

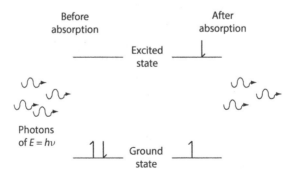

Figure 4.4 Promotion to an excited state by the absorption of a photon.

This loss of a photon can be detected as a dimming of the light passing through a sample. The number of photons absorbed will depend on the number of absorbers, which provides the basis for analytical measurements using absorption spectroscopy. Having more absorbing matter in the path of a light beam, allows less light to pass through that sample. Look at a white wall through a piece of colored glass or plastic. Now, repeat the measurement with several pieces of the colored material—less light gets through.

Emission of EM radiation is the reverse of the process in Figure 4.4. The requirement to use emission as a measurement tool requires that matter be excited to higher energy levels. These metastable levels will return to the ground state eventually by converting the excited state energy to thermal energy (heat) or by emitting a photon. When a photon is emitted, it provides a mechanism to detect and quantitate the matter that is emitting the photon.

Scattering does not require a real transition, but the photon will be redirected from its initial path because of its interaction with matter. Emission and scattering of EM radiation can both be measured by placing an optical detector near the sample to record photons emitted or scattered by the sample. As with absorption, emission and scattering of EM radiation are dependent on the density or concentration of the matter, making measurement of light intensity an analytical tool.

Energy must be conserved in the process in Figure 4.4 and for any type of spectroscopic transition. The resonance condition for this absorption process requires that the energy of the photon equal the difference in energy between the ground and excited states:

$$E_{\text{photon}} = h\nu = E_{\text{excited state}} - E_{\text{ground state}} \tag{4.7}$$

When the interaction with matter is strong, the photon is in resonance with a transition from one energy level to another. If my young daughter is on a swing and swinging back toward me once every 2 s, but I am moving my arms to push once every 7 s, there is not much interaction (there is some unhappiness which gets me back into resonance quickly). Similarly, the sun emits light across a wide band of energies. We know that many wavelengths, most notably near-UV through near-IR, penetrate to the surface. These photons have energies that do not interact strongly with the components of the atmosphere. Figure 4.5 shows the light power (y-axis) reaching the earth from the sun as a function of wavelength (x-axis).[2] Such a plot is called a *spectrum*. The figure labels spectral bands where the energy levels of certain species are in resonance with light. The result is a drop in the light of these wavelengths that reach the Earth's surface. In some cases, near 1400 and 1900 nm, the light is completely absorbed by the water in the atmosphere.

---

[2]Data from Reference Solar Spectral Irradiance: Air Mass 1.5, https://www.nrel.gov/grid/solar-resource/spectra-am1.5.html; accessed August 2022.

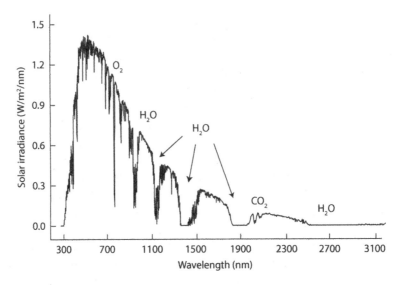

Figure 4.5 The solar spectrum.

You-Try-It 4.A
The conversions worksheet in you-try-it-04.xlsx contains two sets of atomic energy levels. From the energy levels, predict the wavelength, frequency, and energy for each transition.

4.4.2 Atomic Energy Levels

Gas-phase atoms have discrete energy levels that are often widely separated, so absorption and emission transitions between atomic levels have very specific photon energies. Figure 4.6 shows several energy levels of the hydrogen atom. The energy unit is in inverse centimeters, cm^{-1}, or wavenumbers (see Example 4.2). The energies are labeled as negative values because zero energy is assigned to the ionization potential, the energy necessary to remove the electron. Since a transition energy is the difference between two energy levels, it does not matter how we label the individual levels.

Example 4.2 Hydrogen Spectrum Calculation. Hydrogen has transitions in the visible region at 656.28, 486.13, 434.05, 410.17, and 397.01 nm.[3] Use these numbers to improve the precision of the energy labels of the levels above $n = 2$ in Figure 4.6. You can take the energy of the $n = 2$ level to be exactly $-27,420.0$

---

[3]Wavelength data from the NIST Atomic Spectra Database for air, https://www.nist.gov/pml/atomic-spectra-database; accessed August 2022.

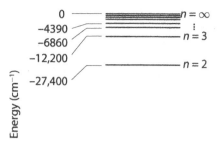

Figure 4.6 Allowed energies for the electron in the hydrogen atom.

cm^{-1}. One example is shown here (note that your calculations might give slightly different values than the figure shows for $n = 3$ and $n = 4$)

$$\text{transition energy} = \frac{1}{656.28\,\text{nm}}\left(\frac{1\,\text{nm}}{1\times10^{-9}\,\text{m}}\right)\frac{1\,\text{m}}{100\,\text{cm}}$$

$$\text{transition energy} = 15237\,\text{cm}^{-1}$$

$$\text{energy position for } n = 3 \text{ is } 15{,}237\,\text{cm}^{-1} - 27{,}420.0\ \text{cm}^{-1} = -12183\ \text{cm}^{-1}$$

Figure 4.7 shows the emission spectrum of a Hg vapor lamp. In this example, the strongest peak at 254 nm is off scale to show the weaker peaks. The

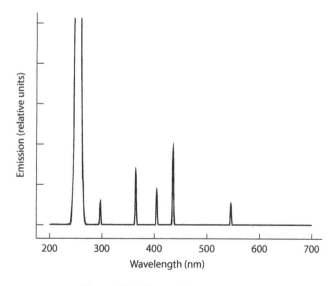

Figure 4.7 Hg emission spectrum.

main point is that the spectral lines of a gas-phase atom are narrow and easily separated. Atomic spectra can be recorded with high resolution so that the absorption or emission lines from different elements can be measured separately from each other. The details of atomic spectroscopy and more advanced instrumentation are found in Chapter 10.

4.4.3 Molecular Energy Levels

To illustrate molecular spectroscopy with the simplest case, Figure 4.8 shows an idealized energy-level diagram for the ground state and the first excited state of a diatomic molecule. The x-axis, r, is the distance between the two atoms. At small distances, the energy increases rapidly because the nuclei repel each other. At large distances the atoms are too far apart to interact. The repulsive and attractive forces balance at an equilibrium distance, which is 0.74 Å for this example. We call the resulting curve a potential well for the two electrons that form the bond between the atoms. When the molecule is in the ground electronic state, the two electrons are in the lower potential well. Providing sufficient energy to the molecule so that one electron is placed in the upper potential well places the molecule in an excited electronic state. Only one excited-state potential well is shown in the figure, but a molecule can have a large number of excited states for the unfilled electron orbitals. For our purposes, the transition of an electron between one state and another can be taken to be instantaneous.

The key difference between atoms and molecules is the horizontal lines, which represent vibrational levels for each of the electronic states. A series of rotational energy levels also exist for each vibrational level. Figure 4.8 does not show the rotational energy levels for clarity. The vertical arrows show only a few of the possible absorption transitions from the electronic ground state to the

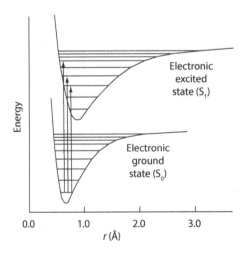

Figure 4.8 Ground and excited electronic states of a diatomic molecule.

first electronic excited state. The multiple transitions that are possible in molecules for each electronic transition lead to broad absorption bands in the UV/Vis absorption spectra. The broad bands are very different from the sharp-line spectra of an atom, making it difficult to measure overlapping molecular spectra.

IR radiation, typically 4000–400 cm^{-1} (2.5–25 μm), is in the energy range that can excite molecular vibrations to higher vibrational energy levels. This energy range is much lower than that of UV and visible photons 50,000–12,500 cm^{-1}. When an IR photon of the same energy as a vibrational mode passes by a molecule, the molecule can absorb that IR photon, converting the photon energy to a greater vibrational amplitude. Eventually, this energy is transferred to the surroundings, for example, by collisions with solvent molecules, resulting in an increase in the temperature of the sample. (If you do not believe me go stand in the sun for a while!) An IR absorption transition (not shown in Figure 4.8) would occur from the ground vibrational state to another vibrational level of the electronic ground state.

UV and visible photons do not excite vibrations directly because the photon energies are far from resonance with the vibrational transitions. Only IR photons can be in resonance with vibrational transitions to cause vibrational excitation. You might think, based on how warm you feel when out in the sun, that there is vibrational energy absorbed. Your skin, clothing, and most of anything that is not extremely reflective absorb UV and visible photons. What happens to that energy? Your skin obviously does not reemit the absorbed energy; otherwise we would all glow. The absorbed energy is converted to heat as the molecules return to the ground state through nonradiative pathways. In fact, nonradiative decay is more common than emission for molecules. Some exceptions are fluorescent organic molecules that are used to brighten paper, clothing, etc.

4.5 UV/VIS ABSORPTION SPECTROSCOPY

The energies of photons in the UV and visible spectral regions are between 50,000 and 14,000 cm^{-1} (200–700 nm). 200 nm is a practical short wavelength limit of UV/Vis instruments due to absorption by atmospheric gases at shorter wavelengths. Many instruments can measure to longer wavelength, depending on the specific detector in the instrument. Photons in this energy range can promote an electron from a filled orbital to an unfilled orbital in an atom or molecule. Types of molecular electronic absorption transitions and their approximate energies are listed in Table 4.4 and diagrammed very simply in Figure 4.9.

The broad horizontal bands in Figure 4.9 represent the closely spaced vibrational levels of each electronic state. Only one vertical line is shown for each type of transition, but there are many such transitions, creating the broad absorption bands that are characteristic of molecular spectroscopy. As Figure 4.9 suggests, all molecules will have electronic absorptions. In spectroscopic measurements of solutes in solution, the solvent that contains the analyte must be transparent to

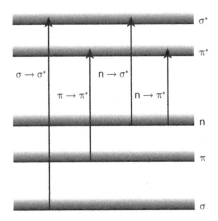

Figure 4.9 Relative energies of different electronic transitions.

TABLE 4.4 Types of Electronic Molecular Transitions

Transition	λ (nm)	$\varepsilon\,(M^{-1}cm^{-1})$	Examples
$\sigma \rightarrow \sigma^{*}$	< 200		Alkanes, C–C and C–H
$n \rightarrow \sigma^{*}$	160–260	10^{2}–10^{3}	–OH, –NH$_2$, –halogen
$n \rightarrow \pi^{*}$	250–600	10–100	–C=O, –COOH, –nitro, –nitrate
$\pi \rightarrow \pi^{*}$	200–500	10^{4}	Alkenes (C=C), alkynes, aromatics

σ orbitals are the in-line single bonds between atoms.
π orbitals are the out-of-plane orbitals that form double and triple bonds.
n are nonbonding orbitals.
* Asterisk specifies antibonding orbitals, that is, the empty excited states.

light. Simple molecular structures are common for spectroscopy solvents so that solvent absorption does not overlap with analyte absorption bands. Table 4.5 lists some common solvents and their UV cutoff.[4] As you can see from the table, these useful solvents have fairly simple structures and therefore little absorption in the near-UV and visible spectral regions. Acetone and benzene are included in the table for illustration, but they are not usually used as solvents for UV/Vis spectroscopy because of their relatively long-wavelength cutoff.

Figure 4.10 illustrates the nature of molecular spectra with the UV/Vis absorption spectrum of $Ru(bpy)_3^{2+}$ (the inset shows a representation of the structure). The central atom of this complex is an octahedrally coordinated Ru^{2+} and the (bpy) are the bipyridine ligands that complex to the metal ion through the nitrogen atoms. Hydrogen atoms and PF_6^{-} counterions are not shown in the structure. The spectrum shows wavelength on the x-axis and molar absorption coefficient, ε on the y-axis. The molar absorption coefficient

[4]Values for Burdick and Jackson high purity solvents. The listed wavelength cutoff is where the solvent transmits only 10% of light (absorbance = 1.0).

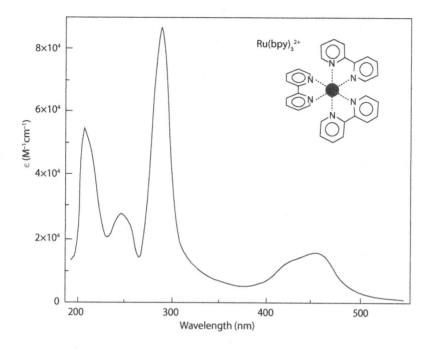

Figure 4.10   UV/Vis absorption spectrum of $Ru(bpy)_3^{2+}$ in acetonitrile.

TABLE 4.5   Onset of UV Absorption for Common Solvents

Solvent	Formula	λ_{cutoff} (nm)
Acetone	CH_3COCH_3	330
Acetonitrile	CH_3CN	190
Benzene	C_6H_6	280
Cyclohexane	C_6H_{12}	200
n-Hexane	C_6H_{14}	195
Ethanol	CH_3CH_2OH	210
Methanol	CH_3OH	205
Tetrahydrofuran (THF)	C_4H_8O	212
Trichloromethane (chloroform)	$CHCl_3$	245
Water	H_2O	190

and absorbance are defined below. For now, it is sufficient to know that a higher value of ε corresponds to greater absorption of light by the sample. Several points to note about the spectrum of $Ru(bpy)_3^{2+}$, which are characteristic of molecular spectra, are as follows:

- The lines are much broader than in atomic absorption spectra.
- There are multiple absorption bands because there are multiple excited states.
- The relative absorption of different bands can vary greatly.

A significant consequence of the broad electronic transitions of molecules and complexes in solution is that it can be very difficult to resolve spectroscopic measurements if more than two or three absorbing components are present in a mixture. The advantage for analytical measurements is that the broader transitions make it easier to obtain reproducible spectroscopic measurements. Small instrumental errors, such as selecting a spectrometer wavelength, will not result in large measurement errors. Having the wavelength-dispersing element set off a peak by a tenth of a nanometer will have almost no consequence for a molecular absorption, compared to possibly being off resonance completely for an atomic transition.

4.6 UV/VIS INSTRUMENTATION

Equipment for making spectroscopic measurements can range from simple visual comparators to elaborate computer-controlled instruments. You have probably done a simple spectroscopic comparison using pH paper, which contains a molecular indicator that changes color depending on pH. Fancier comparators will add a reagent or reagent mixture that forms a colored complex in the sample solution. Figure 4.11 shows an example. The intensity of the color of the test portion is compared to reference colors to determine the concentration of the analyte. Besides pH, visual comparators are common for water analysis of analytes such as ammonia, chlorine, fluoride, alkalinity (as mg/L $CaCO_3$), and metals. These simple instruments work well for repetitive measurements of samples that do not vary in composition, as they can be susceptible to interferences in very complex samples.

An instrument that disperses and detects light is called a *spectrometer*. A spectrometer that includes a light source for making absorption measurements is called a spectrophotometer. A more elaborate version of a comparator is a simple colorimeter, as illustrated in Figure 4.12. This simple absorption spectrometer consists of a light source, a cell or cuvette to hold the sample, and a light detector. The instrument housing and various apertures, optics, and electronic components are not shown.

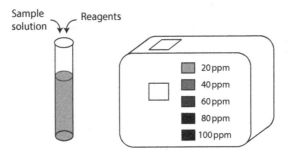

Figure 4.11 Schematic of a visual comparator.

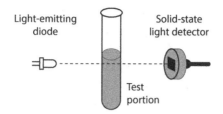

Figure 4.12 Schematic of a simple colorimeter.

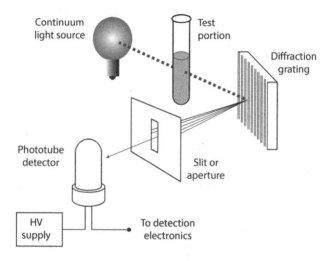

Figure 4.13 Schematic of a grating spectrometer.

The detector in this instrument will measure the attenuation of the light power due to absorption by the sample. From this attenuation, we can determine the concentration of an absorbing analyte. By converting the light intensity to a voltage readout, much finer gradations are achievable in the measurement compared to a visual comparator. Although simple and suitable for many dedicated measurements, this colorimeter has the disadvantage that the wavelength is fixed at one, or in some commercial instruments, only a few wavelengths.

More flexible instruments use continuum light sources, such as an incandescent lamp, which emits light of a broad range of wavelengths. A dispersing element, such as a prism or a diffraction grating, redirects the light at different angles depending on the wavelength. One narrow band of wavelengths can be selected by dispersing the continuum light and blocking most of the dispersed radiation with a slit or aperture. Because these instruments select one wavelength or color, they are often called monochromators.

Figure 4.13 shows a simplified schematic of an absorption spectrophotometer with light passing through a test portion in a sample cuvette. The light is dispersed by a grating and an aperture allows only "one" wavelength to reach the detector. The other wavelengths emitted by the light source are diffracted at different angles and blocked by the aperture. The specific wavelength that passes through the sample is selected by rotating the grating. Simple instruments have a knob and a wavelength readout on the top of the spectrometer to set wavelength. This selected wavelength is actually a range of wavelengths, called the *bandpass*, that depends on the grating dispersion and the physical width of the opening. The bandpass should be smaller than the width of spectral features, which is easy to achieve for the broad bands of molecules. The "HV supply" is a high voltage source to power the detector, which converts the light power striking it to an electrical signal.

The schematic in Figure 4.13 illustrates an older but very common spectrometer design for making measurements at one select wavelength. Figure 4.14 shows a simplified schematic of an alternate design that uses an array-based detector in place of the slit and phototube. The array detector is similar to the sensor chip in a phone camera but with only a linear array rather than the two-dimensional array needed to create an image. The grating disperses the different wavelengths of light onto the detector, where each pixel in the array functions as an individual detector. The main advantage of this design is that all wavelengths are recorded at once. Obtaining a spectrum with a phototube-based instrument requires rotating the grating to scan the wavelength. By making simultaneous measurements of all wavelengths with the array detector, the time is reduced from minutes to seconds. The array-based instrument is also easier to miniaturize, and commercial systems including light source have a footprint similar to that of a laptop computer.

The choice of spectrometer design is governed by the application. For a spectroscopic measurement that must be field-portable, a simple comparator or colorimeter is desirable. For recording routine spectra in the laboratory, the rapid array-based spectrometer is common. For high resolution or more general spectral measurements, the design in Figure 4.13 allows flexibility to change the light source, aperture size, and other parameters. The disadvantage of the simple design in Figure 4.13 is that recording an absorbance spectrum actually requires two measurements at each wavelength. We'll look at a modification to this design after explaining absorbance in the next section.

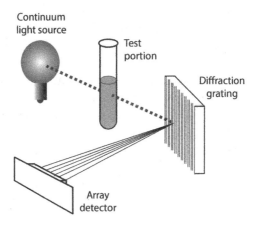

Figure 4.14 Schematic of an optical spectrometer with an array detector.

4.7 BEER–LAMBERT LAW

4.7.1 Transmittance and Absorbance

The detector in an absorption spectrophotometer outputs an electrical current that is proportional to the power of the light that strikes it. This current is converted to a voltage and displayed on a readout or recorded and stored by a computer. Experimental measurements are reported in terms of transmittance, T, or absorbance, A, which are defined as

$$T = \frac{P}{P_0} \quad \text{and} \quad A = -\log(T) \tag{4.8}$$

where P is the power of light after it passes through the sample and P_0 is the reference power. A quantitative absorption measurement thus requires two measurements, P and P_0. When measuring an analyte in solution, the sample container and solvent might absorb or scatter some of the incident light. The reference power, P_0, is measured by placing the sample holder containing only pure solvent in the light beam. P is then measured using the sample holder with the analyte present. The schematic in Figure 4.15 illustrates the measurement. P_s is the total power of the light source incident on the sample. The two measurements made to obtain transmittance are P_0, the light power passing through the sample holder and solvent, and P, the light power passing through the sample (holder, solvent, and analyte). Note that the narrower arrow for P in the figure represents a smaller amount of light passing through the cuvette when the absorbing analyte is present.

To determine the relationship between transmittance and the amount of an absorbing analyte, consider light passing through slices of an absorbing sample. For a constant absorber concentration, as expected for an analyte in solution, the amount of absorber is directly proportional to distance through the sample. For a given thickness of the sample, there is some fixed probability for a photon to be absorbed. The result is a fixed fraction of light making it through the slice. If we take a hypothetical case where 10% of photons are absorbed in a 0.1-cm slice of sample, a light beam will be attenuated to 90% of its initial value after 0.1 cm. As this light beam passes through another

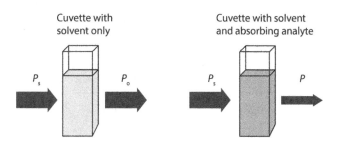

Figure 4.15 Light power measurements to determine transmittance.

0.1-cm slice, it is again attenuated by 90%, resulting in 81% of the light making it through a total path length of 0.2 cm. For each additional 0.1-cm slice that the light beam passes through, it is further attenuated by another 90%.

Plotting the intensity of a light beam versus distance as it is attenuated by 90% for every 0.1 cm of distance results in an exponential decay. This exponential dependence, due to a fixed probability, is the same as observed for radioactive decay and other kinetic phenomena. Figure 4.16 shows the attenuation of light, labeled as "# photons" on the y-axis, as a function of distance, labeled as "path length" on the x-axis. The total distance that the light travels through the sample is 1.0 cm. The two curves are for the same absorber but at two different concentrations. The solid line replicates our hypothetical case of a sample that attenuates the light by 90% for each 0.1-cm slice. The dashed line is for the same absorber present at twice the concentration of the solid line. For both absorber concentrations, the attenuation of photons follows an exponential decay as a function of distance through the sample. Path length is equivalent to the amount of absorber, so we can also deduce that attenuation of light will also follow an exponential dependence on analyte concentration. Comparing the value of P for the two absorber concentrations at any given path length does not follow the factor of 2 difference in concentration.

The overall result is that the measured transmittance is *not* linearly proportional to the distance that the light beam travels through the sample or to absorber concentration. We prefer linear relationships when making quantitative measurements. We can make the curves in Figure 4.16 linear by

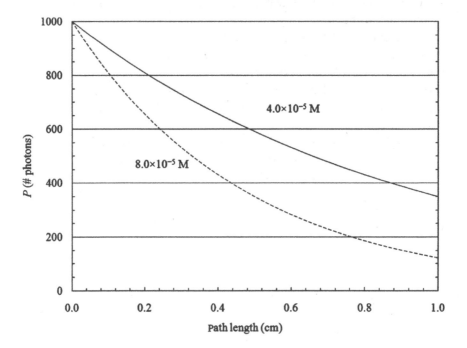

Figure 4.16 Light power as a function of concentration and distance.

plotting the logarithm of the y values. The concentration of an analyte is related directly to $-\log T$, which is called the absorbance, A:

$$A = -\log T = -\log \frac{P}{P_o} \tag{4.9}$$

Table 4.6 contains some of the values from the curves in Figure 4.16, where the solid and dashed subscripts refer to the 4.0×10^{-5} M and 8.0×10^{-5} M curves, respectively. Comparing values of P for a given concentration versus path length, b, shows nonlinear dependence. Likewise, P at a given path length does not vary by the factor of 2 difference in the concentrations, c. The data in Table 4.6 verifies the nonlinear dependence of transmittance on path length, b, and concentration, c. If you make the same comparisons for the absorbance values, A is directly proportional to both b and c.

The linear relationship between absorbance, A, path length, b, and concentration, c, of an absorber is called the Beer–Lambert law.[5] The general Beer–Lambert law is usually written as

$$A = a(\lambda) \times b \times c \quad \text{or,} \quad \text{simply,} \quad A = abc \tag{4.10}$$

A is the measured absorbance, $a(\lambda)$ is a wavelength-dependent absorption coefficient, b is the path length, and c is the analyte concentration. $a(\lambda)$ is a proportionality factor that is related to the probability of a photon being absorbed, as for the 90% factor that we used in the example above. This proportionality constant depends on the specific absorber, the measurement wavelength, and the specific solvent. When working in concentration units of molarity, the Beer–Lambert law is written as

$$A = \varepsilon bc \tag{4.11}$$

TABLE 4.6   **Attenuation of Photons as a Function of Distance and Concentration**

Distance (cm)	P_{solid}	P_{dashed}	A_{solid}	A_{dashed}
0.0	1000	1000	0.0	0.0
0.2	810	656	0.0915	0.183
0.4	656	430	0.183	0.366
0.6	531	282	0.275	0.549
0.8	430	185	0.366	0.732
1.0	348	121	0.458	0.915

[5]Also called simply Beer's law or the Beer–Lambert–Bouguer law after the researchers who determined the dependencies on b and c.

where ε is the wavelength-dependent molar absorption coefficient with units of $cm^{-1} M^{-1}$. Strong absorptions in molecules will have ε values of $10^4 - 10^5 cm^{-1} M^{-1}$.

Example 4.3 *A*, *T*, and %-*T*. Express the measurement of the transmitted light for the 4.0×10^{-5} M solution (solid curve) at a path length of 1.0 cm in terms of transmittance, %-transmittance, and absorbance.

The upper curve has 348 photons transmitted at 1.0 cm. Since there were 1000 photons at the start, 0.0 cm, the transmittance is

$$T = \frac{348}{1000} = 0.348$$

%-transmittance is simply T expressed as a percentage:

$$\%\text{-}T = 0.348 \times 100\% = 34.8\%$$

Absorbance is $-\log$ of T:

$$A = -\log(0.348) = 0.458$$

Modern spectrophotometers can usually display data as transmittance, T, %-T (%-transmittance), or absorbance, A (see Example 4.3). T and %-T are sometimes more useful to describe a sample when a physical characteristic is of interest rather than an analyte concentration. Examples include the transmittance of optical windows and filters, where the % provides a more intuitive measure of how much light gets through the optic. Having said that, engineers and photographers use a logarithmic scale for neutral density filters that attenuate light intensity. Neutral density refers to a flat or constant wavelength dependence throughout the visible spectrum. An attenuating filter that had a wavelength dependence would give a color cast to photographs. Neutral density, or optical density (OD), is the same as absorbance and is related to photographic exposure by the factors in Table 4.7. Using OD is convenient when light intensity varies by orders of magnitude.

Now that we understand absorbance, we can revisit a detail of spectrophotometer designs. For repetitive sample measurements at one wavelength in the same solvent, the P_o measurement is made once for any number of subsequent measurements of P. This approach works well for quantitative absorbance measurements using a colorimeter or the single-beam instrument design shown in Figure 4.13. Likewise, in an array-based spectrophotometer, a reference spectrum of the cuvette and pure solvent is recorded and stored in the software that controls the instrument. The software then generates T or A over the full wavelength range for subsequent samples.

TABLE 4.7   Relationship Between Photographic f-stop, Optical Density, and %-Transmittance

f-stops	OD	%-T
1/3	0.1	79
2/3	0.2	63
1	0.3	50
4/3	0.4	40
5/3	0.5	32
2	0.6	25
3	0.9	12.5
3 1/3	1.0	10
6 2/3	2.0	1.0
10	3.0	0.1

It is possible to record an absorption spectrum with a single-beam scanning spectrophotometer (Figure 4.13). However, doing so is quite tedious as it requires measurement of P and P_o at each wavelength as the grating is rotated in steps. Figure 4.17 shows an addition to the spectrophotometer that allows simultaneous measurement of P and P_o. The solid arrows represent the light path of the continuum source. The dashed arrows show the light path of the selected wavelength, which is scanned by rotating the diffraction grating. This light is split into two beams with an optical chopper that alternately reflects the light through the sample cuvette or allows it to pass through the reference cuvette with pure solvent. A 50% mirror or beamsplitter allows both beams to

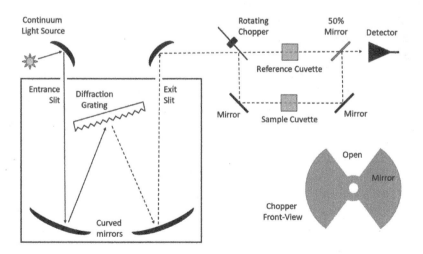

Figure 4.17   Top-view schematic of a scanning dual-beam UV/Vis spectrophotometer. The graphic to the right shows the design of the optical chopper.

reach the detector.[6] The detector alternately records the values of P and P_0 much faster than the time at which the grating scans the wavelength. Accurate measurements do require a matched pair of identical cuvettes. The values of P and P_0 are ratioed to display T or A versus λ to obtain a spectrum. This *double-beam* design has the advantage that it is insensitive to drift in light intensity or other experimental parameters.

Analytical chemists often stress the limits of detection for a method, but it is equally important to know the upper limit of a given measurement technique. If measuring the purity of a substance, there is an obvious upper limit of having a 100% pure material. For an analyte in solution, there might be a concentration where the analyte forms dimers, micelles, or a separate phase. These upper limits are analyte specific and can depend on the solvent or sample matrix. Figure 1.6 showed an example of the measurement range and the linear range for a fluorescence measurement. In addition to chemical upper limits, UV/Vis absorption spectroscopy has an upper limit due to an instrumental factor called stray light.

In UV/Vis absorption, we expect a linear relationship based on the Beer-Lambert law. The following discussion explains why real measurements deviate from this linear relationship. The upper limit in an absorbance measurement depends on the design and quality of the instrument. Typical benchtop UV/Vis spectrophotometers have an absorbance measurement range of 0–4. Compact spectrophotometers will have a smaller linear range, and specialized instruments can extend this range higher. The difference arises due to the diffraction grating not being perfect. It nominally diffracts each wavelength of light at a distinct angle. However, there is also a very small fraction of a given wavelength that scatters at all angles. This scattered "background" light affects all wavelengths and is called *stray light*. It is an intrinsic specification of an instrument and the effect varies with wavelength. The deviations from linearity are much less severe at longer wavelength because the continuum light source is much brighter.

The effect of stray light can be understood taking the extreme case of placing an object in the sample cuvette that absorbs all UV light. No UV light transits the sample cuvette and the measurement of P should be zero (an absorbance of infinity in principle). The reality is that some amount of visible stray light is incident on the detector. The measured light power, P, is not zero and the instrument records an absorbance value. Table 4.8 shows the deviation at high absorbance for the stray light specification at 220 nm of two commercial spectrophotometers. Values of %-T and T are listed for scale. At a sample absorbance of 4.0, only 0.0100% of the lamp power at 220 nm reaches the detector. Since the lamp power is weak at shorter wavelength, the stray light is significant. The measured absorbance at 220 nm has a significant error in both spectrophotometers.

---

[6]Some designs use two synchronized choppers to achieve the same effect.

TABLE 4.8   Effect of Stray Light on Absorbance

Absorbance (true)	%-Transmittance	Transmittance	Absorbance (0.007% stray light)	Absorbance (0.05% stray light)
0.0	100.0	1.0000	0.00	0.00
1.0	10.0	0.1000	1.00	1.00
2.0	1.0	0.0100	2.00	1.98
3.0	0.10	0.0010	2.97	2.82
4.0	0.010	0.0001	3.77	3.22

You-Try-It 4.B

The absorbance worksheet in you-try-it-04.xlsx contains light measurements for a sample and blank. Determine T, %-T, and A from the measurements. Use the standard to find the unknown concentration. Use the repetitive measurements to estimate the LOD and LOQ.

4.7.2   Using the Beer–Lambert Law

Nonabsorbing species can be measured by UV/Vis absorption spectroscopy if they can be converted to an absorbing form. There are a large number of complexing agents that form absorbing complexes with metal ions. One example is the addition of ammonia to a copper-containing solution to form the stable tetraamminecopper(II) complex (spectator anions not shown):

$$Cu^{2+}(aq) + 4NH_3(aq) \rightleftharpoons Cu(NH_3)_4^{2+}(aq)$$

This complex produces a bright blue solution with a maximum absorbance, λ_{max}, near 600 nm. You will see λ_{max} often in analytical protocols; it is simply the wavelength where the absorbing species has its maximum absorbance or peak for a specific absorption band.

Table 4.9 lists the absorbance at 600 nm of a series of standard solutions of $Cu(NH_3)_4^{2+}$ and an unknown solution. Figure 4.18 shows the trend line and best-fit equation for the absorbance values of the standards in the table. The trend line shows that the absorption measurement is linear over the full range of the standards, that is, it follows Beer's law. The data is in milligrams per milliliter rather than in molar, so we write Beer's law as $A = abc$ using the general absorption coefficient, a. Comparing the unknown measurement to the standards, we see that it falls between standards 4 and 5, so we can predict that the unknown solution contains a Cu^{2+} concentration slightly less than 0.800 mg/ml. Example 4.4 shows the exact calculation.

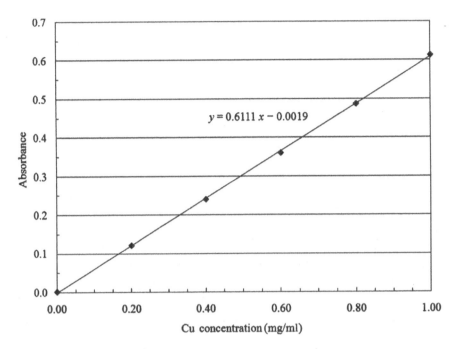

Figure 4.18   Plot of absorbance versus concentration.

TABLE 4.9   Sample Absorbance Data

Sample	Cu Concentration (mg/ml)	Absorbance
Standard 1	0.000	0.001
Standard 2	0.200	0.121
Standard 3	0.400	0.240
Standard 4	0.600	0.360
Standard 5	0.800	0.487
Standard 6	1.000	0.613
Unknown	?	0.482

Example 4.4 Unknown Determination using Absorbance.   What is the copper concentration of the unknown solution given in Table 4.9?

The calibration data shows that the absorbance is linear between 0 and 1.000 mg/ml. Since absorbance is unitless, the slope of the calibration function has units of 1/mg/ml. Plugging the unknown absorbance into the equation of the line from Figure 4.18 gives us the concentration of the unknown:

$$0.482 = (0.611 \text{ ml/mg}) \left[Cu^{2+} \right] - 0.0019$$

$$\left[Cu^{2+} \right] = 0.792 \text{ mg/ml}$$

You-Try-It 4.C

The Beers-law worksheet in you-try-it-04.xlsx contains spectral data for solutions of a riboflavin standard and an unknown. Plot the data and calculate the molar absorption coefficient from the spectrum of the standard. Determine the riboflavin concentration of the test portion.

4.7.3 An Example Illustrating Several Reaction Types

Measuring the concentration of glucose in urine and blood is a common medical analysis. Glucose absorbs in the UV portion of the EM spectrum. Unfortunately, many other biological compounds also absorb in the UV and interfere in a direct absorption measurement of glucose. Glucose contains an aldehyde that can be oxidized to a carboxylic acid, and the glucose concentration can be determined by an indirect analysis. Cu^{2+} will oxidize glucose to gluconic acid. The copper is reduced to Cu^+ in the reaction and precipitates as Cu_2O:

$$\text{glucose(aq)} + Cu^{2+}(aq) \rightarrow \text{gluconic acid(aq)} + Cu_2O(s)$$

Benedict's and Fehling's solutions are two common reagent mixtures for such analyses. Fehling's solution is an alkaline solution containing Cu^{2+} as the tartrate complex (Figure 4.19), which absorbs strongly in the red. A high pH keeps the tartrate unprotonated and speeds up the redox reaction. Complexation with the tartrate prevents the copper from precipitating as $Cu(OH)_2(s)$. In a typical glucose analysis, a known amount of Cu^{2+} in the Fehling's solution is added to a solution containing glucose. The molar amount of Cu^{2+} must be larger than the amount of glucose, that is, in excess. This same concept was introduced in Chapter 3 for performing a back titration.

Since the Cu^+ that forms reacts to form a precipitate, it is removed from solution, driving the reaction to the right. This strong driving force is why I use the right arrow in the reaction rather than the right and left harpoons of an equilibrium. The assumption that the reaction goes to completion is valid

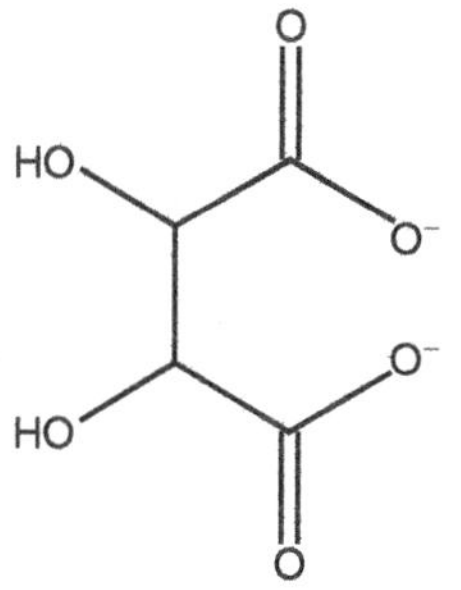

Figure 4.19 Tartrate ligand.

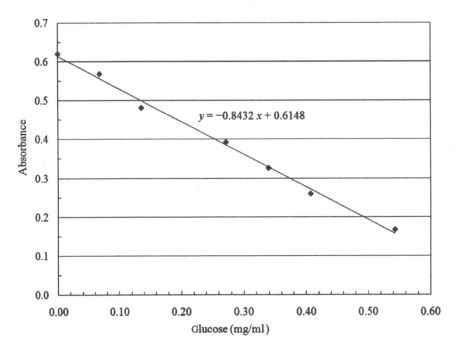

Figure 4.20 Calibration curve for glucose using Fehling's solution measured at 680 nm.

given that the precipitation removes the Cu^+ product from solution (recall Le Chatelier's principle).

The precipitate will scatter light, so it is removed by filtering or centrifugation. The amount of Cu^{2+}-tartrate remaining in solution is then measured by absorption spectroscopy. This indirect method using red light has the additional advantage of reducing interferences from other solution components, which are more likely to absorb in the UV and blue part of the spectrum. An unknown glucose concentration is determined from a calibration curve created from a series of glucose standards. Figure 4.20 shows an example. Note that the slope decreases with increasing glucose concentration because we are measuring the remaining copper tartrate in solution after some of the Cu^{2+} is removed by reaction with glucose. The maximum absorbance at 0.0 mg/ml of glucose is due to the full amount of the copper tartrate in the Fehling's solution. Even though we diluted the test solutions by adding the Fehling's solution, we don't need to correct the standard glucose concentrations on the x-axis if we keep the ratio of Fehling's solution to all test solutions constant.

4.8 MOLECULAR FLUORESCENCE

Atoms, molecules, or solids that are excited to higher energy levels can return to lower levels by emitting radiation. For atoms excited by a high-temperature energy source such as a flame or plasma, this light emission is called atomic or optical emission. Light emission from a molecule is called *fluorescence* if there is no change in electron spin during the transition.[7] In this case, the molecule remains in a singlet state and we label the ground state as S_0 and the excited state as S_1. If the spin of an electron flips in an excited state, the molecule is then a triplet state, T_1. Light emission can still occur with the electron spin flipping during the radiative process, and this emission is called *phosphorescence*. Triplet states are uncommon for organic molecules but do occur in transition-metal complexes. Light emission from inorganic solids is called luminescence, although fluorescence, phosphorescence, and luminescence are often used interchangeably for materials.

Molecular fluorescence is usually generated by exciting a molecule to a higher electronic energy level by absorption of EM radiation. Analytical applications include quantitative measurements of molecules in solution and fluorescence detection in liquid chromatography. The main advantage of fluorescence compared to absorption measurements is the greater sensitivity. The fluorescence signal is measured on essentially a zero background. Absorption relies on measuring the difference in two light powers, so sensitivity is limited by the noise of the light source. Since most molecules do not fluoresce in solution, fluorescence is also selective for those molecules that do fluoresce or that can be derivatized or "tagged" selectively to a fluorescent form. This aspect can help avoid interferences compared to absorption measurements. Fluorescent tags are often employed in fluorescence microscopy to visualize certain biomolecules selectively.

Figure 4.21 is a simplified energy-level diagram for an absorbing molecule. The two potential wells each have a series of vibrational sublevels. These levels are labeled vibronic levels in the excited state to indicate that they are a combined electronic + vibrational excited state. Molecules will have additional electronic states at energies higher than S_1, which are not shown in the figure. Molecules can be excited to these higher electronic states, but decay to S_1 is

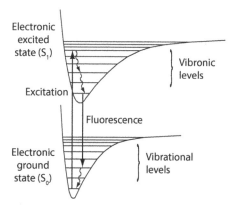

Figure 4.21   Molecular energy-level diagram showing excitation and fluorescence.

[7]Recall that a pair of electrons in an orbital will exist as one with spin up and one with spin down.

rapid. The net result is that only one fluorescence band is observed, originating from S_1. The other simplification in Figure 4.21 is the number of vibrational levels in each electronic state. I drew a limited number of horizontal lines to make it easier to show transitions between the levels. The number of vibrational states increases rapidly as molecules get larger. A more realistic figure will completely fill the potential wells with horizontal lines. The dense vibrational structure leads to smooth, featureless spectra for most molecules (see Figure 4.22).

The upward transition labeled "Excitation" in Figure 4.21 is the same process as absorption. The energy of a resonant photon moved an electron from the ground to the excited state. The wavy arrows represent nonradiative decay to the lowest vibrational energy level of an electronic state. The energy difference in this nonradiative decay is dissipated to the solvent or surroundings as heat. This nonradiative decay is much faster than the probability of fluorescence occurring, and excited vibronic levels quickly decay to the lowest level of the excited state. Most molecules are not fluorescent because the excited state can convert to the ground state by a nonradiative process called internal conversion. In these cases, all of the energy of the absorbed photon is converted to heat in the system.

For molecules in which the internal conversion is relatively slow, the excited state persists long enough for fluorescence to occur. We say that the excited state is metastable, but keep in mind that the excited-state lifetime is on the order of 1–10 ns. The downward arrow labeled "Fluorescence" in Figure 4.21 represents the process of light emission to return the molecule to the electronic ground state. The figure shows only one specific fluorescence transition, but there are a large number of transitions to other ground-state vibrational levels. The result is a fluorescence spectrum that appears as a smooth, broad band (see Figure 4.22). Unlike absorption spectra, which can have multiple bands

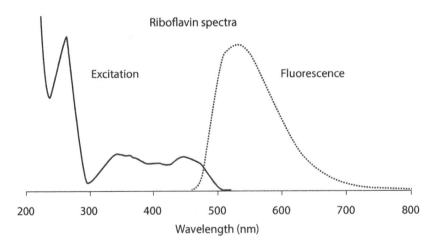

Figure 4.22 Excitation and fluorescence spectra of riboflavin.

due to transitions to multiple excited states, there is only one fluorescence band due to rapid decay to the lowest vibronic level of S_1. Figure 4.22 illustrates these characteristics for the excitation and fluorescence spectra of riboflavin (vitamin B_2).[8] Note that the absorption and fluorescence spectra in this figure were recorded on two different instruments and combined graphically. No y scale is shown, and the two spectra were scaled vertically to be comparable.

Figures 4.21 and 4.22 illustrate another common feature of fluorescence. The energy of the emitted light will always be less than the energy of the excitation light. In spectral terms, the fluorescence spectrum is always at longer wavelengths than the excitation or absorption spectrum. This "red shift" compared to the excitation wavelength is called the *Stokes shift*. A practical advantage of a large Stokes shift is that it makes it easier to block scattered excitation light with an optical filter and prevent it from reaching the detector. Any light other than the actual fluorescence that reaches the detector increases the baseline and reduces the sensitivity of a measurement.

Instruments for recording a fluorescence signal or a fluorescence spectrum are called spectrofluorometers, fluorometers, or fluorimeters. The simplest fluorometer consists of an excitation light source, a sample cell, optical filters, a light detector, and some type of electrical recording or readout. Figure 4.23 shows a schematic of this simple design (detector readout omitted). The long dark arrow represents excitation light at a wavelength that is absorbed by the analyte. The lighter colored arrows represent the fluorescence that is emitted in all directions. This simple schematic leaves out several common elements. Curved mirrors or lenses are usually used to focus the excitation light through the sample cuvette and to collect the fluorescence. As the figure shows, only a fraction of the fluorescence reaches the detector. The actual fraction of fluorescence is not as important as measuring standards and unknowns under identical conditions to maintain an equivalent calibration. Commercial fluorometers are modular, allowing LED excitation sources and filters to be swapped in and out of an instrument to use the most appropriate wavelengths for different analytes. When exciting with UV

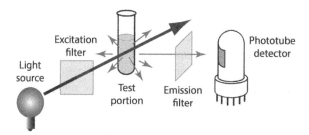

Figure 4.23 Schematic of a simple fluorometer.

[8]Du, H.; Fuh, R. A.; Li, J.; Corkan, A.; Lindsey, J. S. "PhotochemCAD: A computer-aided design and research tool in photochemistry," *Photochem. Photobiol.* **1998**, *68*, 141–142; https://omlc.org/spectra/PhotochemCAD/html/004.html; accessed August 2022.

light, the sample cuvette must be high-quality quartz to transmit the UV wavelength. This simple fluorometer design is used for quantitative measurements, but it does not have the ability to scan wavelength to record a spectrum.

The filters in Figure 4.23 are critical to achieving a sensitive fluorescence measurement. They are optical quality glass that block one spectral region and transmit another. The "Excitation filter" in the excitation light beam blocks light of longer wavelengths from reaching the sample cuvette and possibly scattering to the detector. The "Emission filter" blocks the shorter wavelength excitation light but passes the longer wavelength fluorescence (refer to the Stokes shift defined above). The 90° geometry between the excitation light and fluorescence collection is typical to optimize the blocking of scattered light. Scattered light that reaches the detector is indistinguishable from fluorescence. Trying to measure a small fluorescence signal on a large baseline is difficult. Making the baseline as small as possible, by filtering scattered light, increases sensitivity of the instrument.

High-resolution fluorometers use monochromators with rotatable diffraction gratings in place of, or in addition to, optical filters. The monochromators serve the same function to discriminate fluorescence from background light, but they also allow the wavelength to be scanned to obtain spectra. A xenon discharge lamp is a common light source to provide a continuum of excitation wavelengths from the UV through the visible. The light from the xenon lamp passes through an excitation monochromator to select one specific excitation wavelength. The fluorescence excited by this light is focused into a second emission monochromator, which selects the specific wavelength of fluorescence that reaches the detector. Quantitative measurements are made by placing the excitation and emission monochromators at the optimum wavelengths, the same as with a filter fluorometer. An excitation spectrum (left spectrum in Figure 4.22) is obtained by recording fluorescence at one specific wavelength and scanning the excitation monochromator. A fluorescence spectrum (right spectrum in Figure 4.22) is obtained by exciting the sample at one specific wavelength and scanning the emission monochromator.

4.8.1   Quantitative Fluorometry

The light intensity of an emissive sample is linearly proportional to concentration at low analyte concentration. The linear dependence and sensitivity make atomic emission and molecular fluorescence very useful for quantitating emitting species. The full relationship between fluorescence intensity, F, and analyte concentration is

$$F = k\Phi(\lambda_{em})P_0(\lambda_{exc})(1 - 10^{-\varepsilon bc}) \qquad (4.12)$$

F will depend on both the excitation, λ_{exc}, and the emission, λ_{em}, wavelengths that are selected for the measurement. In the absence of spectral interferences, λ_{exc} and λ_{em} are set at the peaks of the absorption and fluorescence spectra, respectively.

The k is an instrumental factor for the fraction of emitted light that is collected and ultimately converted to an electrical signal. It depends on the geometric arrangement, transmission of optical components, detector efficiency, and adjustable parameters such as slit width. Several of these factors will depend on the measurement wavelengths, λ_{exc} and λ_{em}, that are selected. We generally do not try to determine k. If all instrument settings and measurement wavelengths are maintained the same, k is simply one component in the proportionality constant in Equation (4.12). As always, keeping experimental conditions the same when measuring standard and unknown test solutions keeps the calibration function appropriate for the unknowns. $\Phi(\lambda_{em})$ is the *quantum yield*, the number of photons emitted at λ_{em} divided by the number of photons absorbed. It is a fraction between 0 and 1, where a high quantum yield indicates a very efficient fluorophore. This factor will be constant for an analyte in a given solvent and temperature. Some analytes are sensitive to interferences that reduce the quantum yield, which we call *quenching*. The sensitivity of an analyte to quenching is something to determine in validation experiments.[9] $P_o(\lambda_{exc})$ is the radiant power of the excitation source at λ_{exc}. A brighter excitation light will get you more emitted photons. ε is the molar absorption coefficient at λ_{exc}, b is the path length of the excitation light to get to the emissive analyte, and c is the analyte concentration. These quantities are the same as those used in the Beer–Lambert law.

The $(1 - 10^{-\varepsilon bc})$ term is present because F will depend on the amount of the excitation light that is absorbed as it transits the sample. If the analyte concentration is low so that εbc is 0.01 or less, the power of the excitation light will not vary significantly through the sample. Converting this exponential term to base e, $10^{-\varepsilon bc} = e^{-2.303\varepsilon bc}$, and expanding in a power series gives

$$F = k\Phi(\lambda_{em})P_o(\lambda_{exc})\left\{1-\left[1+(-2.303\varepsilon bc)+\frac{(-2.303\varepsilon bc)^2}{2!}+\cdots\right]\right\}$$

On dropping the higher terms, we have the linear expression

$$F = k\Phi(\lambda_{em})P_o(\lambda_{exc})2.303\varepsilon bc \qquad (4.13)$$

This relationship is valid at low concentrations, typically $<10^{-5}$ M, depending on ε for a given fluorescent analyte. The key point is that we have a linear relationship where $F \propto c$. We can use this linear relationship to determine unknown concentrations after calibrating the fluorometer with one or more standards. The process is the same as we have done for other types of analyses, and examples are found in the you-try-it and end-of-chapter exercises.

A key aspect of using fluorescence is recognizing when data is nonlinear. Figure 1.6 showed a calibration curve for a set of fluorescence measurements

---

[9]Numerous techniques in biology actually use fluorescence quenching to determine protein conformations and other properties of biomolecules.

where the curve appears to become nonlinear at concentration >1.0 ppm. We can use this data to calculate the attenuation of the excitation light, which is often called an *inner filter effect*. The molar absorption coefficient at the excitation wavelength of 266 nm is 33,000 $M^{-1}cm^{-1}$. I use a path length of 0.5 cm to calculate the absorbance at the center of the cuvette:

$$A = (33,000 \ M^{-1}cm^{-1}) \ (0.5 \ cm)c$$

This calculation gives an approximate result because the fluorometer will collect light from some volume of excited molecules in the cuvette. Table 4.10 lists the results, where I've converted ppm to M concentration and converted A to %-T. What I see from the result is that the attenuation is quite gradual with concentration, and there is probably some error introduced at 1.0 ppm. If an analyst finds that a set of solutions have a high enough concentration to be in the nonlinear fluorescence region there are two simple solutions. All solutions can be diluted by a factor that gets them into the linear range. Another approach, which can take less effort, is to measure all test and standard solutions at an excitation wavelength that has a lower absorption. For the case of riboflavin, setting λ_{exc} to one of the weaker absorbing peaks at 370 or 450 nm will reduce the fluorescence intensity by about a factor of 3. Given that concentrations are relatively high, the loss of signal is not a problem and all measurements will be in the linear range.

Nonlinearity due to the inner filter effect is easy to predict because we can measure the absorbance of test solutions. There can be other contributions to nonlinearity, and the measurement range should always be determined experimentally. If the Stokes shift is small, unexcited fluorophore in the cuvette might reabsorb some of the fluorescence that is emitted from the center of the cuvette. Some fluorophores are known to aggregate, which can change the quantum yield. These issues are concentration dependent and can be corrected by dilution.

TABLE 4.10 Light Transmittance at Center of Fluorometer Cell

Concentration (ppm)	Concentration (M)	Absorbance	Transmittance
0.0	0.0	0.000	1.00
0.2	5.31×10^{-7}	0.009	0.98
0.4	1.06×10^{-6}	0.017	0.96
0.6	1.59×10^{-6}	0.026	0.94
0.8	2.13×10^{-6}	0.035	0.92
1.0	2.66×10^{-6}	0.044	0.90
1.2	3.19×10^{-6}	0.052	0.89
1.4	3.72×10^{-6}	0.061	0.87

The presence of interferences that absorb the excitation light, emit fluorescence at the wavelength being measured, or absorb the analyte fluorescence will all introduce errors in a measurement. As noted above, some fluorophores are susceptible to quenching by sample matrix components. If interferences are suspected, calibration with matrix-matched standards or validating a test solution with a standard addition spike can identify problems (see Example 4.5). The point of discussing these nonlinearity issues is not to say fluorescence doesn't work. It is an extremely sensitive analytical tool. The point is that the analyst must be aware of issues leading to bias to be able to take corrective actions to make accurate measurement.

Example 4.5 Comparing Sensitivity of Absorption and Fluorescence. Use Figure 1.7 to estimate the limit of detection of riboflavin using UV/Vis absorption.

The sensitivity of a fluorescence measurement depends on the details of the fluorometer. Commercial instruments can achieve a LOQ for riboflavin of 0.2 ppb.[10] Specialized fluorometers with laser excitation and low-noise detectors can achieve much less. Here we will calculate LOQ for absorbance. Table 1.11 lists an absorbance at 266 nm of 0.013 with S/N of 13. Using the criterion of S/N = 10 for the LOQ, I divide the signal by 1.3 (0.013/1.3) to estimate the smallest quantifiable peak at 266 nm to have an absorbance of 0.010. Now calculating the concentration equivalent to that absorbance in a 1.0-cm cuvette

$$0.010 = \left(33000 \text{ M}^{-1}\text{cm}^{-1}\right)(1.0 \text{ cm})c$$

$$c = 3.0 \times 10^{-7} \text{M or 0.11 ppm (110 ppb)}$$

This measurement assumes that there are no interferences, turbidity, or other factors affecting the absorbance measurement. Even in this ideal case, the LOQ is approximately 500 times lower for fluorescence compared to absorption spectroscopy.

You-Try-It 4.D
The fluorescence worksheet in you-try-it-04.xlsx contains measurements of a series of riboflavin standards and of an unknown solution. Construct a calibration curve and determine the concentration of the unknown.

---

[10]Upstone, S. *Measurement of LOD and LOQ of Vitamin B2 Using the FL 6500 and FL 8500 Fluorescence Spectrometer*, Perkin-Elmer Application Note, 2019.

Chapter 4. What Was the Point? This chapter introduced the concepts and instrumentation for using spectroscopic transitions to make analytical measurements. Our spectroscopic tools of absorption and fluorescence provide a linear dependence for analysis of molecules and complexes.

PRACTICE EXERCISES

1. Disk players use laser diodes to read the information encoded on the disks. DVD players operate at 650 nm and Blu-ray players use 405 nm light.
   (a) What is the frequency of the light from these laser sources?
   (b) What is the energy of one photon for these wavelengths in joules and inverse centimeters?

2. Order the following types of EM radiation from the lowest energy to the highest energy: γ-ray, IR, microwave, radio, UV, visible, X-ray.

3. An article by a national news organization described an astronomer's discovery of previously undetected matter as "The astronomers think the missing matter exists as highly charged hydrogen between galaxies, ..." What is the highest charge of a hydrogen ion? What might the author of the article have meant rather than charge?

4. The absorption of IR light by a carbon–hydrogen stretching vibration occurs at an energy of 3100 cm^{-1}. What is the wavelength of this light? How many photons at the energy of this vibration are necessary to provide the total energy of one photon of an electronic absorption band at 280 nm?

5. A molecular absorption transition can occur from the ground state to
   (a) a filled antibonding orbital
   (b) an empty antibonding orbital
   (c) either of these
   Explain your answer.

6. The schematic in Figure 4.4 illustrated an electronic absorption transition. Sketch your own diagrams for the emission and scattering of photons.

7. The bond length of carbon–halogen bonds varies in the order of C–F < C–Cl < C–Br < C–I. Which carbon–halogen bond will have the highest IR absorption energy for the stretching vibration? Which carbon–halogen bond will have the longest wavelength absorption band for this stretching vibration?

8. Using the calibration curve in Figure 4.20, what is the glucose concentration in a sample that gives an absorbance of 0.521?

9. Which one of the following compounds will have its lowest energy UV/ Vis absorption band at a shorter wavelength than the other compounds? Explain your rationale.

- Acetone, CH_3COCH_3
- Benzene, C_6H_6
- Hexanol, $C_6H_{13}OH$
- *n*-Hexane, C_6H_{14}

10. A sample in a 1-cm cuvette has a measured absorbance of 0.333 AU (absorbance units).
    (a) What will be the absorbance of this same sample in a 5-cm cell?
    (b) What will be the transmittance of this same sample in a 5-cm cell?
    (c) What will be the percent-transmittance of this same sample in a 2-cm cuvette?

11. Calculate the transmitted light for the 4.0×10^{-5} M solution data in Table 4.6 (solid line in Figure 4.17) at a path length of 2.0 cm. Answer in terms of number of photons, transmittance, %-transmittance, and absorbance.

12. A single-beam UV/Vis absorption spectrophotometer is calibrated accurately to 0 and 100-% T using a clean cuvette and pure solvent. If the cuvette becomes stained with the analyte during sample measurements, what will happen to subsequent results?

13. Typical instructions for UV/Vis absorption measurements suggest keeping the absorbance of the samples between 0.1 and 1.0 absorbance units. Calculate T for these two limits and explain these recommendations.

14. Discuss the relative advantages and disadvantages of array versus scanning UV/Vis absorption spectrophotometers.

15. Using the data in Table 4.6, what is ε for the absorbing species? Do the data values you use make a difference in the result?

16. Use the calibration curve in Figure 4.20 for this question.
    (a) What is the glucose concentration of an unknown sample that has an absorbance of 0.271?
    (b) After completing a measurement, you notice some cloudiness in the sample solution. You run a blank and find an absorbance of 0.009. Correct your answer in (a) for this new blank measurement.

17. Fluorescence is intrinsically more sensitive than absorption because you are measuring a light power in the absence of any background. What factor limits the sensitivity in an absorption measurement?

18. Figure 1.8 shows the fluorescence from riboflavin standards as a function of concentration. What is the concentration of an unknown sample that produces a fluorescence signal of 14.5?

19. Iron concentrations can be measured with UV/Vis absorption spectroscopy by forming complexes of iron(II)-1,10-phenanthroline, $Fe(phen)_3^{2+}$. Refer to the following procedure for the next questions:

Procedure to Measure Iron in Solution via UV/Vis Spectroscopy.
1. To a 100.0-ml volumetric flask add:
   - 10.00 ml of sample solution.
   - 10 ml of 1,10-phenanthroline solution.
   - 10 ml of a redox and pH buffer.
2. Dilute to mark and mix.
3. Rinse a 1.00-cm cuvette with a small
   amount of this solution and discard.
4. Refill the cuvette and measure the
   absorbance at 508 nm.

(a) A standard solution of $4.00 \times 10^{-5}\,M\,\mathrm{Fe(phen)}_3^{2+}$ has an absorbance of 0.549. What is the molar absorption coefficient of $\mathrm{Fe(phen)}_3^{2+}$ at 508 nm?

(b) A water sample from strip mine runoff, prepared by the above procedure, has an absorbance of 0.455. What is the $\mathrm{Fe(phen)}_3^{2+}$ concentration in the measured solution?

(c) What is the Fe concentration in the original water sample?

(d) Another water sample from the strip mine, prepared by the above procedure, has an absorbance of 3.421. How will you analyze this sample?

(e) Predict the effect on an iron measurement if the redox buffer (hydroxylamine) is omitted and some of the iron persists as Fe^{3+}? The molar absorption coefficient of $\mathrm{Fe(phen)}_3^{3+}$ at 508 nm is lower than ε for $\mathrm{Fe(phen)}_3^{2+}$.

Part I. What Was the Point?

Part I of the text introduced core concepts of analytical chemistry that you will see repeatedly: calibration, interferences, and proper data reporting. The techniques described for sample preparation and classical analysis are common laboratory manipulations that are useful in analytical procedures and other purposes.

Most of the chemical examples discussed in Part I relied on reactions going to completion:

- complete neutralization of acid and base at the titration end point
- complete precipitation and collection of an insoluble salt in a gravimetric analysis
- complete formation of an absorbing complex in an absorbance measurement
- complete extraction of an analyte from a sample matrix

Processes that occur at less than 100% efficiency lead to inaccurate results. For extraction of analytes, and other sample processing steps, we use spikes and control samples to monitor that we achieve complete extraction to ensure accuracy. In the measurement step of an analysis, we rely on chemical controls to ensure that our analyte is in 100% of the form that we are measuring. It is up to the analyst to use these aspects of good laboratory practice to ensure reliable and accurate results.

The wet chemical methods, based on reactions that go to completion, fill the critical role of providing methods to validate primary standards. The classical methods are also quite suitable, with fewer potential problems, when quantitating analytes at relatively high concentrations. Having reiterated the importance of classical methods, you will find that most analytical procedures now involve at least some instrumentation.

Part II. What Is Next?

Aqueous-phase species often exist in multiple forms. The tartrate complexing agent shown for the Cu^{2+} analysis has two COO^- functional groups, which supply the unbonded electron pairs to form the absorbing complex. At some pH, the tartrate becomes protonated and exists as tartaric acid, $C_2H_2(OH)_2(COOH)_2$. If these COO^- groups were protonated and existed in solution as COOH, the absorbing complex might not form. Fehling's and Benedict's solutions are adjusted to alkaline pH to prevent this interference.

Part II of this text provides a fundamental description of chemical equilibria to help us recognize these types of interferences. We look at acid–base, complexation, and precipitation equilibria. The chapters build in complexity to describe the types of competing equilibria that occur in industrial waste streams, natural aquatic systems, body fluids, and in your careers.

REACTIONS THAT DO NOT GO TO "COMPLETION." EQUILIBRIA IN AQUEOUS SOLUTIONS

CHAPTER 5

ACID–BASE EQUILIBRIA AND ACTIVITY

Learning Outcomes

- Classify acids and bases as strong or weak.
- Calculate ionic strength and activity coefficients.
- Write equilibrium constant expressions for weak acids and weak bases.
- Correct thermodynamic equilibrium constants using activity coefficients.
- Predict equilibrium concentrations in solutions of weak acids and weak bases.

5.1 ACIDS AND BASES

We saw in Chapter 3 that we can use a neutralization reaction in an acid–base titration, which is a common type of analysis. An assumption in a titration is that the neutralization reaction goes to completion. In practice, we use strong acids or strong bases as titrants so that this condition is true. To understand the chemistry of aqueous systems, it is more common that we are dealing with equilibrium mixtures of multiple forms of a given substance. This chapter and the next discuss how to describe such equilibrium mixtures of acids and bases.

Basics of Analytical Chemistry and Chemical Equilibria: A Quantitative Approach, Second Edition. Brian M. Tissue.
© 2023 John Wiley & Sons, Inc. Published 2023 by John Wiley & Sons, Inc.
Companion Website: www.wiley.com/go/tissue/analyticalchemistry2e

For most of the situations that we encounter in analyzing aqueous solutions, we can use the Brønsted–Lowry definitions for acids and bases:

Acid: Any chemical species that can provide a proton, H^+, to a base.
Base: Any chemical species that can remove a proton, H^+, from an acid.

These definitions are not the most general. They do not describe how metal salts affect acidity, which we discuss with complexation in Chapter 7. Also, they do not describe the chemistry of many compounds in nonaqueous solution. In these cases, the Lewis acid and Lewis base descriptions are much better. Lewis acids and Lewis bases are described as electron-pair acceptors and electron-pair donators, respectively. These more general definitions encompass the Brønsted–Lowry acids and bases. A species that can provide an H^+ can accept an electron pair. Likewise, a Brønsted–Lowry base can donate an electron pair. Noting these limits of the Brønsted–Lowry definitions, we will use them because of their simplicity and suitability for acids and bases in aqueous equilibria.

Most Brønsted–Lowry acids have a hydrogen atom bound to a more electronegative atom, making the bond ionic. An atom in an ionic bond is more likely to dissociate in water than one that is bound covalently. We write this reaction as

$$HBase(aq) \rightleftharpoons H^+(aq) + Base^-(aq)$$

where the double-arrow harpoons indicate that we have an equilibrium between the reactant on the left and the products on the right. When dealing with an equilibrium, it is equivalent to write the reaction as the reverse:

$$H^+(aq) + Base^-(aq) \rightleftharpoons HBase(aq)$$

In practice, the convention for weak acids and bases is to use the first description. In this qualitative discussion, we have said nothing about the solubility of HBase in water or the degree to which HBase dissociates. We will deal with these issues when we start using equilibrium constants to describe equilibria quantitatively. Our first concern is to be able to recognize acids, bases, and mixtures of these species that lead to neutralization reactions versus equilibrium mixtures.

Table 5.1 lists Pauling electronegativities for the first three rows of the periodic table.[1] These values provide the first approximation for predicting when a hydrogen atom in a molecule will be acidic. As an extreme example, molecular hydrogen, H_2, forms a highly covalent bond. It will not provide an

---

[1]Adapted with permission from Allen, L. C. "Electronegativity is the average one-electron energy of the valence-shell electrons in ground-state free atoms," *J. Am. Chem. Soc.* **1989**, *111*, 9003–9014. Copyright 1989 American Chemical Society.

TABLE 5.1 Pauling Electronegativities

H 2.20								He –
Li 0.98	Be 1.57		B 2.04	C 2.55	N 3.04	O 3.44	F 3.98	Ne –
Na 0.93	Mg 1.31		Al 1.61	Si 1.90	P 2.19	S 2.58	Cl 3.16	Ar –

acidic proton when bubbled through water.[2] If we could move hydrogen in the periodic table to be closer to elements with similar electronegativities, we would slide it over to be above boron and carbon. Hydrogen atoms bound to atoms with similar electronegativities, such as C or N, tend to be covalent, and these hydrogen atoms are rarely acidic protons (an exception is the weak acid HCN). The most common weak acids have an H atom bound to an electronegative atom such as O or S, and the –COOH group occurs often. The acidity will further be affected by the local chemical structure. Alcohols usually do not provide an acidic proton in water, but phenol, C_6H_5OH, is an exception.

Brønsted–Lowry bases have nonbonded electrons, usually a lone pair of electrons, that can form a coordinate covalent bond with a proton. The interaction between a base and a proton is completely analogous to a ligand forming a complex with a metal ion, M^+:

$$H^+(aq) + Base(aq) \rightleftharpoons HBase^+(aq)$$

$$M^+(aq) + Ligand(aq) \rightleftharpoons MLigand^+(aq)$$

We see in Chapter 7 that we often need to worry about protons competing with metal ions for the lone pairs on ligands. Many weak bases have a negative charge, but the charge is not a requirement for a species to act as a base. We will encounter many neutral amines that can bind a proton with their unbonded pair of electrons (see the weak base structures in Figure 5.2).

5.1.1 Strong Acids and Bases

Strong acids and strong bases are a subset of the ionic compounds referred to as *strong electrolytes* (Table 2.13). A strong acid is a strong electrolyte anion associated with a proton, H^+. A strong base is a strong electrolyte cation associated with the hydroxide ion, OH^-. The most common strong acids and strong bases are listed in Table 5.2. $Ca(OH)_2$ is a borderline case because it is less

---

[2]We can convert between H_2 and H^+ via redox chemistry (see Chapter 9), but H_2 will not dissociate spontaneously in water to form H^+.

soluble in water than the other strong bases, and it will precipitate when the concentrations of Ca^{2+} and OH^- are high.

When dissolved in water, the strong acids and bases dissociate completely. For example,

$$HNO_3(l) + H_2O \rightarrow NO_3^-(aq) + H_3O^+(aq)$$

$$NaOH(s) + H_2O \rightarrow Na^+(aq) + H_2O + OH^-(aq)$$

where the right arrow indicates complete dissociation. For consistency with the Brønsted–Lowry definitions, the dissociation of HNO_3 in water shows H_2O acting as a base. H_2O removes the proton from HNO_3 because NO_3^- and H_3O^+ are thermodynamically more stable than neutral HNO_3 in water.

The dissociation of NaOH in water is similar in that H_2O acts as an acid by providing a proton to the NaOH salt.

$$NaOH + H^+ \cdot \cdot OH^- \rightarrow Na^+(aq) + OH^- \cdot \cdot H^+ + OH^-(aq)$$

The two dots, $\cdot \cdot$, represent the bond that breaks in a water molecule (left side of the reaction) and the new bond that forms when the OH^- from the NaOH reacts with the H^+ (right side of the reaction). This description and terminology is more extensive than necessary to understand the strong bases, but it does show that the Brønsted–Lowry theory is sufficiently general to describe all of the strong acids, strong bases, weak acids, and weak bases that we will discuss. This example also highlights the two ways that we often refer to a hydrogen ion: H^+ and H_3O^+. I will use H^+ when discussing reactions that involve a proton, such as neutralization or redox reactions. I will use the hydronium ion, H_3O^+, when discussing the equilibrium concentration of the hydrogen ion in water. This convention gives an indication that what exists in solution is a proton surrounded by water molecules.[3]

TABLE 5.2   Common Strong Acids and Bases

Strong Acids		Strong Bases	
HCl	Hydrochloric acid	LiOH	Lithium hydroxide
HBr	Hydrobromic acid	NaOH	Sodium hydroxide
HI	Hydroiodic acid	KOH	Potassium hydroxide
HNO_3	Nitric acid	RbOH	Rubidium hydroxide
$HClO_3$	Chloric acid	$Ca(OH)_2$	Calcium hydroxide
$HClO_4$	Perchloric acid	$Sr(OH)_2$	Strontium hydroxide
H_2SO_4	Sulfuric acid	$Ba(OH)_2$	Barium hydroxide

[3]The structure of water is a current research topic. The actual degree of hydration probably involves two to four water molecules that form hydrogen bonds with the proton.

Owing to the complete dissociation that occurs when a strong electrolyte ion is involved, the concentration of H_3O^+ is equal to the concentration of the strong acid that is added to water multiplied by the number of H^+ per acid molecule. Similarly, the concentration of OH^- when a strong base is added to water is equal to the concentration of the added base multiplied by the number of OH^- per base molecule. If 0.010 mol of nitric acid is added to 1.0 l of water, the solution is 0.010 M H_3O^+ and 0.010 M NO_3^-. We label this solution as 0.010 M HNO_3, understanding that there is no "undissociated" nitric acid present in solution.

If 0.010 mol of $Ca(OH)_2$ is added to 1.00 l of water, the solution is 0.010 M Ca^{2+} and 0.020 M OH^-. In this case, there are two OH^- for each $Ca(OH)_2$. We label this solution 0.010 M $Ca(OH)_2$, again recognizing that we are dealing with strong electrolytes and that there is essentially no undissociated $Ca(OH)_2$ molecules in solution. This naming convention for solution concentration is the same as for any strong electrolyte salt; for example, 0.1 M NaCl contains 0.1 M Na^+ and 0.1 M Cl^-, but essentially no molecular NaCl.[4]

These examples explain the difference between a formal concentration and an equilibrium concentration. When preparing solutions of soluble compounds that are not strong electrolytes, we usually do not know the equilibrium concentrations of the solutes. However, we do know what we have put into the solution and if we expect any reaction to occur. If a reaction does occur, we can calculate the resulting concentrations as we do for any limiting reagent calculation. We call these "obvious" concentrations *formal concentrations* and we use the symbol *c*. Formal concentrations are a convenient means to describe a solution, label a bottle, and determine what reactions might take place if other solutes are added to a solution. The *equilibrium concentration*, denoted by [ ], is used for the concentration of the specific forms that actually exists in solution. With the exception of the strong electrolytes, it takes the formal concentration and the equilibrium constant to predict the form and equilibrium concentrations of most chemical species in solution.

> *Formal concentration*: The concentration that we use to describe or label a solution. Often the total concentration of a chemical species, regardless of its true form in solution. It is specified with the notation *c*, for example, c_{Na^+} or c_{OH^-}.
>
> *Equilibrium concentration*: The actual concentration in solution of any given chemical species or form of a solute. It is specified with the notation [ ], for example, $[Na^+]$ or $[OH^-]$.

As we begin using more complicated expressions, the variables with chemical symbols as subscripts can get confusing. Be careful that you do not interpret a

---

[4]When we discuss insoluble salts in Chapter 8, we will see that we will have small equilibrium concentrations of the "ionic molecule" in solution.

plus or minus in a subscript as a mathematical operator. Where this typo-graphical condition can occur, I will often use parentheses to keep things clear.

Example 5.1 Strong Acid.   What is the H_3O^+ concentration of a solution pre-pared by adding 0.0250 mol of $HClO_4$ to water and diluting to 1.00 l?

$HClO_4$ is a strong electrolyte and dissociates completely in water, that is, it is a strong acid:

$$HClO_4 + H_2O \rightarrow ClO_4^-(aq) + H_3O^+(aq)$$

Thus, the equilibrium and formal concentrations are equal

$$[H_3O^+] = c_{H_3O^+} = \frac{0.0250\,mol}{1.00\,l} = 0.0250\,M$$

Example 5.2 Strong Base.   What is the concentration of OH^- when 0.0050 mol of $Ca(OH)_2$ is added to 1.00 l of water? (This amount of $Ca(OH)_2$ is solu-ble in water.)

When $Ca(OH)_2$ is dissolved in water,

$$1\ mol\ Ca(OH)_2 \rightarrow 1\ mol\ Ca^{2+} + 2\ mol\ OH^-$$

and the solution will contain 0.0050 mol Ca^{2+} and 0.0100 mol OH^-.

As for the case of a strong acid, we would refer to this solution as 0.0050 M $Ca(OH)_2$. However, there is no undissociated $Ca(OH)_2$ present in the solution. As in Example 5.1, the formal and equilibrium concentrations are equal:

$$[Ca^{2+}] = c_{Ca^{2+}} = \frac{0.0050\,mol}{1.00\,l} = 0.0050\,M\,Ca^{2+}$$

$$[OH^-] = c_{OH^-} = \frac{0.010\,mol}{1.00\,l} = 0.010\,M\,OH^-$$

As shown in Examples 5.1 and 5.2, the formal and equilibrium ion concen-trations are equal for a strong electrolyte in a solution. This will not be true for many other types of solutes in water. When we wish to discuss equilibrium concentrations, we must refer to formal or equilibrium concentrations explicitly; hence, the different notations.

As we saw in Chapter 3, a strong acid will react with any strong or weak base that is added to solution. Likewise, a strong base will react with any strong or weak acid that is present. This statement is true for a strong base neutral-izing all of the protons of a polyprotic acid or for all of the protons from a mixture of different acids in solution. We will defer discussion of neutraliza-tion reactions involving weak acids and weak bases to Chapter 6, but reactions between strong acids and bases are simply limiting reagent calculations.

Example 5.3 illustrates this type of calculation. Similar limiting reagent calculations are often necessary when working with weak acids and bases to determine the species that will be present at equilibrium.

Example 5.3 Limiting Reagent.   What are the concentrations of all ions on mixing 10.0 ml of 0.20 M HCl with 200.0 ml of 0.010 M $Ca(OH)_2$ and diluting to 1.00 l of water?

The mole amounts of ions that we are mixing in solution are

- (0.0100 l)(0.20 M HCl) = 0.0020 mol H^+
- (0.0100 l)(0.20 M HCl) = 0.0020 mol Cl^-
- (0.2000 l)(0.010 M $Ca(OH)_2$) = 0.0020 mol Ca^{2+}
- (0.2000 l)(0.010 M $Ca(OH)_2$) $\left(\dfrac{2 \text{ mol } OH^-}{1 \text{ mol } Ca(OH)_2} \right)$ = 0.0040 mol OH^-

Since Ca^{2+} and Cl^- are ions of a strong electrolyte, concentrations are simply

$$\left[Ca^{2+} \right] = c_{Ca^{2+}} = \frac{0.0020 \text{ mol}}{1.00 \text{ l}} = 2.0 \text{ mM}$$

$$\left[Cl^- \right] = c_{Cl^-} = \frac{0.0020 \text{ mol}}{1.00 \text{ l}} = 2.0 \text{ mM}$$

The H^+ and the OH^- will react on mixing:

$$
\begin{array}{ccccc}
H^+(aq) & + & OH^-(aq) & \rightarrow & H_2O \\
0.0020\,\text{mol} & & 0.0040\,\text{mol} & & \\
\approx 0\,\text{mol} & & 0.0020\,\text{mol} & & 0.0020\,\text{mol}
\end{array}
$$

In this reaction, H^+ is the limiting reagent and the net reaction shows that we have excess OH^-. The OH^- concentration is

$$[OH^-] = c_{OH^-} = \frac{0.0020\,\text{mol}}{1.00\,\text{l}} = 2.0\,\text{mM}$$

In terms of formal concentrations, c_{H^+} = 0 M. We will see in Section 5.3 how to calculate the H^+ equilibrium concentration, $[H^+]$.

As a check, we can tabulate the results in Example 5.3 and make sure that our results make sense. A solution must have an overall neutral charge, so if your math is correct above, the positive and negative charges will be equal. Table 5.3 shows the concentration of each ion in solution and the concentration of charge. Because each Ca^{2+} has a charge of +2, the "concentration" of

TABLE 5.3 Checking Charge Balance

Ca^{2+}	2 mM	Positive charge	4 mM
H^+	≈ 0 M	Positive charge	≈ 0 M
Cl^-	2 mM	Negative charge	2 mM
OH^-	2 mM	Negative charge	2 mM

charge due to Ca^{2+} is $(c_{Ca^{2+}})(+2)$. We can see that we have a total of 4 mM of positive charge and a total of 4 mM of negative charge, which balance to give us a neutral solution overall.

You-Try-It 5.A

The neutralization worksheet in you-try-it-05.xlsx contains a table of strong acid and strong base solutions. Determine the limiting reagent when each pair of solutions is mixed. From this result, you can also determine if the resulting solution is acidic or neutral and calculate the pH (pH = $-\log[H^+]$). Use Example 5.3 as a guide.

5.2 WEAK ACIDS AND WEAK BASES

5.2.1 Identifying Weak Acids and Weak Bases

Unlike the complete dissociation that occurs when strong acids or strong bases are added to water, weak acids and weak bases react with water only to a limited extent. The extent to which they react varies and is quantified by an equilibrium constant. These constants for the weak acid and weak base hydrolysis reactions are labeled K_a and K_b, respectively. Later in the chapter, we will see how to use these constants to determine equilibrium concentrations. First, we must be able to identify weak acids and weak bases in order to predict when reactions will occur. How do we identify weak acids and weak bases? It is not always obvious, so for compounds with which you are not familiar see if they are listed in tables of K_a or K_b values. Useful reference sources include the *CRC Handbook of Chemistry and Physics*, the *Merck Index*, and *Lange's Handbook of Chemistry*. If you find them in these sources, they are weak acids or weak bases.

There are a large number of inorganic and organic weak acids (see Tables 5.7 and 5.8 for examples). Inorganic weak acids include the protonated forms of fluoride, F^-; sulfide, S^{2-}; and oxyanions such as sulfite, SO_3^{2-}, and nitrite, NO_2^-. Most organic acids have one or more carboxylic acid groups, –COOH. There are some exceptions of organic weak acids without the carboxylic acid group, such as HCN; phenol, C_6H_5OH; and cysteine, which has an –S–H group. Figures 5.1 and 5.2 show some examples of weak acids and weak bases.

The weak acids in Figure 5.1 range from, left to right, the simple organic acetic acid, the diprotic organic phthalic acid, and the diprotic and charged inorganic acid hydrogen sulfite ion. The hydrogen sulfite ion example is shown with one proton having been neutralized. The amino acid alanine is shown in Figure 5.2 (rightmost structure).

Most weak bases are molecules with an amine group or a nitrogen atom in a cyclic ring. The nitrogen atom in these compounds has a nonbonded pair of electrons, shown as two dots in Figure 5.2, that can form a coordinate covalent bond with a proton. The weak bases in Figure 5.2, from left to right, are ammonia, imidazole, and the amino acid alanine. The amino acid is shown in the neutral form, but it actually exists in neutral pH solutions with the amine being protonated and the carboxylic acid being unprotonated. Draw the structures and try to identify the acidic and basic groups in the cases in Example 5.4 before reading the answers.

Example 5.4 Weak Acids and Weak Bases.   Identify the acidic protons and basic groups in each of the following molecules. (Refer to Tables 5.7 and 5.8 for help and look up structures as necessary.)

- Fluoroacetic acid
- Nitrous acid
- Ethanol
- Glutamic acid

Fluoroacetic acid and nitrous acid each have one K_a value listed in Tables 5.7 and 5.8, respectively, so they each have one acidic proton. Neither of these molecules have an atom that can be protonated, that is, there are no basic groups.

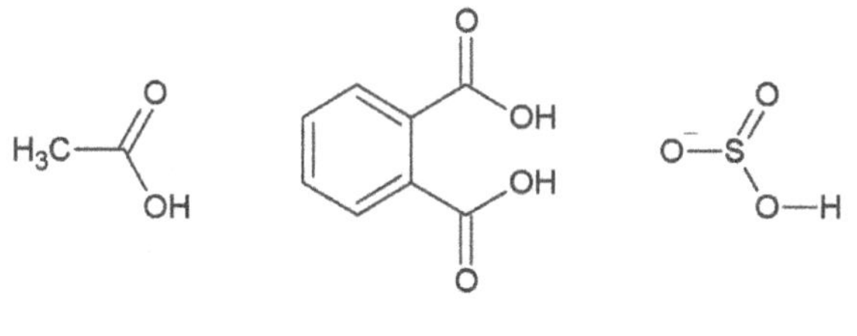

Figure 5.1   Examples of structures of weak acids.

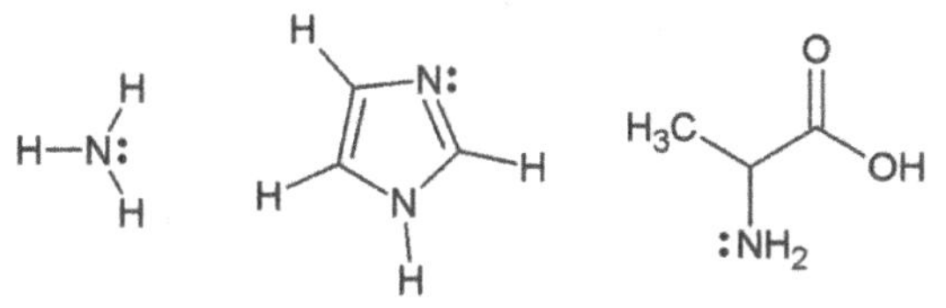

Figure 5.2   Examples of structures of weak bases.

Ethanol is not listed in Table 5.8, so the proton on the hydroxyl group is not an acidic proton in water.

Glutamic acid is an amino acid. In addition to the usual acidic and basic end groups of an amino acid, it has an additional carboxylic acid in the R group.

5.2.2 Weak Acid and Weak Base Equilibria

Owing to the complete dissociation of strong acids and strong bases in water, solutions of these compounds contain only the ions that made up the strong acids and bases. For example, dissolving HCl in water gives us a solution of H^+ and Cl^- (and a very small amount of OH^-, which we will discuss later). This situation is not true for weak acids and weak bases. When a weak acid dissolves in water, both the undissociated acid and the dissociated form are present. For the example of dissolving acetic acid, CH_3COOH, in water, we have a solution containing CH_3COOH, CH_3COO^-, and H^+ (as well as a very small amount of OH^-).

We need a description of both the acidic and basic forms, so we label the base that is produced the "conjugate base." Similarly, when a weak base dissolves in water, it produces some amount of hydroxide ion and the protonated form of the base. We call this protonated form the "conjugate acid." In Example 5.5, ammonia is a weak base that, when dissolved in water, produces some of the conjugate acid ammonium ion. Very often, we simply refer to a solution as containing a conjugate weak acid–base pair.

Example 5.5 Weak Base Equilibrium. What equilibrium occurs in solution when ammonia is added to water?

You have to know a little about ammonia to answer this example. Ammonia is $:NH_3$, where the two dots represent a lone pair of electrons. The lone pair provides electrons that can attract and bond a proton, H^+, forming the ammonium ion, NH_4^+. As water is a source of protons, the equilibrium is

$$NH_3\,(aq) + H_2O \rightleftharpoons NH_4^+\,(aq) + OH^-\,(aq)$$

Following the rules for writing K expressions listed in Section 2.5, the equilibrium constant expression for this equilibrium is

$$K_b' = \frac{[NH_4^+][OH^-]}{[NH_3]}$$

K_b' for ammonia is $\approx 1.8 \times 10^{-5}$. This value tells us qualitatively that the majority of ammonia in solution remains in the NH_3 form. A small amount exists as NH_4^+. The calculation of the actual equilibrium concentrations is left until later in the chapter.

As shown in Example 5.5, solutions will contain both acidic and basic species on reaching equilibrium. Keeping track of what is in solution can be confusing after taking account of neutralization and limiting reagent calculations. We refer to the solution in Example 5.5 as ammonia solution, but keep in mind that in addition to NH_3, it also contains NH_4^+, OH^-, and H_3O^+. Example 5.6 shows the case of creating a mixture of a weak acid and its conjugate base by partial neutralization.

Example 5.6 Weak Acid Equilibrium. What weak acid–weak base equilibrium is present when 0.05 mol of NaOH is added to 0.2 mol of acetic acid and diluted to 1.0 l?

An acid and a base are being mixed, so we must first consider the initial neutralization reaction and determine if there is a limiting reagent:

$$CH_3COOH(aq) \ + \ OH^-(aq) \ \longrightarrow \ CH_3COO^-(aq) \ + \ H_2O$$

0.20 mol	0.05 mol		
0.15 mol	≈ 0 mol	0.05 mol	0.05 mol

The strong base is the limiting reagent and it is neutralized completely. The result is that a portion of the acetic acid is converted to acetate ion. The predominant species in this solution after reaction are Na^+, acetic acid, and acetate ion. We typically write the equilibrium as the weak acid going to the conjugate base:

$$CH_3COOH(aq) + H_2O \rightleftharpoons CH_3COO^-(aq) + H_3O^+(aq)$$

(Writing the weak base going to the conjugate acid is equally correct.)

5.2.3 Salt Examples

For more practice, let us look at the ion concentrations and acid–base equilibria that occur in some salt solutions. Note that in these examples, I use a single arrow to indicate that the salt dissolves completely. Salts of strong electrolytes do not affect the acid–base properties of water directly. Adding NaCl to water does not make the solution acidic or basic; it remains a neutral solution. (When we get to activity, we will see that the total concentration of all ions can affect acid–base equilibria by affecting the ionic strength of a solution.) Some salts when added to water make the solution acidic or basic, and it is a common error to not recognize these salts. To know the effect of the salt

- determine if the cation can participate in acid–base equilibria, that is, it is not a strong electrolyte (see Example 5.7);
- determine if the anion can participate in acid–base equilibria, that is, it is not a strong electrolyte (see Example 5.8);

- if both the cation and anion can react with water, determine the predominant effect by comparing K_a' and K_b'. The ion with the larger K' will determine the acidity or alkalinity of the solution.

Now we will illustrate the first two of these cases with examples. Later in the chapter, we will consider the third case with amphiprotic species after we see how to convert between K_a' and K_b'.

Example 5.7 An Acidic Salt. What equilibrium is present when the soluble salt ammonium chloride is added to water? Write the equilibrium constant expression.

The dissolution reaction is

$$NH_4Cl(s) + H_2O \rightarrow NH_4^+(aq) + Cl^-(aq) + H_2O$$

Cl^- is a strong electrolyte and does not react with water. NH_4^+ has an acidic proton that can be provided to a base, so it is a weak acid. Overall, the solution will be acidic with the following equilibrium present:

$$NH_4^+(aq) + H_2O \rightleftharpoons NH_3(aq) + H_3O^+(aq)$$

Following our general rules, the equilibrium constant expression is

$$K_a' = \frac{[NH_3][H_3O^+]}{[NH_4^+]}$$

Example 5.8 A Basic Salt. What equilibrium is present when the soluble salt sodium acetate is added to water? Write the equilibrium constant expression.

The dissolution reaction is

$$CH_3COONa(s) + H_2O \rightarrow CH_3COO^-(aq) + Na^+(aq) + H_2O$$

Na^+ is a strong electrolyte and does not react with water. CH_3COO^- has a lone pair of electrons that can bind a proton, so it is a weak base. Overall, the solution will be basic

$$CH_3COO^-(aq) + H_2O \rightleftharpoons CH_3COOH(aq) + OH^-(aq)$$

and the equilibrium constant expression is

$$K_b' = \frac{[CH_3COOH][OH^-]}{[CH_3COO^-]}$$

5.3 WATER AND K_W

5.3.1 Self-Dissociation of Water

For strong electrolytes, in the absence of any precipitation or complexation reactions, the formal and equilibrium concentrations are equivalent, for example, $c_{Na^+} = [Na^+]$. We make the same assumption for the simple cases of adding strong acids or strong bases to water when we consider the dominant species that affects pH. The relationship between $[H_3O^+]$ and $[OH^-]$ in aqueous solution depends on the following equilibrium:

$$H_2O(aq) + H_2O(aq) \rightleftharpoons H_3O^+(aq) + OH^-(aq)$$

The presence of this equilibrium reaction indicates that H_3O^+ and OH^- will always exist together in aqueous solutions. The reason for water self-dissociation is entropy. Statistically, it is too improbable that all water molecules will be perfect at a temperature above 0 K. This statistical probability corresponds to a certain amount of energy, so that a small fraction of the water molecules self-ionize.

The presence of ions in pure water can be determined by measuring the electrical conductivity of a solution. Table 5.4 lists the solution conductivity at 25°C for several water samples. The units for solution conductivity are expressed as siemens per centimeter (S/cm), where siemens, S, is the unit of conductance. You will also see deionized water described by solution resistivity with the unit megohm-centimeters (MΩ-cm), which is the inverse of conductivity. Even though the conductivity of "pure water" is much smaller than the other water samples, it is a measurable value and arises from the presence of the H_3O^+ and OH^- ions.

The amount of the H_3O^+ and OH^- depends on the equilibrium constant expression and the value of the equilibrium constant, K'_w:

$$K'_w = [H_3O^+][OH^-] = [H^+][OH^-] = 1.01 \times 10^{-14} \quad \text{(at 25°C)} \quad (5.1)$$

where the subscript w indicates that the equilibrium constant is for water. As a reminder, the prime, ′, is used when working with concentrations to indicate that K' is a "formal equilibrium constant." Later in the chapter, we will see how to

TABLE 5.4 Conductivity of Water

Water Sample	Conductivity
Seawater	53 mS/cm
Tap water	≈50 μS/cm
Pure water exposed to atmospheric $CO_2(g)$	≈1 μS/cm
Pure water	0.055 μS/cm

correct thermodynamic equilibrium constant, K, as tabulated in reference sources, to obtain formal equilibrium constant, K', for working with nonideal solutions.

K_w is called the *ion product, dissociation constant,* or *ionization constant of water*. It is a measured thermodynamic quantity that is temperature dependent. Table 5.5 lists some values of K_w between 0 and 100°C.[5]

For pure water at 25°C, the resulting equilibrium concentrations are

$$[H_3O^+] = [OH^-] = \sqrt{K'_w} = 1.0 \times 10^{-7}\,M$$

Given the value of K'_w at 25°C, $[H_3O^+]$ and $[OH^-]$ can vary from ≈ 1 to 1×10^{-14} M; that is, if $[H^+] = 1$ M, using Equation (5.1) gives $[OH^-] = 1 \times 10^{-14}$ M. Describing the acidity or basicity of water over this range is rather inconvenient, so we use a logarithmic scale. "p" is a shorthand notation for $-\log$. Thus, $p[H_3O^+]$ is a shorthand notation for $-\log[H_3O^+]$, and $p[OH^-]$ is a shorthand notation for $-\log[OH^-]$. Taking the $-\log$ of the above expression,

$$p[H_3O^+] = p[OH^-] = 7.00$$

and

$$p[H_3O^+] + p[OH^-] = 14.00$$

When discussing calculated or estimated concentrations, I will use $p[H_3O^+]$ and $p[OH^-]$. The terms pH and pOH are used for actual measurements obtained from a pH meter. In the literature you will see pH and pOH used for both calculations and measurements.[6]

Solutions are called

- neutral when $[H_3O^+] = [OH^-]$;
- acidic when $[H_3O^+] > [OH^-]$;
- basic when $[H_3O^+] < [OH^-]$.

TABLE 5.5 K_w Versus Temperature

Temperature (°C)	K_w
0	1.15×10^{-15}
25	1.01×10^{-14}
50	5.31×10^{-14}
75	1.94×10^{-13}
100	5.43×10^{-13}

[5]Adapted with permission from Marshall, W. L.; Franck, E. U. "Ion product of water substance, 0–1000°C, 1–10,000 bars new international formulation and its background," *J. Phys. Chem. Ref. Data* **1981**, *10*, 295–304. Copyright 1981 American Institute of Physics.

[6]Owing to limitations in labeling in spreadsheets, many figures use pH in place of $p[H_3O^+]$.

For room temperature, in the absence of any other ionic species, we can predict

- neutral when pH = 7.0;
- acidic when pH < 7.0;
- basic when pH > 7.0.

Note that these relationships are not definitions; they are the result of K_w' being 1.01×10^{-14} at 25°C.

5.3.2 Amphiprotic Species

The self-dissociation of water is an example of a solute acting as both an acid and a base. One water molecule provides a proton that another water molecule can bind with a lone pair of electrons. We call a solute that can act as both an acid and a base an *amphiprotic species*. Identifying amphiprotic species is no different than identifying any other acid or base, as discussed at the beginning of the chapter. The main difference is to recognize the multifunctional nature of one or more groups in the structure. In the case of water, we have seen that on adding an acid to water, a water molecule accepts a proton to form H_3O^+. On adding a base, a water molecule can provide a proton to leave OH^-.

A soluble salt that consist of electrolytes that can react with water is the most straightforward case of an amphiprotic species. On dissolving, such a salt will place a weak acid and a weak base in solution. To the first approximation, the "stronger" component of the salt will dominate. For the case where the cation is acidic and the anion is basic, the solution will be acidic if K_a' of the acid is greater than K_b' of the base. For the reverse case, the solution will be basic, $K_b' > K_a'$. This concept is best illustrated by Example 5.9. Before considering this case, we must have a means to determine the relative strength of acids and bases, that is, to compare K_b values with K_a values.

Example 5.9 Using K_w to Find K_b.   Derive the relationship between K_a and K_b.

Using the generic formulas HA and A^- to represent a weak acid and its conjugate base, respectively, we can write the following equilibria for these species with water:

$$HA(aq) + H_2O \rightleftharpoons A^-(aq) + H_3O^+(aq)$$
$$A^-(aq) + H_2O \rightleftharpoons HA(aq) + OH^-(aq)$$
$$\overline{HA(aq) + H_2O + A^-(aq) + H_2O \rightleftharpoons A^-(aq) + H_3O^+(aq) + HA(aq) + OH^-(aq)}$$

When we sum the equilibria, the equilibrium constant for the overall reaction is the product of the individual equilibria, which is, in this case, $K' = K_a' K_b'$.

Simplifying the overall reaction, we see that it is the water self-dissociation equilibrium:

$$H_2O + H_2O \rightleftharpoons H_3O^+(aq) + OH^-(aq)$$

which has the equilibrium constant K_w'.

Thus,

$$K_a' \times K_b' = K_w'$$

and we can use the K_a' of any weak acid to find the K_b' of its conjugate base:

$$K_b' = \frac{K_w'}{K_a'}$$

The same result is obtained by manipulating the K_a' and K_b' expressions for the equilibria given earlier:

$$K_a' = \frac{[A^-][H_3O^+]}{[HA]}$$

$$K_b' = \frac{[HA][OH^-]}{[A^-]}$$

Multiplying these expressions together,

$$K_a'K_b' = \frac{[A^-][H_3O^+]}{[HA]}\frac{[HA][OH^-]}{[A^-]}$$

and simplifying,

$$K_a'K_b' = [H_3O^+][OH^-] = K_w'$$

Now that we can interconvert the K_a' and K_b' values, let us consider the acidity or basicity of a solution when an amphiprotic salt is present. Keep in mind that the prediction in Example 5.10 is for the case of adding the salt to pure water. If strong acid or strong base is also present, the acidity or basicity will be controlled by a limiting reagent condition and any resulting equilibrium.

Example 5.10 An Amphiprotic Salt. Predict if the resulting solution will be acidic or basic when the soluble salt ammonium fluoride is added to water.

On dissolving this salt in water, we have

$$NH_4F(s) + H_2O \rightarrow NH_4^+(aq) + F^-(aq) + H_2O$$

NH_4^+ can donate a proton to the solution and F^- can accept a proton. The following two equilibria are possible:

$$NH_4^+(aq) + H_2O \rightleftharpoons NH_3(aq) + H_3O^+(aq)$$

$$F^-(aq) + H_2O \rightleftharpoons HF(aq) + OH^-(aq)$$

In the first equilibrium, NH_4^+ is a weak acid and NH_3 is its conjugate base. In the second equilibrium, F^- is a weak base and HF is its conjugate acid.

The K_a' of NH_4^+ is 6×10^{-10} and the K_b' of F^- is 1.4×10^{-11}. $K_a' > K_b'$, so we predict the solution to be acidic.

Another example of an amphiprotic species is a polyprotic acid that has lost some acidic protons. We will consider polyprotic acids in greater detail in Chapter 6, including an expression for finding pH of solutions of amphiprotic species. Example 5.11 shows that the process of comparing K_a' and K_b' values works for these amphiprotic species as it does for a mixture of acidic and basic ions.

Example 5.11 Amphiprotic Species.   Predict if the resulting solution will be acidic or basic when the soluble salt sodium hydrogen carbonate ($NaHCO_3$, also called *sodium bicarbonate*) is added to water.

The dissolution reaction is

$$NaHCO_3(s) + H_2O \rightarrow Na^+(aq) + HCO_3^-(aq) + H_2O$$

Na^+ is a strong electrolyte ion and does not affect the acid–base equilibrium. HCO_3^- can act as either an acid or a base, and such compounds are called amphiprotic. HCO_3^- can react with water to take part in the following two equilibria:

$$HCO_3^-(aq) + H_2O \rightleftharpoons CO_3^{2-}(aq) + H_3O^+(aq) \quad K_a' = 4.7 \times 10^{-11}$$

$$HCO_3^-(aq) + H_2O \rightleftharpoons H_2CO_3(aq) + OH^-(aq) \quad K_b' = ?$$

Given the relationship between K_a' and K_b',

$$K_a' \times K_b' = K_w'$$

we can use the K_a' of the conjugate acid of HCO_3^- to find K_b' for the second equilibrium:

$$H_2CO_3(aq) + H_2O \rightleftharpoons HCO_3^-(aq) + H_3O^+(aq) \quad K_a' = 4.45 \times 10^{-7}$$

$$K_a' \times K_b' = K_w'$$

$$K_b' = \frac{(1.01 \times 10^{-14})}{(4.45 \times 10^{-7})}$$

$$K_b' = 2.25 \times 10^{-8}$$

$K_b' > K_a'$, so the second equilibrium predominates. In practice we use a simpler expression discussed in Section 6.4.

You-Try-It 5.B
The amphiprotic-salts worksheet in you-try-it-05.xlsx contains a table of salt solutions. For each salt determine if the solution will be neutral, acidic, or basic. For the weak bases, calculate K_b values from the K_a of the conjugate acid. Use Examples 5.9–5.11 as guides.

5.4 ACID STRENGTH

The acidity or basicity (also called *alkalinity*) of a solution depends on both the concentration and on the intrinsic "strength" of the added weak acid or weak base. "Strength" is not an issue for strong acids and strong bases, which dissociate completely in water. $p[H_3O^+]$ and $p[OH^-]$ of a solution of strong acid or strong base depend only on the concentration of strong acid or strong base in the solution after any neutralization reaction has occurred. Predicting $p[H_3O^+]$ and $p[OH^-]$ for solutions of weak acids and weak bases is more involved than for those of the strong acids and strong bases because the weak electrolytes react only partially with water. However, keep in mind that solutions of weak acids and weak bases are not necessarily less acidic or less basic than solutions of strong acids and strong bases. For example, it is possible for a weak acid at a high concentration to produce a more acidic solution (a lower $p[H_3O^+]$) than a strong acid at a low concentration.

Our measure of the relative strength of different weak acids is the equilibrium constant, K_a. By acid "strength," we refer to the degree to which a weak acid dissociates in water to produce H_3O^+ and its conjugate base. The larger the K_a value, the more readily the acid gives up its proton to water. Table 5.6 provides K_a and pK_a ($-\log K_a$) values for selected inorganic and organic weak acids. All of the examples in Table 5.6 are monoprotic weak acids, that is, they have only one acidic proton. Tables 5.7 and 5.8 provide more complete lists of K_a and

TABLE 5.6 Relative Strength of Selected Weak Acids

Weak Acid	Formula	Conjugate Base	K_a	pK_a
Trichloroacetic acid	CCl_3COOH	CCl_3COO^-	3.0×10^{-1}	0.52
Dichloroacetic acid	$CHCl_2COOH$	$CHCl_2COO^-$	5.5×10^{-2}	1.26
Chloroacetic acid	$CH_2ClCOOH$	CH_2ClCOO^-	1.36×10^{-3}	2.867
Acetic acid	CH_3COOH	CH_3COO^-	1.75×10^{-5}	4.756
Hypochlorous acid	$HClO$	ClO^-	2.90×10^{-8}	7.537
Ammonium ion	NH_4^+	NH_3	5.68×10^{-10}	9.246

pK_a values for monoprotic and polyprotic weak acids.[7] A note on the chemical formulas: I use the older COOH and COO$^-$ notation to provide some indication of the structure. You will also see simpler formula such as $C_7H_6O_2$ for benzoic acid (C_6H_5COOH). I use the simpler formula for some of the anionic forms for compactness. The values in these tables are thermodynamic K_a values, and K_a' values are approximately equal to these values only when the concentration of ions in a solution is very low.

The significance of acid strength is most easily observed for solutions of different weak acids at equal concentrations. A weak acid with a larger K_a will produce a solution with a lower pH. Similarly, if a mixture of weak acids is titrated with a strong base, the strongest of the weak acids is neutralized completely before any of the other acids will be neutralized.

We will delay calculations to predict $p[H_3O^+]$ until Section 5.6, after we have learned how to correct the K_a constant to obtain K_a' values. What we will do is compare relative acid strength by calculating the percentage dissociation of different acids using K_a values (see Example 5.12). Any difference in these calculations compared to using K_a' values will be small for low acid concentrations.

Example 5.12 Percentage Acid Dissociation. Calculate the percent dissociation of acetic acid and dichloroacetic acid for solutions that are 0.010 M. You may assume that using K_a in place of K_a' creates a negligible error.

The percentage dissociation is the product of the fraction that dissociates and 100%:

$$\frac{[A^-]}{c_{HA}} \times 100\%$$

where HA and A$^-$ are generic labels for the weak acid and conjugate base, respectively. The equilibrium is

$$HA(aq) + H_2O \rightleftharpoons A^-(aq) + H_3O^+(aq)$$

To find the ratio ($[A^-]/c_{HA}$), we set up the K_a expression and simplify

$$K_a = \frac{[A^-][H_3O^+]}{c_{HA} - [H_3O^+]}$$

As the only source of A$^-$ and H_3O^+ is the dissociation of HA (neglecting the H_3O^+ from water dissociation), $[A^-] = [H_3O^+]$ and our K_a expression simplifies to

$$K_a = \frac{[A^-]^2}{c_{HA} - [A^-]}$$

---

[7]Values are for 25°C. Adapted with permission from *Lange's Handbook of Chemistry*, 16th ed., Speight, J. G., Ed.; McGraw-Hill: New York, 2005.

We can now rearrange this expression for solving with the quadratic equation:

$$[A^-]^2 + K_a[A^-] - K_a c_{HA} = 0$$

Recall that a quadratic equation will have two roots given by,

$$x = \frac{-b \pm \sqrt{b^2 - 4ac}}{2a}$$

Entering the values of K_a and c_{HA} for acetic acid gives

$$[A^-]^2 + 1.75 \times 10^{-5}[A^-] - 1.75 \times 10^{-5}(0.010\,M) = 0$$

so in our case, we have $a = 1$, $b = 1.75 \times 10^{-5}$, and $c = -1.75 \times 10^{-7}$. Solving with the quadratic equation gives the results -0.000427 and 0.000410. In our case, the negative root has no physical meaning, and we take the positive root as our answer:

$$[CH_3COO^-] = 0.00041\,M$$

Converting this result to the fraction dissociated and the percentage dissociation gives

$$\text{Fraction dissociated} = \frac{0.00041\,M}{0.010\,M} = 0.041$$

$$\%\,\text{Dissociation} = 0.041 \times 100\% = 4.1\%$$

Repeating the calculation for dichloroacetic acid gives

$$[CHCl_2COO^-] = 0.0084\,M$$

$$\text{Fraction dissociated} = \frac{0.0084\,M}{0.010\,M} = 0.84$$

$$\%\,\text{Dissociation} = 0.84 \times 100\% = 84\%$$

As you can see, the dichloroacetic acid is quite a strong "weak acid." From the K_a values in Table 5.6, we can predict the percentage dissociation of trichloroacetic acid to be even higher and the percentage dissociation of chloroacetic acid to be between that of acetic acid and dichloroacetic acid. For completeness we will calculate $p[H_3O^+]$ in Example 5.13, assuming that $K_a' = K_a$.

Example 5.13 Weak Acid Example. What is $p[H_3O^+]$ for the two solutions in Example 5.12?

Most of the work has been completed in the previous calculation. The important point to recognize, for a solution made by dissolving a weak acid in water with nothing else, is that $[H_3O^+] = [A^-]$. Repeating from above, the only source of A^- and H_3O^+ is the dissociation of HA (neglecting the H_3O^+ from water dissociation). To answer this question, the two weak acid solutions contain

$$[H_3O^+]=[CH_3COO^-]=0.00041\,M$$

and

$$[H_3O^+]=[CHCl_2COO^-]=0.0084\,M$$

Now taking the negative log of each result gives

$$p[H_3O^+]=-\log(0.00041\,M)=3.39$$

for 0.010 M acetic acid, and

$$p[H_3O^+]=-\log(0.0084\,M)=2.08$$

for 0.010 M dichloroacetic acid.

TABLE 5.7 K_a **Values of Inorganic Weak Acids**

Acid	Acid Formula	fw (g/mol)		K_a	pK_a
Ammonium ion	NH_4^+			5.68×10^{-10}	9.246
Hydrobromic acid	HBr	80.912		*Strong*	
Hypobromous acid	HBrO	96.911		2.8×10^{-9}	8.55
Hydrocyanic acid	HCN	27.025		6.2×10^{-10}	9.21
Hydrochloric acid	HCl	36.461		*Strong*	
Hypochlorous acid	HClO	52.460		2.90×10^{-8}	7.537
Chloric acid	$HClO_3$	84.459		*Strong*	
Perchloric acid	$HClO_4$	100.459		*Strong*	
Hydrofluoric acid	HF	20.006		6.3×10^{-4}	3.20
Hydroiodic acid	HI	127.912		*Strong*	
Iodic acid	HIO_3	175.911		1.57×10^{-1}	0.804
Periodic acid	HIO_4	191.910		2.3×10^{-2}	1.64
Nitrous acid	HNO_2	47.013		7.2×10^{-4}	3.14
Nitric acid	HNO_3	63.013		*Strong*	
Phosphoric acid	H_3PO_4	97.995	K_{a1}	7.11×10^{-3}	2.148
Dihydrogen phosphate ion	$H_2PO_4^-$		K_{a2}	6.34×10^{-8}	7.198
Hydrogen phosphate ion	HPO_4^{2-}		K_{a3}	4.8×10^{-13}	12.32
Dihydrogen sulfide	H_2S	34.081	K_{a1}	1.1×10^{-7}	6.97
Hydrogen sulfide ion	HS^-		K_{a2}	1.3×10^{-13}	12.90
Sulfurous acid	H_2SO_3	82.079	K_{a1}	1.3×10^{-2}	1.89
Hydrogen sulfite ion	HSO_3^-		K_{a2}	6.24×10^{-8}	7.205
Sulfuric acid	H_2SO_4	98.078	K_{a1}	*Strong*	–
Hydrogen sulfate ion	HSO_4^-		K_{a2}	1.0×10^{-2}	1.99

TABLE 5.8 K_a Values of Organic Weak Acids

Acid	Acid Formula	fw (g/mol)		K_a	pK_a
Acetic acid	CH_3COOH	60.052		1.75×10^{-5}	4.756
Bromoacetic acid	$CH_2BrCOOH$	137.940		1.25×10^{-3}	2.902
Chloroacetic acid	$CH_2ClCOOH$	93.489		1.36×10^{-3}	2.867
Dichloroacetic acid	$CHCl_2COOH$	128.942		5.5×10^{-2}	1.26
Trichloroacetic acid	CCl_3COOH	164.395		3.0×10^{-1}	0.52
Fluoroacetic acid	CH_2FCOOH	77.034		2.59×10^{-3}	2.586
Iodoacetic acid	CH_2ICOOH	184.941		6.68×10^{-4}	3.175
Acetylsalicylic acid	$C_8H_7O_2COOH$	180.157		3.3×10^{-4}	3.48
Ascorbic acid	$C_6H_8O_6$	176.124	K_{a1}	6.8×10^{-5}	4.17
Hydrogen ascorbate ion	$C_6H_7O_6^-$		K_{a2}	2.7×10^{-12}	11.57
Benzoic acid	C_6H_5COOH	122.121		6.25×10^{-5}	4.204
Butyric acid	C_3H_7COOH	88.105		1.52×10^{-5}	4.817
Carbonic acid	H_2CO_3	62.025	K_{a1}	4.45×10^{-7}	6.352
Hydrogen carbonate ion	HCO_3^-		K_{a2}	4.69×10^{-11}	10.329
Citric acid	$C_3H_4OH(COOH)_3$	192.124	K_{a1}	7.45×10^{-4}	3.128
Dihydrogen citrate ion	$C_6H_7O_7^-$		K_{a2}	1.73×10^{-5}	4.761
Hydrogen citrate ion	$C_6H_7O_7^-$		K_{a3}	4.02×10^{-7}	6.396
Ethylenediaminetetraacetic acid	H_4EDTA		K_{a1}	1.02×10^{-2}	1.99
	H_3EDTA^-		K_{a2}	2.14×10^{-3}	2.67
	H_2EDTA^{2-}		K_{a3}	6.92×10^{-7}	6.16
	$HEDTA^{3-}$		K_{a4}	5.50×10^{-11}	10.26
Formic acid	$HCOOH$	46.025		1.77×10^{-4}	3.751
Fumaric acid	$C_4H_2(COOH)_2$	140.094	K_{a1}	7.9×10^{-4}	3.10
Hydrogen fumarate ion	$C_4H_3O_4^-$		K_{a2}	2.5×10^{-5}	4.60
Hexanoic acid	$C_5H_{11}COOH$	116.158		1.42×10^{-5}	4.849
Lactic acid	$C_2H_4OHCOOH$	90.078		1.39×10^{-4}	3.858
Maleic acid	$C_2H_2(COOH)_2$	116.072	K_{a1}	1.23×10^{-2}	1.910
Hydrogen maleate ion	$C_4H_3O_4^-$		K_{a2}	4.7×10^{-7}	6.33
Malonic acid	$CH_2(COOH)_2$	116.072	K_{a1}	1.49×10^{-3}	2.826
Hydrogen malonate ion	$(COOH)COO^-$		K_{a2}	2.01×10^{-6}	5.696
Oxalic acid	$(COOH)_2$	90.035	K_{a1}	5.36×10^{-2}	1.271
Hydrogen oxalate ion	$C_2HO_4^-$		K_{a2}	5.35×10^{-5}	4.272
Pentanoic acid	C_4H_9COOH	102.132		1.44×10^{-5}	4.842
Phenol	C_6H_5OH	94.111		1.0×10^{-10}	9.99
o-Phthalic acid	$C_6H_4(COOH)_2$	166.131	K_{a1}	1.12×10^{-3}	2.950
Hydrogen phthalate ion	$C_8H_5O_4^-$		K_{a2}	3.91×10^{-6}	5.408
Pyruvic acid	C_2H_3OCOOH	88.062		3.24×10^{-3}	2.49
Succinic acid	$C_2H_4(COOH)_2$	118.088	K_{a1}	6.21×10^{-5}	4.207
Hydrogen succinate ion	$C_4H_5O_4^-$		K_{a2}	2.32×10^{-6}	5.635
Tartaric acid	$C_2H_2(OH)_2(COOH)_2$	134.087	K_{a1}	9.20×10^{-4}	3.036
Hydrogen tartrate ion	$C_4H_5O_5^-$		K_{a2}	4.31×10^{-5}	4.366

Note that 0.010 M of a strong acid would produce a solution pH of 2.0, so again we see that even though we classify some acids as weak acids, their behavior can be very close to that of a strong acid.

These types of calculations should be familiar from general chemistry. We will delay doing more examples until after we discuss ionic strength and see how to correct K_a' values from K_a values using activity coefficients. You will see that our purpose in analytical chemistry is to calculate equilibrium concentrations and to determine the factors that affect these first approximation calculations.

You-Try-It 5.C
The acidity worksheet in you-try-it-05.xlsx contains a table with pairs of acid solutions. For each pair determine which solution will be the most acidic, that is, which will have the lowest pH. First try to deduce which solution will be the most acidic based on concentration and K_a values. For the cases that are not easily deduced, calculate p[H_3O^+] of the acid solutions. Finally, calculate the percent dissociation for each weak acid. Use Examples 5.12 and 5.13 as guides.

5.5 THE CONCEPT OF ACTIVITY

We have mentioned the constants K and K', and this section explains the concepts that lead to these two different versions of an equilibrium constant. Although we are introducing the concept of activity with weak acid and weak base equilibria, the concept applies to all types of aqueous equilibria that involve ions. The thermodynamic equilibrium constant, K, that we can look up in tables is for equilibria that occur in ideal, that is, dilute, solutions. Solutions that have a total ion concentration greater than approximately 0.001 M cannot really be treated as ideal solutions. When ion concentrations become high enough that ionic interactions affect solution equilibria, the thermodynamic equilibrium constant, K, no longer describes the equilibrium concentrations well. In these cases, we correct K values to find the formal equilibrium constant, K'. Equilibria are affected not only when the concentrations of the ions involved in an equilibrium become high but also when the concentrations of strong electrolytes not involved in the equilibrium become high. Spectator ions are not involved directly in equilibria, but they affect the environment of the ions that are part of an equilibrium.

Consider two containers that are connected with a tube that is closed initially (Figure 5.3). The container on the left holds 0.001 M KBr and that on the right holds distilled water. Opening the valve in the tube between the containers creates a system that is not at equilibrium. Intuitively, we know that there is a driving force that causes the K^+ and Br^- to diffuse throughout the whole solution. This diffusion is not necessarily fast, especially in the absence of stirring or shaking, but we know that it will occur. We can also describe this system from a statistical viewpoint. There is essentially zero chance for all of the K^+ and Br^- to remain in only one-half of the solution, with the other half containing pure water.

We can quantify the driving force, which we call activity, from the concentration difference, which is 0.001 M. Once the concentration of K^+ and Br^- are equal throughout the system, that is, 0.0005 M, the system is at equilibrium and there is no driving force for further change. Ions are still moving throughout the solution, but the ion concentrations will remain uniform.

Now consider the case when 1 M NaCl is also present in each container (Figure 5.4). When the connecting tube is opened in this case, do you think the driving force is the same as in the earlier case? Your intuition tells you that it is not, because the K^+ and Br^- from the KBr are "shielded" by the large amount of Na^+ and Cl^-. The activity of the KBr in this solution is not the same as the concentration, 0.001 M; it is somewhat less than 0.001 M. The nonequilibrium situation will go to equilibrium, but the conditions are different. In dealing with ionic equilibria, we must consider the ion-dependent activity rather than only concentration.

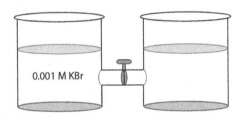

Figure 5.3 Diffusion driving force.

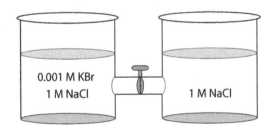

Figure 5.4 Diffusion driving force with spectator ions.

5.5.1 An Activity Analogy

To use an analogy, let us say that you and a date plan to meet five other couples at an ice rink. You are expecting a nice evening, with the concentration of couples on the rink being six couples per rink (cpr). As you and your date skate onto the ice, you see the other couples, but they are not always skating together as couples. (The topic of why couples dissociate and recombine is best left for a physical chemistry class.) Thus, you refine your thinking a bit and say to yourself that the "formal concentration" is 6 cpr, but the equilibrium concentration of couples per rink is somewhat less than six. (If you are on your first date, you probably do not want to share these thoughts with your date on the chance that you might want to have a second date with him or her.) Using AB for a couple and A and B for single skaters, we can write a dissociation equilibrium reaction and an equilibrium constant expression as

$$AB \rightleftharpoons A + B$$
$$K_c = \frac{[A][B]}{[AB]}$$

As you skate around, you notice that everyone skates in pairs most of the time, but on average, one couple is "dissociated" at any given time. You could describe the equilibrium concentrations as five undissociated couples and one dissociated couple per rink and even determine an equilibrium constant for couple dissociation (temperature dependent, as for any K):

$$K_c = \frac{(1)(1)}{5} = 0.2$$

where the subscript "c" indicates "couple dissociation."

Now, what happens if a pair of school buses pull up to the ice rink and 100 little kids skate onto the ice? The little kids are rather hyperactive and will not skate as couples themselves or with the adult couples, but they do tend to cluster around any individual adult skater. The rate at which couples dissociate is the same as before (for the usual reasons), but as soon as they dissociate, the individual skaters are immediately surrounded by little kids. The little kids shield individual skaters from other individual skaters, making it more difficult for two skaters to recombine to form a couple. Now you might look around and see that there are four undissociated and two dissociated couples per rink and that the individual skaters tend to be surrounded by little kids. Now, our equilibrium constant is

$$K_c' = \frac{(2)(2)}{4} = 1.0$$

and we put a prime on the equilibrium constant to indicate that there are new, nonideal conditions.

The presence of the little kids has affected couple equilibrium drastically. You can imagine that if there were 500 little kids on the skating rink, the shielding between isolated adult skaters would be even greater. Similarly, if there were only 10 little kids scattered around the rink, they probably would not affect the equilibrium concentrations of the undissociated and dissociated couples very much. When there were no or a few little kids on the skating rink, we could call the condition as dilute or ideal, and in such cases, the formal equilibrium constant is nearly equal to the thermodynamic equilibrium constant, $K_c' \cong K_c$.

5.5.2   Ionic Strength

In the aforementioned analogy, the undissociated and dissociated couples represented solutes in an equilibrium and the little kids represented spectator ions. Note that for the little kids to affect the equilibrium, there had to be an attraction between the little kids and the individual adult skaters. Ions in solution have electrostatic interactions with other ions. Ions of opposite charge experience an attractive force and ions of like charge experience a repulsive force. Neutral solutes do not have such interactions. When the concentrations of ions in a solution is greater than approximately 0.001 M, a shielding effect occurs around the ions, similar to the little kids around individual adults in our analogy. This shielding effect becomes quite significant for concentrations of singly charged ions of 0.01 M and greater. Cations tend to be surrounded by nearby anions and anions are surrounded by nearby cations.

Doubly or triply charged ions "charge up" a solution even more than singly charged ions, so we need a standard way to talk about charge concentration in solution. We describe the charge concentration of a solution by the *ionic strength*, I_c, which is calculated from the following expression:

$$I_c = \frac{1}{2}\sum_i z_i^2 c_i \tag{5.2}$$

where the subscript "c" indicates that we are working with molar concentration, z_i is the charge on ion i, and c_i is the formal concentration of ion i. Note that the charge is squared so that positive and negative charges do not cancel. You can think of ionic strength as the overall "ionic character" of a solution. The factor of 1/2 is present because every positive charge is balanced by a negative charge and there is no point in counting the overall charge twice. Example 5.14 illustrates the calculation.

Example 5.14   Ionic Strength.   What is the ionic strength of a 0.050 M Na_2SO_4 solution?

Na_2SO_4 is a strong electrolyte, so this solution contains 0.100 M Na^+ and 0.050 M SO_4^{2-}. The ionic strength is

$$I_c = \frac{1}{2}\left\{(+1)^2(0.100\,M) + (-2)^2(0.050\,M)\right\}$$

$$I_c = \frac{1}{2}\left\{(0.100\,M) + (0.200\,M)\right\}$$

$$I_c = 0.150\,M$$

Recall that we refer to the solution by the formal concentration, 0.050 M Na_2SO_4, which tells us how to make the solution. This concentration does not provide a quantitative description of the ionic nature of the solution directly. Thus, we calculate ionic strength when we need a quantitative measure of the ionic character of the solution.

Note that we neglected two other ions in the solution in the calculation in Example 5.14. They are the H_3O^+ and OH^- ions, which are present in all aqueous solutions. Assuming that no other species were present, the pH of this salt solution is nearly 7 and the $[H_3O^+]$ and $[OH^-]$ are $\approx 1 \times 10^{-7}$ M. Comparing these concentrations with the concentrations in the example, we see that they contribute an insignificant concentration of ions compared to $[Na^+]$ and $[SO_4^{2-}]$.

5.5.3 Activity Coefficients

Given that equilibria depend on the overall ionic character of a solution, how do we predict equilibrium concentrations? One approach is to tabulate equilibrium constants at different ionic strengths, and you can find tables in some reference sources. Doing so for all aqueous equilibria would be a lot of work and it would create huge tables. A more general approach is to use the thermodynamic K values, which are tabulated for the ideal case of zero ionic strength and 25°C (see Tables 5.7 and 5.8 for the K_a of weak acids). These values can be corrected for the ionic strength of a specific solution. We call a corrected equilibrium constant the *formal equilibrium constant* and add a prime to the symbol, K'. These K' values better predict equilibria at higher ionic strength; that is, when the overall concentration of ions is high.

To correct K values and obtain K', we define the "effective concentration" of ions in solution as the *activity*. We then relate the activity to the equilibrium concentration of ion i by an activity coefficient, γ_i:

$$a_i = \gamma_i[i] \qquad (5.3)$$

Using these activity coefficients as correction factors, we can look up thermodynamic equilibrium constants measured for ideal solutions and calculate the formal equilibrium constant, K'. The K' is then used to predict equilibrium concentrations in our specific, nonideal solution.

Using the solubility of an AgCl precipitate as an illustrative example, the expression for the thermodynamic equilibrium constant, K_{sp}, is

$$K_{sp} = a_{Ag}a_{Cl}$$

where a_{Ag} and a_{Cl} are the solution activities of Ag^+ and Cl^-, respectively. Using Equation (5.3) to substitute for activity:

$$K_{sp} = \gamma_{Ag}\left[Ag^+\right]\gamma_{Cl}\left[Cl^-\right]$$

On rearranging this expression, we have a formal equilibrium constant that better replicates equilibrium behavior as a function of ionic strength:

$$K'_{sp} = \left[Ag^+\right]\left[Cl^-\right] = \frac{K_{sp}}{\gamma_{Ag}\gamma_{Cl}}$$

Since γ_i values are less than one, a high ionic strength due to spectator ions will always increase the solubility of a precipitate. The impact of activity coefficients will vary for other types of equilibrium constant expressions depending on the extent to which γ_i values cancel in the numerator and denominator. The general process to obtain formal equilibrium constants is the same.

The activity coefficients are unitless numbers that can be calculated from several expressions, depending on ionic strength. For low to moderate ionic strength, the Debye–Hückel equation provides good predictions compared to experimental measurements:

$$\log \gamma_i = \frac{-Az_i^2\sqrt{I_c}}{1+Bd_i\sqrt{I_c}} \tag{5.4}$$

where

- I_c is the ionic strength of the solution;
- z_i is the charge of the ion i;
- d_i is the ion size parameter (Tables 5.9 and 5.10);[8]
- A and B are constants that depend on the solvent and temperature.

The ion size parameter is also called the *effective diameter of the hydrated ion* or the *distance of closest approach*.[9] Interaction between ions depends on their

---

[8]Adapted with permission from Kielland, J. "Individual activity coefficients of ions in aqueous solutions," *J. Am. Chem. Soc.* **1937**, *59*, 1675–1678. Copyright 1937 American Chemical Society.

[9]de Levie, R. *Aqueous Acid-Base Equilibria and Titrations*, Oxford University Press: Oxford, 1999.

TABLE 5.9   Ion Size Parameter for Cations in Aqueous Solution at 25°C

Ion	d_i, nm
Ag^+, Cs^+, Rb^+, Tl^+, NH_4^+	0.25
K^+	0.3
Na^+, $CdCl^+$	0.4–0.45
Hg_2^{2+}	0.4
Pb^{2+}	0.45
Sr^{2+}, Ba^{2+}, Ra^{2+}, Cd^{2+}, Hg^{2+}	0.5
Li^+, Ca^{2+}, Cu^{2+}, Co^{2+}, Fe^{2+}, Mn^{2+}, Ni^{2+} Sn^{2+}, Zn^{2+}	0.6
Be^{2+}, Mg^{2+}	0.8
H^+, Al^{3+}, Cr^{3+}, Fe^{3+}, La^{3+}, Y^{3+}	0.9
Ce^{4+}, Sn^{4+}, Th^{4+}, Zr^{4+}	1.1

TABLE 5.10   Ion Size Parameter for Anions in Aqueous Solution at 25°C

Ion	d_i, nm
Br^-, Cl^-, I^-, CN^-, NO_2^-, NO_3^-	0.3
F^-, OH^-, HS^-, SCN^-, BrO_3^-, ClO_3^-, ClO_4^-, IO_4^-, MnO_4^-	0.35
HCO_3^-, IO_3^-, HSO_3^-, $H_2PO_4^-$	0.4–0.45
CrO_4^{2-}, SO_4^{2-}, PO_4^{3-}, HPO_4^{2-}	0.4
CO_3^{2-}, SO_3^{2-}, $C_2O_4^{2-}$, CH_3COO^-, chloroacetate, hydrogen citrate	0.45
S^{2-}, citrate, malonate, succinate, tartrate, trichloroacetate	0.5
Benzoate, chlorobenzoate, hydroxybenzoate, phenylacetate, phthalate	0.6
Diphenylacetate	0.8

charges and the distance between them. Ions that are hydrated strongly, that is, having a sphere of water molecules clustered around them tightly, cannot approach other ions as closely. We'll see in Chapter 12 that this effect allows ions of the same charge to be separated from each other using an ion-exchange stationary phase.

For aqueous solutions, at 25°C, $A = 0.509$ M$^{-1/2}$ and $B = 3.28$ M$^{-1/2}$ nm^{-1}.[10] The Debye–Hückel equation can then be written as

$$\log \gamma_i = \frac{-0.509 z_i^2 \sqrt{I_c}}{1+(3.28 d_i \sqrt{I_c})} \tag{5.5}$$

As the Debye–Hückel equation contains the charge and effective hydrated diameter of a specific ion, each ion in a solution can have a different calculated

[10]Values of A and B at other temperatures are in Grenthe, I.; Wanner, H.; Östhols, E. "TDB-2: Guidelines for the extrapolation to zero ionic strength," (OECD Nuclear Energy Agency, 2000); Available at https://www.oecd-nea.org/jcms/pl_37492/tdb-2-guidelines-for-the-extrapolation-to-zero-ionic-strength. Accessed August 2022.

activity coefficient. Measuring the ionic nature of a solution gives only an average activity coefficient for the ions that are present, so the tabulated values are extracted from a large number of experiments. The ionic strength is a description of the overall solution, and the same value of I_c is used for calculating the activity coefficients of each ion in solution. Example 5.15 shows a sample calculation for activity coefficients.

For ionic strength of approximately 0.1–0.5 M, the Davies equation adds an empirical correction to a simplified form of the Debye–Hückel equation:

$$\log \gamma_i = -0.509 z_i^2 \left(\frac{\sqrt{I_c}}{1 + \sqrt{I_c}} - 0.3 I_c \right) \tag{5.6}$$

This equation produces equivalent activity coefficients for all ions of the same charge. More sophisticated approaches use specific ion interaction models to determine activity coefficients. For examples see:

- Pitzer, K. S., "Electrolyte theory—improvements since Debye and Hueckel," *Acc. Chem. Res.* **1977**, *10*, 371–377.
- IUPAC Project 2000-003-1-500: Ionic Strength Corrections for Stability Constants.[11]

The Debye–Hückel equation does a good job of describing experimental data for $I_c \approx 0.1$ M or less. At higher ionic strength, the Davies equation or other theories provide better descriptions of real solutions. Why might these other theories be relevant? Think of the ionic strength of seawater ($I_c \approx 0.7$ M) or biological fluids and the importance of equilibria in environmental and biological processes. We will use the Debye–Hückel equation for most of our calculations, but keep in mind that we are making theoretical predictions, and our predictions will have varying degrees of accuracy depending on the real behavior in the solutions we consider.

Example 5.15 Activity Coefficients. What are the activity coefficients for H_3O^+, benzoate ($C_6H_5COO^-$), and SO_4^{2-} for a 0.050 M Na_2SO_4 solution that has a small amount of benzoic acid added? The solution pH is measured to be pH = 4.

The measured pH tells us that $[H_3O^+]$ and $[C_6H_5COO^-]$ = 1×10^{-4} M. These concentrations are much smaller than the 0.050 M Na_2SO_4 and we can use the ionic strength calculated in Example 5.14, $I_c = 0.150$ M. Inserting the ion size parameters for H_3O^+, benzoate, and SO_4^{2-} from Tables 5.9 and 5.10 in the Debye–Hückel equation (A and B units dropped for clarity) gives

---

[11]https://iupac.org/project/2000-003-1-500/. Accessed August 2022.

$$\log \gamma_{H_3O^+} = \frac{-0.509(+1)^2\sqrt{0.150\,M}}{1+3.28(0.9\,nm)\sqrt{0.150\,M}}$$

$$\log \gamma_{H_3O^+} = -0.092$$

$$\gamma_{H_3O^+} = 0.81$$

$$\log \gamma_{C_6H_5COO^-} = \frac{-0.509(-1)^2\sqrt{0.150\,M}}{1+3.28(0.6\,nm)\sqrt{0.150\,M}}$$

$$\log \gamma_{C_6H_5COO^-} = -0.112$$

$$\gamma_{C_6H_5COO^-} = 0.77$$

$$\log \gamma_{SO_4^{2-}} = \frac{-0.509(-2)^2\sqrt{0.150\,M}}{1+3.28(0.4\,nm)\sqrt{0.150\,M}}$$

$$\log \gamma_{SO_4^{2-}} = -0.522$$

$$\gamma_{SO_4^{2-}} = 0.30$$

You can see that at this relatively high ionic strength, the values are quite different from the ideal case of $\gamma = 1$.

Now let us use the activity coefficients from Example 5.15 with Equation (5.3) to find the activity of each ion in Example 5.16.

Example 5.16 Activity Values. What are the activities of H_3O^+, benzoate ($C_6H_5COO^-$), and SO_4^{2-} in a 0.050 M Na_2SO_4 solution containing benzoic acid. The concentrations of H_3O^+ and $C_6H_5COO^-$ are 0.0001 M based on pH measurement.

For the H_3O^+ concentration of 1.0×10^{-4} M, the activity is

$$a_{H_3O^+} = (\gamma_{H_3O^+})[H_3O^+]$$
$$a_{H_3O^+} = (0.81)(1.0\times10^{-4}\,M) = 0.81\times10^{-4}\,M$$

The calculation is the same for the 1.0×10^{-4} M $C_6H_5COO^-$, using the activity coefficient for benzoate:

$$a_{C_6H_5COO^-} = (0.77)(1.0\times10^{-4}\,M) = 0.77\times10^{-4}\,M$$

The SO_4^{2-} concentration is $[SO_4] = 0.050$ M, and the activity is

$$a_{SO_4^{2-}} = (\gamma_{SO_4^{2-}})[SO_4]$$
$$a_{SO_4^{2-}} = (0.30)(0.050\,M) = 0.015\,M$$

These calculations show the degree to which ionic strength will impact the behavior of ions in solution. Sulfate is a spectator ion in the benzoic acid equilibrium, but note how ionic strength impacts the doubly charged SO_4^{2-} much more than the singly charged ions.

You can see from Example 5.16 that a, activity, what I have called an *effective concentration*, can be quite different from the nominal concentration. The difference can be especially pronounced for ions with multiple charges. The actual impact of ionic strength on equilibrium concentrations will depend on the nature of the equilibrium expression. As an example, Table 5.11 lists the activity coefficients calculated with the Debye–Hückel equation for hydronium and benzoate ions as a function of ionic strength (at 25°C). The K_a' column of Table 5.11 shows the effect of ionic strength on K_a' (the thermodynamic K_a is listed in the first row). The data starts at 0.00022 M ionic strength because that is the intrinsic ionic strength for 0.001 M benzoic acid, $[H_3O^+] = [C_6H_5COO^-] = 0.00022$ M (we will see how to do this calculation in Section 5.6). The last column shows the effect of ionic strength on $p[H_3O^+]$ for a 0.001 M benzoic acid solution.

The 1.0 M listing in Table 5.11 is calculated with the Davies equation:

$$\log \gamma_i = -0.509 \left(\frac{\sqrt{I_c}}{1 + \sqrt{I_c}} - 0.3 I_c \right) \tag{5.7}$$

You can see that at such a high ionic strength, there is a deviation between the Debye–Hückel equation and the more appropriate Davies equation. This difference highlights that although we can calculate activity coefficients and equilibrium concentrations to two or three significant figures, the actual predictive reliability is only as good as our constants. Choosing the Davies correction versus the Debye–Hückel equation results in a 25% difference in the value of K_a'. This level of uncertainty, or even higher, is not uncommon when working with real solutions of importance in biological and environmental systems. The bottom line is that we should not take the precision of our calculations too seriously, but to use our predictions to help guide our understanding of chemical systems.

As the ionic strength increases, the K_a' value increases, leading to a greater degree of dissociation for the acid and a lower $p[H_3O^+]$ of the solution. The change is relatively small, K_a' differs from K_a by approximately a factor of 2, resulting in a decrease of approximately 0.1 of a pH unit between the ideal and high ionic strength cases. The significance of such a change is a different matter and must be evaluated in the context of a specific chemical system. Will a pH change of 0.1 affect the taste of drinking water? Not likely. In some chemical systems, a pH change of 0.1 might be enough to disturb other equilibria or change reaction rates.

TABLE 5.11 Activity Coefficients Versus Ionic Strength

I_c	$\gamma_{H_3O^+}$	$\gamma_{benzoate^-}$	Product of γ	K_a'	$p[H_3O^+]$
0	1.0	1.0	1.0	6.25×10^{-5}	3.66
0.00022	0.983	0.983	0.978	6.47×10^{-5}	3.65
0.001	0.967	0.966	0.934	6.69×10^{-5}	3.64
0.01	0.914	0.907	0.829	7.54×10^{-5}	3.62
0.1	0.826	0.796	0.658	9.51×10^{-5}	3.58
1.0	0.744	0.674	0.502	1.25×10^{-4}	3.53
1.0 (Davies)	0.791	0.791	0.626	9.99×10^{-5}	3.57

As an example of the significance of ionic strength effects, the South-eastern United States experienced very dry weather during the summer of 2007. The lack of precipitation resulted in a high level of fertilizer salts remaining in the soil and lower than typical measurements of soil pH.[12] Farmers use measurement of soil pH to determine the amount of lime to spread on fields to maintain optimal pH for different crops. Another agricultural report recommends making soil pH measurements in 0.01 M $CaCl_2$ rather than distilled water.[13] Soil pH values are approximately 0.6 pH units lower than those measured in water; however, the measurement will be less susceptible to variable salt concentrations year to year.

Example 5.17 demonstrates the general process for finding K' values from the thermodynamic K constant.

Example 5.17 Activity Values. Use the activity coefficients in Table 5.11 to calculate K_a' for benzoic acid in a solution with an ionic strength of 0.10 M.

The thermodynamic equilibrium constant expression is

$$K_a = \frac{(a_{C_6H_5COO^-})(a_{H_3O^+})}{a_{C_6H_5COOH}}$$

Substituting the activity coefficient and concentration for each activity gives

$$K_a = \frac{(\gamma_{C_6H_5COO^-})[C_6H_5COO^-](\gamma_{H_3O^+})[H_3O^+]}{(\gamma_{C_6H_5COOH})[C_6H_5COOH]}$$

[12]Hardy, D. H. "Effect of fertilizer salts on soil pH," North Carolina Department of Agriculture and Consumer Services Agronomic Division, January 2008; Available at http://www.ncagr.gov/agronomi/pdffiles/fsalts.pdf. Accessed August 2022.
[13]Kissel, D. E.; Vendrell, P. F. "Soil testing: soil pH and salt concentration," University of Georgia Cooperative Extension, Circular 875; Available at http://www.caes.uga.edu/publications. Accessed August 2022.

Rearranging to collect similar terms gives

$$K_a = \left(\frac{(\gamma_{C_6H_5COO^-})(\gamma_{H_3O^+})}{(\gamma_{C_6H_5COOH})} \right) \left(\frac{[C_6H_5COO^-][H_3O^+]}{[C_6H_5COOH]} \right)$$

The concentration terms in this equation are equal to the K_a' expression:

$$K_a = \left(\frac{(\gamma_{C_6H_5COO^-})(\gamma_{H_3O^+})}{\gamma_{C_6H_5COOH}} \right) K_a'$$

Finally rearranging for the quantity that we wish to find,

$$K_a' = \left(\frac{\gamma_{C_6H_5COOH}}{(\gamma_{C_6H_5COO^-})(\gamma_{H_3O^+})} \right) K_a$$

All of the terms on the right side of this expression are quantities that we can calculate or look up. Inserting the values for K_a and the activity coefficients (remember that the activity of a neutral species is 1.0),

$$K_a' = \frac{1.0}{(0.826)(0.796)} 6.25 \times 10^{-5}$$

$$K_a' = 9.51 \times 10^{-5}$$

For any equilibrium reaction that consists of a neutral species going to two ions, K' will always be K divided by the activity coefficients. In case of equilibria that have different numbers of ions in the denominator or numerator of the equilibrium constant expression, there will be a different relationship between K' and K. Although Example 5.17 shows a relatively small effect, the effect on equilibrium constant calculations will depend on the form of the equilibrium constant expression. In other types of problems, such as many acid–base buffer solutions (Chapter 6), the deviation between ideal and real behavior will not be great because the activity coefficients will partially cancel in the calculation of equilibrium concentrations. In contrast, solubility problems that have ions with multiple charges can have much greater divergence between ideal and real behavior than acid–base equilibria. The general approach of setting up the thermodynamic K expression and substituting activity coefficients and concentrations for the activities will always produce the correct relationship and is more reliable than trying to memorize the outcome for every individual case.

You-Try-It 5.D
Write formulas in the ionic-strength worksheet in you-try-it-05.xlsx
to find ionic strength and activity coefficients. Use Example 5.17 as a
guide.

To summarize the concept of activity, in dilute solutions ions that are surrounded only by water molecules and electrolytes in solution behave "ideally," that is, thermodynamic equilibrium constants provide good predictions of equilibrium concentrations. When ionic strength is larger than approximately 0.001 M, the shielding of ions affects equilibria significantly. To use an ideal thermodynamic equilibrium constant, K, we multiply concentrations by correction factors called activity coefficients to generate a formal equilibrium constant, K'.

5.6   ACID–BASE EQUILIBRIUM CALCULATIONS

We calculated equilibrium concentrations for weak acids in terms of percentage dissociation in Section 5.4. Before continuing with similar calculations for weak acids and weak bases, let us think broadly about all of the steps that we should consider when presented with a problem. Figure 5.5 shows some of the factors that are possible in neutralization or acid–base equilibrium reactions. Many errors in equilibrium calculations are made in setting up the problem correctly rather than in the actual calculation of equilibrium concentrations.

Determining the species that are present and their formal concentrations correctly must precede setting up an equilibrium calculation. Once the formal concentrations of all ionic species are known, we can predict the ionic strength of a solution and correct K_a values with activity coefficients to find the K_a' constant. For most monoprotic weak acid calculations, ionic strength effects are not significant for $I_c \leq 0.001$ M (Table 5.11). For solutions of neutral weak acids or neutral weak bases that have no additional spectator ions, a quick approximate calculation can determine if the acid or base reaction will generate ion concentrations that affect equilibrium concentrations. For example, assuming that $[HA] = c_{HA}$ and rearranging the K_a expression leads to

$$[H_3O^+] = [A^-] \approx \sqrt{K_a c_{HA}}$$

From this expression, we can see that adding a weak acid of moderate acidity, such as acetic acid, $K_a = 1.75 \times 10^{-5}$, to water generates an ionic strength of 0.001 M only for $c_{HA} \geq 0.06$ M. Weak acids or weak bases that are charged will have counterions present, and correcting for ionic strength effects will often be necessary for acid or base formal concentrations greater than 0.001 M.

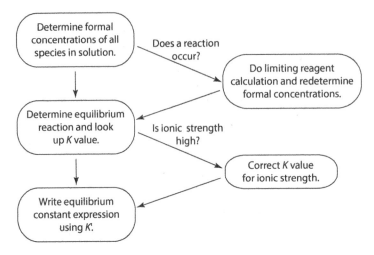

Figure 5.5   Steps before setting up an equilibrium calculation.

Once the steps in Figure 5.5 are completed, predicting equilibrium concentrations is done by setting up the equilibrium constant expression and solving for the unknowns, as for any other algebraic expression. Rather than presenting general rules for solving equations, these types of calculations are illustrated here by a series of examples. Example 5.18 is benzoic acid neglecting ionic strength effects, and Example 5.19 repeats this calculation after correcting K_a to find K_a'. Example 5.20 illustrates a weak base calculation.

Example 5.18 Weak Acid Example.   What is $[H_3O^+]$ when 0.200 mmol of benzoic acid is diluted in 0.100 l of water?
    The equilibrium is

$$C_6H_5COOH(aq) + H_2O \rightleftharpoons C_6H_5COO^-(aq) + H_3O^+(aq)$$

The formal concentration of benzoic acid is $(2.00 \times 10^{-4}$ mol$)/(0.100$ l$) = 0.00200$ M. As benzoic acid is a weak acid, it will exist mostly as C_6H_5COOH in water and the concentration of the ions will be very small. The ionic strength will be low and the activity coefficients will be very close to 1 for the ionic solutes. We will assume that $K_a' = K_a = 6.25 \times 10^{-5}$. You can check this assumption with the approximate expression: $[H_3O^+] = [C_6H_5COO^-] \approx \sqrt{K_a c_{HA}}$. The equilibrium constant expression is

$$K_a' = K_a = \frac{[C_6H_5COO^-][H_3O^+]}{[C_6H_5COOH]}$$

The only source of H_3O^+ and $C_6H_5COO^-$ in this solution is the dissociation of benzoic acid, and therefore, $[H_3O^+] = [C_6H_5COO^-]$. (We do neglect H_3O^+ from water dissociation.) The equilibrium concentration of the benzoic acid will be slightly less than the formal concentration due to this dissociation, so

$$[C_6H_5COOH] = c_{HA} - [H_3O^+]$$

Note that we use HA as the subscript in c_{HA} only as a shorthand notation, it is not something new. Now inserting the K_a value and making the equality substitution,

$$6.25 \times 10^{-5} = \frac{[H_3O^+]^2}{c_{HA} - [H_3O^+]}$$

Inserting 0.00200 M for the benzoic acid formal concentration and rearranging yields

$$[H_3O^+]^2 + 6.25 \times 10^{-5}[H_3O^+] - 1.25 \times 10^{-7} = 0$$

Solving gives the following result:

$$[H_3O^+] = 3.54 \times 10^{-4} \, M$$

and

$$p[H_3O^+] = 3.49$$

Checking our assumption that $K'_a = K_a$ appears valid because the ionic strength is less than 0.001 M.

Before repeating the calculation for benzoic acid in a salt solution, do you predict the dissociation of benzoic acid to be higher or lower in a high ionic strength solution compared to the extent of dissociation in only water?

Example 5.19 Weak Acid Example at High Ionic Strength. What is $[H_3O^+]$ when 0.200 mmol of benzoic acid is diluted in 0.100 l of a solution of 0.10 M $CaCl_2$?

The equilibrium is the same as in Example 5.18. As the ionic strength is high, we must use activities rather than concentrations. Again, note that the A^- and HA subscripts are simply shorthand notations for $C_6H_5COO^-$ and C_6H_5COOH, respectively.

$$K_a = 6.28 \times 10^{-5} = \frac{(a_{A^-})(a_{H_3O^+})}{a_{HA}}$$

$$6.25 \times 10^{-5} = \frac{(\gamma_{A^-})[C_6H_5COO^-](\gamma_{H_3O^+})[H_3O^+]}{\gamma_{HA}[C_6H_5COOH]}$$

Looking at this expression, you can recognize the same procedure as before, but we now have activity coefficients in place. These coefficients correct the thermodynamic K_a that we look up to get the K_a' constant. We can rewrite the expression to highlight the similarity with the previous calculation:

$$K_a' = \frac{[C_6H_5COO^-][H_3O^+]}{[C_6H_5COOH]} = \frac{(6.25 \times 10^{-5})\gamma_{HA}}{(\gamma_{A^-})(\gamma_{H_3O^+})}$$

Now to find the activity coefficients, we need the ionic strength, which we calculate using the charge and concentration of Ca^{2+} and Cl^- ($[Ca^{2+}] = 0.1$ M and $[Cl^-] = 0.2$ M). As was the case for the water solution, the ion concentration from the acid dissociation is small and can be neglected.

$$I_c = \frac{1}{2}\sum_i z_i^2 c_i$$

$$I_c = \frac{1}{2}\{(+2)^2(0.100\,M) + (-1)^2(0.200\,M)\} = 0.30\,M$$

With the ionic strength and the d_i values from Tables 5.9 and 5.10, we can calculate the activity coefficients for H_3O^+ and $C_6H_5COO^-$:

$$\log\gamma_{H_3O^+} = \frac{-0.509(+1)^2\sqrt{0.30}}{1+3.28(0.9)\sqrt{0.30}}$$

$$\log\gamma_{H_3O^+} = -0.106$$

$$\gamma_{H_3O^+} = 0.783$$

$$\log\gamma_{A^-} = \frac{-0.509(-1)^2\sqrt{0.30}}{1+3.28(0.6)\sqrt{0.30}}$$

$$\log\gamma_{A^-} = -0.134$$

$$\gamma_{A^-} = 0.735$$

Now substituting the activity coefficients into the K_a' expression and simplifying as before,

$$K_a' = \frac{6.25\times10^{-5}(1)}{(0.735)(0.783)} = \frac{[C_6H_5COO^-][H_3O^+]}{[C_6H_5COOH]}$$

$$K_a' = 1.09\times10^{-4} = \frac{[C_6H_5COO^-][H_3O^+]}{[C_6H_5COOH]}$$

$$K_a' = 1.09\times10^{-4} = \frac{[H_3O^+]^2}{c_{HA} - [H_3O^+]}$$

Rearranging as before and solving using the quadratic equation:

$$[H_3O^+]^2 + 1.09 \times 10^{-4}[H_3O^+] - 2.17 \times 10^{-7} = 0$$
$$[H_3O^+] = 4.16 \times 10^{-4} M$$
$$p[H_3O^+] = 3.38$$

We see that the shielding effect of the strong electrolyte ions increases the amount of acid dissociation by nearly 30% in this high ionic strength solution compared to the ideal case. The H_3O^+ and benzoate ions are stabilized in solution by nearby strong electrolytes. We can also think about the shielding in the high ionic strength case as lowering the attraction for the ions to recombine, just like the individual skaters were shielded by little kids from recombining into couples.

To round out these examples, we will complete a sample calculation for a weak base.

Example 5.20 Weak Base Example.   Predict the $p[H_3O^+]$ of the resulting solution when we mix 50.0 ml of 0.010 M HClO with 50.0 ml of 0.010 M NaOH?

First, looking at the compounds that are being mixed. We see that we have an acid and a base, so we can expect a neutralization reaction to occur. Find the molar amounts of each reactant and determine the limiting reagent:

$$HClO(aq) \quad + \quad OH^-(aq) \quad \rightarrow \quad ClO^-(aq) \quad + \quad H_2O$$

5.0×10^{-4} mol	5.0×10^{-4} mol	
≈ 0 mmol	≈ 0 mmol	5.0×10^{-4} mol

On consuming the limiting reagents, both the HClO and OH^- in this case, the result is a solution of sodium hypochlorite. The hypochlorite ion, ClO^-, has a formal concentration of 5.0×10^{-4} mol/0.100 l = 5.0×10^{-3} M, where the 0.100 l is the total for the two solutions that we mixed. Now we can think about what equilibrium will be established. We can see from the formula of ClO^- that it can react with water to be protonated, that is, it is a weak base. The equilibrium reaction is

$$ClO^-(aq) + H_2O \rightleftharpoons HClO(aq) + OH^-(aq)$$

The ionic strength is 5.0×10^{-3} M, from the hypochlorite and sodium ions. On the basis of our previous discussion, we can expect ionic strength effects to be measurable, but probably not large. To correct the thermodynamic equilibrium constant in the same way as in Example 5.19, first write the equilibrium constant expression:

$$K_b = \frac{(a_{HClO})(a_{OH^-})}{a_{ClO^-}}$$

Replacing activities as before gives

$$K_b = \frac{(\gamma_{HClO})[HClO](\gamma_{OH^-})[OH^-]}{(\gamma_{ClO^-})[ClO^-]}$$

and on collecting terms,

$$K_b = \frac{[HClO][OH^-]}{[ClO^-]} \frac{(\gamma_{HClO})(\gamma_{OH^-})}{\gamma_{ClO^-}}$$

Now before going on let us look at this expression. The γ_{HClO} will be 1. The γ_{OH^-} in the numerator and the γ_{ClO^-} in the denominator will cancel, at least partially if not completely. The result is that we can skip correcting for ionic strength effects:

$$K_b' = K_b = \frac{[HClO][OH^-]}{[ClO^-]}$$

Looking up K_a for hypochlorous acid and using that to find K_b gives

$$K_b = \frac{1.01 \times 10^{-14}}{2.90 \times 10^{-8}} = 3.48 \times 10^{-7}$$

You may have set up weak acid and weak base equilibrium calculations in previous courses in a table, so I will show one here. Note that this process is no different than simply starting with a K_a' or K_b' expression and solving. As the only source of HClO and OH$^-$ is the reaction between ClO$^-$ and water, [HClO] = [OH$^-$].

	ClO$^-$	HClO	OH$^-$
c	0.0050 M	≈ 0 M	≈ 0 M
Δc	$-[OH^-]$	$+[OH^-]$	$+[OH^-]$
[ ]	0.0050 M $- [OH^-]$	$[OH^-]$	$[OH^-]$

where c has its usual meaning of formal or initial concentration, Δc is the change in concentration to get to equilibrium, and [ ] symbolizes the resulting equilibrium concentrations.

The equilibrium constant expression is

$$3.48 \times 10^{-7} = \frac{[OH^-]^2}{0.0050 - [OH^-]}$$

Rearranging this expression to solve with the quadratic equation and then converting ultimately to $p[H_3O^+]$:

$$[OH^-]^2 + 3.48 \times 10^{-7}[OH^-] - 1.74 \times 10^{-9} = 0$$
$$[OH^-] = 4.16 \times 10^{-5} \, M$$
$$p[OH^-] = -\log(4.16 \times 10^{-5} M) = 4.39$$
$$p[H_3O^+] = 14.00 - 4.39 = 9.61$$

In Example 5.20, we did all of the steps highlighted in Figure 5.5. We determined the resulting species that were present after a neutralization reaction and we determined ionic strength. On correcting K_b to find K_b', we found that activity coefficients canceled, and we could use the thermodynamic K_b value for the calculation.

You-Try-It 5.E
Write formulas in the equilibrium-calculations worksheet in you-try-it-05.xlsx to predict the $p[H_3O^+]$ of the given solutions. Use Examples 5.18, 5.19, and 5.20 as guides.

Chapter 5. What Was the Point? This chapter reviewed acids, bases, and the equilibria of weak acids and weak bases in water. It introduced the concepts of ionic strength and activity. Activity coefficients are used to correct the thermodynamic equilibrium constant, K, to generate the formal equilibrium constant, K'. The K' values provide a more realistic description of real solutions. In the monoprotic cases that we considered, we did a lot of math for relatively small differences in the results, often only 0.1 pH units. We can often neglect ionic strength effects to predict equilibrium concentrations of neutral weak acids and neutral weak bases in water. For other types of equilibria, though, we will often find very significant differences between the ideal case and the behavior in real, that is, nonideal, solutions.

PRACTICE EXERCISES

1. Answer the following questions for a conceptual review:
   (a) What is in a bucket of deionized water?

(b) What is present if we add a small amount of HNO_3?

(c) What is present if we add a small amount of acetic acid, CH_3COOH, to the HNO_3 solution in the bucket?

2. Determine if the following soluble salts will produce neutral, acidic, or basic solutions when dissolved in water.

(a) $Ba(NO_3)_2$

(b) $Ca(ClO_4)_2$

(c) KI

(d) NaF

(e) NH_4Br

(f) NH_4F

(g) Ammonium acetate

3. Amino acids have the general formula $^+H_3NCHRCOOH$ when fully protonated. For alanine ($R = CH_3$), the pK_a of the –COOH proton is 2.3 and the pK_a of the $-NH_3^+$ proton is 9.9.

(a) What is the net charge of this amino acid when the pH is less than 2.3, between 2.3 and 9.9, and greater than 9.9?

(b) To what pH would you adjust an aqueous solution containing alanine so that you could extract the alanine into a nonpolar organic solvent to separate it from inorganic salts?

4. Predict which solution in each of the following pairs will produce the solution of the lowest pH. Explain your rationale.

(a) 0.01 M HCl, 0.01 M benzoic acid

(b) 0.01 M HClO, 0.01 M $HClO_4$

(c) 0.01 M HClO, 1×10^{-4} M $HClO_4$

5. Predict which solution in each of the following pairs will produce the solution of the highest pH. Explain your rationale.

(a) 0.01 M KOH, 0.001 M KOH

(b) 0.01 M CH_3COONa, 0.01 M KOH

(c) 0.01 M CH_3COONa, 1×10^{-4} M NaOH

6. What is the ionic strength of each of the following solutions?

(a) 0.010 M KI

(b) 0.250 M $Ca(NO_3)_2$

(c) 0.250 M $AlCl_3$ (neglect the reaction of Al^{3+} with H_2O)

(d) 0.250 M $(NH_4)_2SO_4$

(e) 0.250 M CH_3COONa

7. Calculate activity coefficients for all of the ions in the following solutions.

(a) 0.010 M KI

(b) 0.250 M $Ca(NO_3)_2$

8. Calculate K_w' for the following solutions.

(a) 0.010 M KI

(b) 0.250 M $Ca(NO_3)_2$

9. Use the following solutions to answer these questions. (CH_3COOH is acetic acid: $K_a = 1.75 \times 10^{-5}$, CH_3COONa is sodium acetate.)
   - 0.2 M CH_3COOH
   - 0.2 M CH_3COONa
   - 0.2 M CH_3COOH in 0.2 M NaCl
   - 0.2 M CH_3COONa in 0.2 M NaCl
   - 1.0 M CH_3COOH

   (a) Rank the solutions in order from lowest to highest ionic strength.

   (b) In which solution will the activity coefficients have the largest effect on equilibrium concentrations? You may assume that d_i values are the same for all ions.

   (c) Rank the solutions in order from lowest to highest p[H_3O^+].

10. Predict the percent dissociation of the weak acids, HA, in water for the following values of K_a. $c_{HA} = 0.01$ M for each case. You may assume that $K_a' = K_a$.

    (a) $K_a = 1 \times 10^{-5}$

    (b) $K_a = 1 \times 10^{-3}$

    (c) $K_a = 1 \times 10^{-2}$

11. Predict the percent dissociation of the weak acids, HA, in 0.250 M $Ca(NO_3)_2$ for the following values of $K_a \cdot c_{HA} = 0.01$ M. You may assume that the d_i value for the anion, A^-, is 0.4 nm.

    (a) $K_a = 1 \times 10^{-5}$

    (b) $K_a = 1 \times 10^{-3}$

    (c) $K_a = 1 \times 10^{-2}$

12. Predict the p[H_3O^+] of a 0.0100 M solution of acetic acid. (You may neglect activity effects.) The K_a of acetic acid, CH_3COOH, is 1.75×10^{-5}.

13. Use the result from the previous problem to calculate activity coefficients and recalculate p[H_3O^+] for 0.0100 M acetic acid.

14. Predict the p[H_3O^+] of a 0.0100 M solution of acetic acid in a solution that is 0.5 M KCl.

15. Predict the p[H_3O^+] of a 0.0100 M solution of sodium acetate. (You may neglect activity effects.)

16. Predict the p[H_3O^+] of a 0.0100 M solution of sodium acetate. (Include activity effects in your calculation.)

17. Use the pH-calc-weakacid.xlsx spreadsheet to determine the concentration at which the approximate solution, assuming [HA] = c_{HA}, introduces an error of 0.1 pH units for the following weak acids. You may assume that $K_a' = K_a$.

    (a) Acetic acid

    (b) Dichloroacetic acid

18. Use the pH-calc-weakacid.xlsx and activity-coefficients.xlsx spreadsheets to determine at what ionic strength the error in assuming $K_a' = K_a$ results in an error of 0.1 pH units in a calculation of p[H_3O^+] for 1.0 mM

acetic acid. (Hint: First use pH-calculation.xlsx to determine K_a' and the activity coefficients, then use activity-coefficients.xlsx and adjust the ionic strength using different concentrations of NaCl.)

19. A 0.01 M monoprotic acid has a predicted $p[H_3O^+]$ of 3.85. Measured pH of this solution is 3.79. What is the ionic strength of this solution? You may assume $d_i = 0.4$ for the ions that are present.

CHAPTER 6

BUFFER SOLUTIONS AND POLYPROTIC ACIDS

Learning Outcomes

- Identify buffer solutions and predict $p[H_3O^+]$ using the Henderson–Hasselbalch equation.
- Use alpha fraction plots to predict the dominant form of a weak acid as a function of pH.
- Describe the regions of a weak acid titration curve and calculate analyte concentration.
- Identify polyprotic acids and determine the limiting reagent in neutralization reactions.
- Predict $p[H_3O^+]$ for different forms of a polyprotic weak acid.

6.1 BUFFER SOLUTIONS

A buffer in chemistry is anything that counteracts a change to a chemical system. A buffer will be added to a test portion to control one or more aspects of the solution chemistry that might affect a measurement. In redox chemistry, a "redox buffer" can maintain one or more species in a certain oxidation state. Analytical methods for iron will often use a reducing agent, for example, hydroxylamine in the Fe-phenanthroline practice exercise in Chapter 4, as a reagent to retain the iron in the Fe^{2+} form. A *pH buffer* is a solution that can

Basics of Analytical Chemistry and Chemical Equilibria: A Quantitative Approach, Second Edition. Brian M. Tissue.
© 2023 John Wiley & Sons, Inc. Published 2023 by John Wiley & Sons, Inc.
Companion Website: www.wiley.com/go/tissue/analyticalchemistry2e

maintain a nearly constant pH when diluted or when a small amount of strong acid or strong base is added. A buffer solution consists of a mixture of a weak acid and its conjugate base. Describing a buffer as a weak base and its conjugate acid is equivalent. As many analytes and reagents have acidic or basic groups, acid–base buffers are common in analytical applications to control the solution chemistry. You will see from the examples in this chapter that buffer systems also exist in many environmental and biological systems.

How can we make a pH buffer solution? There are several ways to produce a mixture of a weak acid and its conjugate base:

- Mix a weak acid and a salt of its conjugate base in solution.
- Mix a weak acid solution with strong base to neutralize some but not all of the weak acid.
- Mix a weak base solution with strong acid to neutralize some but not all of the weak base.

Any of these procedures will result in a solution that contains a mixture of a weak acid and its conjugate base. The last two cases are limiting reagent calculations, with the strong base or strong acid being the limiting reagent. Adding excess strong base to a weak acid, or excess strong acid to a weak base, will not produce buffer solutions. The pH response of a buffer solution is illustrated in Examples 6.1, 6.2, 6.3, and 6.4.

Example 6.1 Buffer Calculation.   What is the $p[H_3O^+]$ of the resulting solution when we add 0.040 mol of KOH to 1.00 l of 0.10 M carbonic acid? (We will neglect any volume change on adding the KOH.)[1]

We must first determine the result of the neutralization reaction before considering equilibrium. We are adding 0.040 mol of OH^- from the KOH strong base to $(1.00 l)(0.10 M H_2CO_3)$ or 0.10 mol H_2CO_3:

$$H_2CO_3(aq) \ + \ OH^-(aq) \ \rightarrow \ HCO_3^-(aq) \ + \ H_2O$$

| 0.100 mol | 0.040 mol | | |
| 0.060 mol | ≈ 0 mol | 0.040 mol | 0.040 mol |

We have created a mixture of a weak acid, carbonic acid, and its conjugate base, hydrogen carbonate (bicarbonate) ion, which will act as a buffer solution. Looking at the predominant species that are present in solution, we can determine the equilibrium that will take place:

$$H_2CO_3(aq) + H_2O \rightleftharpoons HCO_3^-(aq) + H_3O^+(aq)$$

---

[1]You can't really make 0.10 M carbonic acid due to release of CO_2. I like to use the carbonate system because it is so important in environmental and biological solutions.

Correcting the equilibrium constant from Table 5.8 for ionic strength,

$$K'_a = 6.2 \times 10^{-7} \quad (\text{corrected for } I_c = 0.04 \text{ M})$$

Carbonic acid is a diprotic acid, but we are neglecting the second dissociation to form carbonate ion, CO_3^{2-}. When we discuss polyprotic acids, we will see how we can justify doing so for this problem. Now calculate the equilibrium concentrations and substitute into the K'_a expression:

	H_2CO_3	HCO_3^-	H_3O^+
c	0.060 M	0.040 M	≈ 0
Δc	$-[H_3O^+]$	$+[H_3O^+]$	$+[H_3O^+]$
[ ]	$0.060 \text{ M} - [H_3O^+]$	$0.040 \text{ M} + [H_3O^+]$	$[H_3O^+]$

$$K'_a = \frac{[HCO_3^-][H_3O^+]}{[H_2CO_3]}$$

$$K'_a = \frac{(0.040 \text{ M} + [H_3O^+])[H_3O^+]}{(0.060 \text{ M} - [H_3O^+])} = 6.2 \times 10^{-7}$$

We can rearrange this expression and solve it using the quadratic equation as we did in Section 5.6, but let us first try neglecting $[H_3O^+]$ compared to the formal concentrations of the carbonic acid and hydrogen carbonate ion:

$$\frac{(0.040 \text{ M})[H_3O^+]}{(0.060 \text{ M})} = 6.2 \times 10^{-7}$$

$$[H_3O^+] = 9.3 \times 10^{-7} \text{ M}$$

$$p[H_3O^+] = 6.03$$

Checking our assumption: 9.3×10^{-7} M is much smaller than 0.040 or 0.060 M, so the following relationships were true:

$$(0.040 \text{ M} + [H_3O^+]) \cong 0.040 \text{ M}$$

$$(0.060 \text{ M} - [H_3O^+]) \cong 0.060 \text{ M}$$

Setting up this problem was no different from what we did in Chapter 5. The difference is that we have significant concentrations of both the weak acid and its conjugate base, which affects how we simplify and solve the K'_a expression. Now let us see what happens when we add some strong acid to this solution.

Example 6.2 pH Change with Buffer. What is the $p[H_3O^+]$ of the resulting solution when we add 0.010 mol of HCl to the pH = 6 carbonic acid buffer solution in Example 6.1?

As we are adding strong acid to the buffer, we must first determine the result of the neutralization reaction before considering equilibrium:

$$HCO_3^-(aq) \;+\; H_3O^+(aq) \;\rightarrow\; H_2CO_3(aq) \;+\; H_2O$$

| 0.040 mol | 0.010 mol | 0.060 mol | |
| 0.030 mol | ≈ 0 mol | 0.070 mol | 0.010 mol |

Now we can consider the equilibrium, and we set up the problem in the same way as above:

$$H_2CO_3(aq) + H_2O \rightleftharpoons HCO_3^-(aq) + H_3O^+(aq)$$

Although a neutralization reaction occurs, the ionic strength does not change:

$$K_a' = 6.2 \times 10^{-7} \;(\text{corrected for } I_c = 0.04 \text{ M})$$

$$K_a' = \frac{(0.030 \text{ M} + [H_3O^+])[H_3O^+]}{(0.070 \text{ M} - [H_3O^+])} = 6.2 \times 10^{-7}$$

Again neglecting $[H_3O^+]$ compared to the formal concentrations of the carbonic acid and carbonate ion,

$$\frac{(0.030 \text{ M})[H_3O^+]}{(0.070 \text{ M})} = 6.2 \times 10^{-7}$$

$$[H_3O^+] = 1.4 \times 10^{-6} \text{ M}$$

$$p[H_3O^+] = 5.84$$

So we predict a small change in $p[H_3O^+]$ after adding this amount of strong acid to the buffer solution. For comparison, what would have happened without the buffer system present?

Example 6.3 pH Change without Buffer.   Adding 0.010 mol of HCl to 1.0 l of pH = 6 water leads to

$$\text{moles } H_3O^+ = 0.010 \text{ mol} + (1.01)\,(1.0 \times 10^{-6} \text{ M}) = 0.010 \text{ mol}$$

$$[H_3O^+] = \frac{0.010 \text{ mol}}{1.01} = 0.01 \text{ M}$$

$$p[H_3O^+] = 2.0$$

In the absence of the buffer, the $p[H_3O^+]$ changes significantly, dropping from 6, a slightly acidic solution, to 2, a highly acidic solution. This example is not unlike the acidification of lakes and streams in the northeastern part of North America that flow over granite rock. Natural waters that are in contact with carbonate rock neutralize added acid, and do not become acidified.[2] Let's

[2]US EPA. "Effects of acid rain"; available at https://www.epa.gov/acidrain/effects-acid-rain. Accessed August 2022.

work through another example, and then in Section 6.2 we will develop a graphical approach to visualize these acid–base equilibria.

Example 6.4 Buffer Calculation. What is the $p[H_3O^+]$ of the resulting solution when we mix 100 ml of 0.20 M benzoic acid and 100 ml of 0.10 M KOH?

Again we must first determine the result of the neutralization reaction before considering equilibrium:

$$C_6H_5COOH(aq) \ + \ OH^-(aq) \ \rightarrow \ C_6H_5COO^-(aq) \ + \ H_2O$$

0.020 mol	0.010 mol		
0.010 mol	≈ 0 mol	0.010 mol	0.010 mol

We now have a mixture of weak acid (benzoic acid) and its conjugate base (benzoate ion). Looking at the formal concentrations of the species in solution (e.g., 0.010 mol/0.200 l = 0.050 M), we can determine the equilibrium that will take place:

$$C_6H_5COOH(aq) + H_2O \rightleftharpoons C_6H_5COO^-(aq) + H_3O^+(aq)$$

and as before, correcting for ionic strength,

$$K_a' = 8.8 \times 10^{-5} \text{ (corrected for } I_c = 0.05\,M, \ K_a = 6.25 \times 10^{-5})$$

$$K_a' = \frac{(0.050\text{ M} + [H_3O^+])[H_3O^+]}{0.050\text{ M} - [H_3O^+]} = 8.8 \times 10^{-5}$$

We can rearrange this expression and solve it using the quadratic equation, but let us first try neglecting $[H_3O^+]$ compared to 0.050 M.

$$\frac{(0.050\text{ M})[H_3O^+]}{(0.050\text{ M})} = 8.8 \times 10^{-5}$$

$$[H_3O^+] = 8.8 \times 10^{-5}\text{ M}$$

Our assumption was valid: $0.050\text{ M} - 8.8 \times 10^{-5}\text{ M} \approx 0.050\text{ M}$
Now taking $-\log$ gives, $p[H_3O^+] = 4.06$

There are several points to note from Example 6.4. First, I did not show the calculation of activity coefficients, but you can verify that the K_a' that I used is correct for the ionic strength of this solution (see Section 5.6 for an example calculation). With equal amounts of the weak acid and its conjugate base, the resulting $[H_3O^+]$ equals K_a'. Similarly, taking the negative log results in $p[H_3O^+] = pK_a'$. We can rearrange the K_a expression to see this more clearly.

$$K_a' = \frac{[H_3O^+][A^-]}{[HA]}$$

$$[H_3O^+] = \frac{K_a'[HA]}{[A^-]}$$

Since we assume that $[H_3O^+] \ll [HA]$ and $[H_3O^+] \ll [A^-]$, $[HA] \approx c_{HA}$ and $[A^-] \approx c_{A^-}$. Replacing equilibrium concentrations with formal concentrations:

$$[H_3O^+] = \frac{K'_a c_{HA}}{c_{A^-}}$$

Taking the $-\log$ of each side of this relationship and rearranging gives us:

$$
\begin{aligned}
-\log[H_3O^+] &= -\log\frac{K'_a c_{HA}}{c_{A^-}} \\
-\log[H_3O^+] &= -\log K'_a - \log\frac{c_{HA}}{c_{A^-}} \qquad (6.1) \\
p[H_3O^+] &= pK'_a + \log\frac{c_{A^-}}{c_{HA}}
\end{aligned}
$$

This expression is called the *Henderson–Hasselbalch equation* and it is very convenient for predicting the $p[H_3O^+]$ of a buffer solution. Note that it does make the assumption that formal and equilibrium concentrations are equal. This approximation is usually valid because concentrations of the weak acid and its conjugate base are usually high to obtain a high buffer capacity. Using the Henderson–Hasselbalch equation is equivalent to starting with a K'_a expression. In either case, be sure to check that any approximations are valid.

6.2 ALPHA FRACTION PLOTS

The calculations in Examples 6.1–6.4 provide good estimates of equilibrium concentrations. However, the concentrations are time consuming to do for all weak acid/conjugate base ratios that are useful for buffer solutions. Plotting the fraction of each form of a conjugate acid–base pair against the pH provides a graphical method to visualize the predominant forms of the species that form a buffer system. We encountered a similar fraction, $\alpha_{S(aq)}$, in the extraction calculations in Section 2.2 and we continue using α as our symbol for a fraction. Chemists usually call a plot of fractional quantities an alpha plot. You will also see them called *distribution diagrams* or *distribution curves*, especially in the earth sciences.

An alpha fraction, α, is the ratio of the equilibrium concentration of one specific form of a solute divided by the total concentration of all forms of that solute. It therefore must be a number between 0 and 1. In all cases, the α is for the system at equilibrium. For a monoprotic acid with just two forms, HA and A^-, the definitions are

$$\alpha_{HA} = \frac{[HA]}{c_{total}} \qquad (6.2)$$

$$\alpha_{A^-} = \frac{[A^-]}{c_{total}} \qquad (6.3)$$

where

$$c_{total} = [HA] + [A^-] \qquad (6.4)$$

Remember, we use the c symbol for formal concentrations and [ ] for equilibrium concentrations. When we refer to a solution, for example, 0.05 M benzoic acid, we are talking about the formal concentration. The actual equilibrium concentration of benzoic acid, $[C_6H_5COOH]$, will depend on the relative concentrations of the two forms that it can take in solution, C_6H_5COOH and $C_6H_5COO^-$ (benzoate ion). The total amount of the two equilibrium forms of benzoic acid must equal the amount that was added to the solution, that is, Equation 6.4.

Figure 6.1 shows the alpha plots for benzoic acid and benzoate ion as a function of pH. These plots are calculated from

$$\alpha_{HA} = \frac{[H_3O^+]}{[H_3O^+] + K_a'} \qquad (6.5)$$

$$\alpha_{A^-} = \frac{K_a'}{[H_3O^+] + K_a'} \qquad (6.6)$$

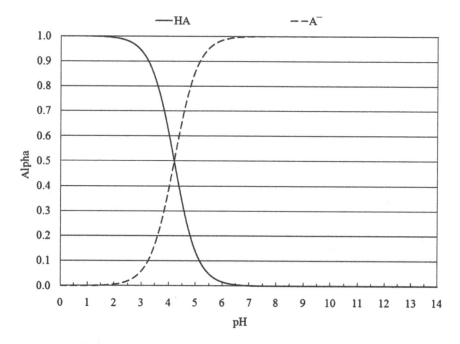

Figure 6.1 Alpha plot for the benzoic acid/benzoate system.

Equations (6.5) and (6.6) are derived from the definitions in Equations (6.2) and (6.3) using the K_a' expression to replace [HA] and [A$^-$] with [H$_3$O$^+$] and K_a', respectively. This new form of the equations allows direct calculation of alpha values as a function of pH.

The two curves in Figure 6.1 designate each equilibrium form of benzoic acid. The solid line is for the protonated acid (from Equation 6.5) and the dashed line is for the deprotonated conjugate base (from Equation 6.6). They each span the full pH range from 0 to 14, although the curves are not easily seen for α values very close to 0 or very close to 1. When we need to see how very small values of α vary with pH, we can plot log(α) to visualize order-of-magnitude changes. We can see that where the two curves cross, that is, [HA] = [HA$^-$], the pH is equal to the pK_a, ≈ 4.1 for benzoic acid. The alpha plots might be less accurate than a calculation because it does not correct K_a' for I_c as pH changes in the plot, but they provide a rapid means to predict the forms that will be present at a given pH. They are also very useful to estimate the relative amounts of a weak acid and its conjugate base for making a buffer at a certain pH (see Examples 6.5 and 6.6).

Example 6.5 Acid–Base Ratio. What ratio of A$^-$ to HA is needed to make a benzoic-acid-based buffer with a pH of 5?

First look at Figure 6.1 to get an approximate ratio. To reach a pH of 5.0, we will need a solution that is roughly 10% HA (solid line) and 90% A$^-$ (dashed line). Now use the Henderson–Hasselbalch equation to check:

$$p[H_3O^+] = pK_a' + \log \frac{c_{A^-}}{c_{HA}}$$

Entering the desired pH and the pK_a for benzoic acid,

$$5.0 = 4.1 + \log \frac{c_{A^-}}{c_{HA}}$$

Rearranging and solving gives,

$$\frac{c_{A^-}}{c_{HA}} = 10^{5.0-4.1} = 7.9$$

Not far from our initial estimate of 9/1. Note that this ratio calculation does not tell us if the assumptions of the Henderson–Hasselbalch equation are valid. If we are making a pH buffer though, the formal concentrations will be relatively high.

You-Try-It 6.A
The Henderson-Hasselbalch worksheet in you-try-it-06.xlsx contains a series of buffer systems. Determine the concentration at which the Henderson–Hasselbalch equation loses accuracy for each buffer.

Example 6.6 Preparing a Buffer Solution. What number of moles of KOH should be added to 1.00 l of a 0.100 M benzoic acid solution to make a pH = 5.0 buffer?

There are multiple approaches to this calculation. Let us use the result from Example 6.5 and find one more equation to solve for the concentration of HA and A⁻. The other thing that we know is that the total amount of benzoic acid plus benzoate must remain 0.100 M.

Thus,

$$c_{A^-} + c_{HA} = 0.100 \text{ M}$$

We already have

$$\frac{c_{A^-}}{c_{HA}} = 7.9$$

so let's rearrange and combine these two equations to solve for one of the concentrations:

$$c_{A^-} = 7.9 c_{HA}$$
$$7.9 c_{HA} + c_{HA} = 8.9 c_{HA} = 0.100 \text{ M}$$
$$c_{HA} = 0.0112 \text{ M}$$

Now we can use this result to find c_{A^-}:

$$c_{A^-} + 0.011 \text{ M} = 0.100 \text{ M}$$
$$c_{A^-} = 0.089 \text{ M}$$

Now knowing the necessary concentrations, we can answer the question of how much KOH to add. As we are working with 1.00 l of solution, we have 0.100 mol of HA and we must neutralize some of that:

$$(1.00 \text{ l})(0.100 \text{ M}) - \text{moles KOH} = 0.011 \text{ mol HA}$$
$$\text{moles KOH} = 0.089 \text{ mol}$$

6.2.1 Buffer Capacity

The *buffer capacity* can be defined as the amount of strong acid or strong base that causes a pH change of 1 in 1.0 l of buffer solution. Buffer capacity depends on both the total concentration of the buffer components and the ratio of the weak acid and its conjugate base. I will not go into the calculation to predict buffer capacity, but as a general rule, a buffer system should be chosen such that the pK_a of the acid is within 1 of the desired solution pH. We can use alpha plots to predict the range over which a certain weak acid/conjugate base pair will be useful. Choosing a monoprotic acid for a given pH is easy to find in a table of pK_a values. The alpha plots can be more useful when predicting the usefulness of a polyprotic acid system.

As an example, if we wanted to use the benzoic acid/benzoate pair as a buffer at a pH of 5, from Figure 6.1 and the sample calculation we can see that we would need 11% of the acid ($\alpha = 0.11$ on the solid curve) and 89% of the conjugate base ($\alpha = 0.89$ on the dashed curve). Given the rule of thumb for useful buffer capacities, the benzoic acid/benzoate system would work. It would not be as useful as a buffer solution for pH greater than 5.

6.3 WEAK ACID TITRATION CURVE

Section 3.5 discussed titrations of strong acids and strong bases to make analytical measurements. As all strong acids are strong electrolytes, the shape of the titration curve will be the same for all strong acids. Weak acids and weak bases have titration curves that are slightly different, which can provide additional qualitative information about the acid–base system. This information is most useful when trying to identify an unknown acid.

Figure 6.2 shows calculated titration curves for 0.10 M strong acid and 0.10 M weak acid (acetic acid). There are several differences in these curves that are instructive to discuss. First, note the initial pH of these solutions. For the strong acid the pH is 1.0, $-\log(0.10\text{ M})$. The initial pH of the 0.1 M acetic acid solution is 2.9 (see Section 5.6 for the calculation). This difference makes sense because the strong acid dissociates completely, but the weak acid dissociates only partially depending on the value of K_a'. As strong base is added to the strong acid solution, the pH changes gradually and then rises rapidly on approaching the

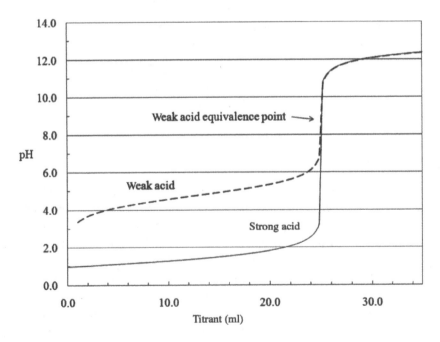

Figure 6.2 Titration curves for strong and weak acids.

equivalence point. At the equivalence point, all the acid is neutralized and the solution contains only ions of strong electrolytes; hence, the pH is 7. On adding more strong base, past the equivalence point, the pH continues upward to alkaline values.

The pH of the weak acid solution responds differently. As strong base is added to the acetic acid solution, there is a slight increase in pH and then it rises more slowly. The fairly flat curve results from being in the buffer region where the solution contains a weak acid and its conjugate base. As nearly all of the acid is neutralized approaching the equivalence point, the pH rises rapidly as it did for the strong acid curve. A key difference is that the equivalence point does not occur at a pH of 7. At the equivalence point, sufficient strong base has been added to neutralize all of the acetic acid, leaving only acetate ion and spectator ions in solution. The pH is ≈ 8.8, due to the base hydrolysis reaction of acetate with water. This value can be confirmed by solving the K'_b expression for the concentration of acetate ion that is present at the equivalence point (see Example 6.7). Past the equivalence point, the situation is the same as for a strong acid and the pH is determined by the concentration of excess strong base.

Example 6.7 Equivalence Point Calculation.   Calculate the pH at the equivalence point for the titration of 0.100 M acetic acid with 0.100 M strong base. The initial volume of the weak acid solution is 25.0 ml.

This calculation is done most directly by setting up the weak base equilibrium and solving for [OH$^-$]. The equilibrium reaction is

$$CH_3COO^-(aq) + H_2O \rightleftharpoons CH_3COOH(aq) + OH^-(aq)$$

From the curve, we see that 25.0 ml of titrant was added to reach the equivalence point. The formal concentration of CH_3COO^- at the equivalence point is

$$c_{A^-} = \frac{(0.100 \text{ M})(0.0250 \text{ l})}{0.0500 \text{ l}} = 0.050 \text{ M}.$$

Unlike the typical titration where we are solving for an unknown, here we know our initial acid concentration and we can calculate formal concentrations of any species at any point on the curve by correcting the concentration for the dilution produced by adding the titrant to the original acid solution.

Now setting up the equilibrium constant expression and solving,

$$K'_b = \frac{K'_w}{K'_a} = \frac{[OH^-]^2}{0.050 \text{ M} - [OH^-]}$$

and

$$[OH^-] = 5.4 \times 10^{-6} \text{ M}$$

Converting to $p[OH^-]$ and then to $p[H_3O^+]$:

$$p[OH^-] = 5.27$$
$$p[H_3O^+] = 14 - 5.27 = 8.72$$

Check the titration curve in Figure 6.2 and you will see that this result matches well.

Before leaving titration curves, we should consider a limitation in using them for analytical measurements. Figure 6.3 shows two titration curves for different concentrations of acetic acid. In both cases, the titrant concentration is the same as the acid concentration. You can see that the equivalence point becomes more difficult to see as the concentration of the acid decreases. Since pure water contains 1×10^{-7} M H_3O^+ and OH^-, titrating weak acids or weak bases near that concentration is not possible. Determining the equivalence point is essentially impossible at an initial acetic acid concentration of 1×10^{-6} M (not shown in the figure).

At this halfway point in the titration, we often find that the $p[H_3O^+]$ equals the pK_a' of the acid. Figure 6.3 shows that obtaining such data from titration curves is also concentration dependent. For acetic acid, the halfway point is equal to pK_a' for an initial concentration of approximately 1×10^{-4} M and higher, but not for lower concentrations.

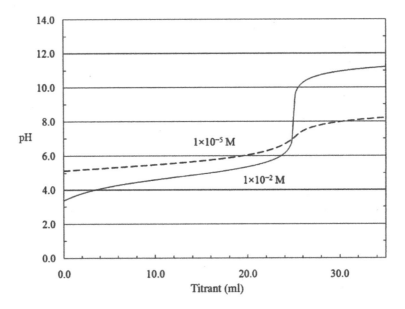

Figure 6.3 Concentration dependence of acetic acid titration.

6.4 POLYPROTIC ACIDS

6.4.1 Multiple Equilibria

Polyprotic acids are acids that possess more than one acidic proton, for example, H_2SO_4, H_2CO_3, H_3PO_4, and $C_2H_4N_2(CH_2COOH)_4$, EDTA. Besides being common in environmental and biological systems, polyprotic acids are quite useful in analytical applications because they can serve as pH buffers over extended pH ranges.

Because there are multiple acidic protons, the K_a and pK_a have an additional numerical subscript to indicate that the constant is for the equilibrium involving the first proton, second proton, etc. The K_a values decrease as protons are removed from a polyprotic acid because each subsequent proton is more difficult to remove as the molecule gains more electronegative character. Figure 6.4 shows the three forms of the diprotic phthalic acid: o-phthalic acid, $C_6H_4(COOH)_2$; hydrogen phthalate ion, $C_8H_5O_4^-$; and phthalate ion, $C_8H_4O_4^{2-}$ (from left to right). The K_a and pK_a values of phthalic acid are

$$K_{a1} = 1.12 \times 10^{-3} \qquad K_{a2} = 3.91 \times 10^{-6}$$
$$pK_{a1} = 2.951 \qquad pK_{a2} = 5.408$$

K_{a1} is the equilibrium constant for phthalic acid; K_{a2}, for the hydrogen phthalate ion. To obtain the K_b for the phthalate ion, $C_8H_4O_4^{2-}$, the appropriate K_a value to use in the calculation is K_{a2}.

When trying to predict equilibrium concentrations of the predominant species of a polyprotic acid in water, there are several different cases that we will encounter. We can usually treat addition of the fully protonated form of a polyprotic acid to water as a weak acid equilibrium problem, as we did for carbonic acid in Example 6.1:

$$H_2A(aq) + H_2O \rightleftharpoons HA^-(aq) + H_3O^+(aq)$$

Similarly, when placing the salt of the fully deprotonated form in water, we can usually treat the problem as a weak base:

$$A^{2-}(aq) + H_2O \rightleftharpoons HA^-(aq) + OH^-(aq)$$

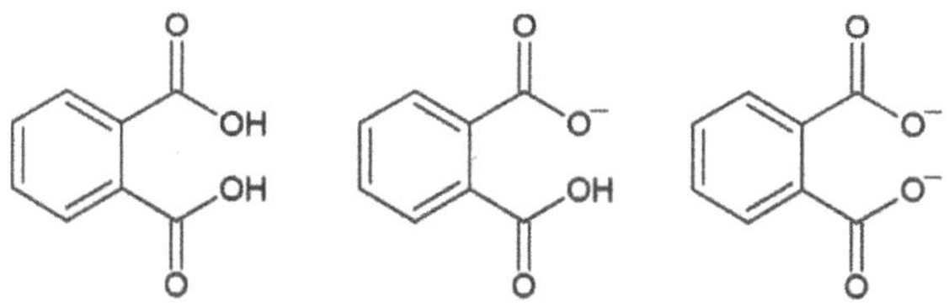

Figure 6.4 Structures of the three forms of o-phthalic acid.

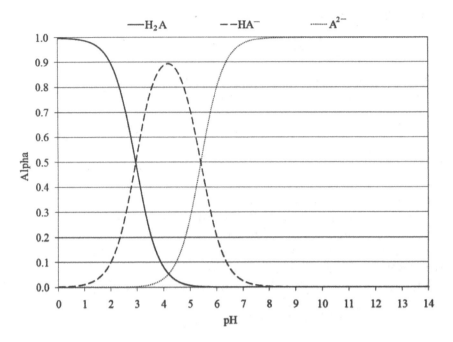

Figure 6.5   Alpha fraction versus pH for the three forms of *o*-phthalic acid.

I say "usually" for these cases because I am assuming that only the one equilibrium, involving two species, is important. Knowing if this is the case is where alpha plots can be very useful. Figure 6.5 shows the alpha plots for the *o*-phthalic acid species as a function of pH. Doing a quick calculation of the expected $p[H_3O^+]$ for a 0.05 M solution of *o*-phthalic acid gives a $p[H_3O^+]$ of 2.2. Looking at the predominant species at this pH, we see that there is a negligible fraction of the phthalate ion, $C_8H_4O_4^{2-}$. We can neglect the second equilibrium and treat a solution of *o*-phthalic acid as a monoprotic weak acid. This case is typical, and only a few weak acids that have very close pK_a values, within ≈ 1, will have appreciable concentration of A^{2-} in addition to H_2A and HA^-.

Similar to the weak acid and weak base end cases, each individual equilibrium can act as a buffer solution. For the example in Figure 6.5, there are two buffer regions:

$$H_2A(aq) + H_2O \rightleftharpoons HA^-(aq) + H_3O^+(aq) \quad pK_a = 2.95$$

and

$$HA^-(aq) + H_2O \rightleftharpoons A^{2-}(aq) + H_3O^+(aq) \quad pK_a = 5.41$$

As before, unless pK_a values are very close, we can consider only a single equilibrium between two forms of the polyprotic acid and neglect the very small concentrations of the other forms. The real power of the alpha plots is to be able to see what pH ranges are feasible for forming buffers for a given weak acid system. For the *o*-phthalic acid example shown in Figure 6.5, buffers will function well for pH ranges of approximately 2.0–4.0 and 4.4–6.4. Our rule of thumb for sufficient buffer action of ±1 from the pK_a is also valid for polyprotic acids that have widely separated pK_a values.

The cases of a weak acid, a weak base, or a buffer solution are all possible for monoprotic and polyprotic acids. Polyprotic acids have the additional case of an amphiprotic species. An *amphiprotic species*, such as hydrogen phthalate ion, $C_8H_5O_4^-$, has two simultaneous equilibria possible and significant concentrations of three forms of the polyprotic acid. The alpha plots in Figure 6.5 show that a solution prepared from a salt of hydrogen phthalate will contain

≈90% hydrogen phthalate, $C_8H_5O_4^-$
≈5% *o*-phthalic acid, $C_6H_4(COOH)_2$, and
≈5% phthalate ion, $C_8H_4O_4^{2-}$.

Because the alpha plot for hydrogen phthalate is symmetric, the expected $p[H_3O^+]$ for a solution of this amphiprotic species will be halfway between the two pK_a values. Numerically, this is the average of pK_{a1} and pK_{a2}:

$$p[H_3O^+] \approx \frac{1}{2}\{pK_{a1} + pK_{a2}\} \qquad (6.7)$$

This approximation works well unless triprotic or higher acids have alpha fraction plots that are asymmetric due to closely spaced pK_a constants. Also note that this approximation is only valid for reasonable concentrations of the amphiprotic species. As in any of the weak acid and weak base calculations, concentrations less than ≈1×10^{-6} M are affected by the pH of water. Note that the example equation is for a diprotic acid. The subscripts might be different for triprotic or higher order acids. As an example, the expression for HPO_4^{2-} will be $p[H_3O^+] \approx \frac{1}{2}\{pK_{a2} + pK_{a3}\}$.

The pK_a values mentioned earlier are the thermodynamic equilibrium constants. If the ionic strength is high, different pK_a' values might be necessary in the calculation. The effect of ionic strength does get canceled to some extent. Using the thermodynamic K_a values is often sufficient for our purposes of predicting equilibrium concentrations in aqueous solution. Example 6.8 shows the difference that a high ionic strength might make in an amphiprotic pH calculation.

Example 6.8 pH of Amphiprotic Species. Predict the $p[H_3O^+]$ of a 0.10 M solution of potassium hydrogen phthalate.

Looking up the pK_a values for phthalic acid and inserting into Equation 6.7 gives

$$p[H_3O^+] \approx \frac{1}{2}\{2.95 + 5.408\} = 4.18$$

As the ionic strength is relatively high, $I_c = 0.105$ M, correcting the pK_a values and using pK_a' values gives

$$p[H_3O^+] \approx \frac{1}{2}\{2.77 + 5.022\} = 3.90$$

(The pK_a' values were found from the equilibrium constant expressions using activity coefficients from the activity-coefficients.xlsx spreadsheet.)

6.4.2 Phosphoric Acid

Phosphoric acid is a triprotic weak acid and thus has four different forms that exist in solution. They all exist in solution simultaneously, but as the pK_a values are widely separated, the concentration of some of them will be very low. The three equilibria are listed below, and Figure 6.6 shows the alpha plots for the four forms of phosphoric acid.[3] Note that the curves cross at pH values equal to the pK_a values. If you need to generate your own curves for high ionic strength solutions, you can enter corrected K_a' values in the alpha-plot-poly-protic-acid.xlsx spreadsheet.

$$H_3PO_4(aq) + H_2O \rightleftharpoons H_2PO_4^-(aq) + H_3O^+(aq) \quad K_{a1} = 7.11 \times 10^{-3}$$
$$pK_{a1} = 2.15$$

$$H_2PO_4^-(aq) + H_2O \rightleftharpoons HPO_4^{2-}(aq) + H_3O^+(aq) \quad K_{a2} = 6.32 \times 10^{-8}$$
$$pK_{a2} = 7.20$$

$$HPO_4^{2-}(aq) + H_2O \rightleftharpoons PO_4^{3-}(aq) + H_3O^+(aq) \quad K_{a3} = 4.49 \times 10^{-13}$$
$$pK_{a3} = 12.35$$

The plots are analogous to the monoprotic case, each individual curve plots the fraction of a specific form of phosphoric acid:

$$\alpha_{H_3A} = \frac{[H_3A]}{c_{total}} \quad \alpha_{H_2A^-} = \frac{[H_2A^-]}{c_{total}} \quad \alpha_{HA^{2-}} = \frac{[HA^{2-}]}{c_{total}} \quad \alpha_{A^{3-}} = \frac{[A^{3-}]}{c_{total}} \quad (6.8)$$

where

$$c_{total} = [H_3A] + [H_2A^-] + [HA^{2-}] + [A^{3-}] \quad (6.9)$$

The alpha fractions for a triprotic acid such as phosphoric acid are calculated from the following equations. These equations are obtained from the above-mentioned definitions by substituting in K_a expressions and rearranging.

---

[3]These alpha fraction plots use the thermodynamic K_a values.

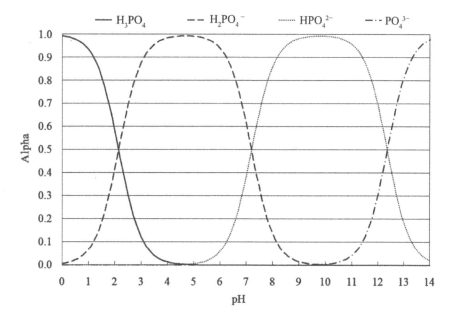

Figure 6.6 Alpha plots for the four forms of phosphoric acid.

$$\alpha_{H_3A} = \frac{[H_3O^+]^3}{[H_3O^+]^3 + K'_{a1}[H_3O^+]^2 + K'_{a1}K'_{a2}[H_3O^+] + K'_{a1}K'_{a2}K'_{a3}} \quad (6.10)$$

$$\alpha_{H_2A^-} = \frac{K'_{a1}[H_3O^+]^2}{[H_3O^+]^3 + K'_{a1}[H_3O^+]^2 + K'_{a1}K'_{a2}[H_3O^+] + K'_{a1}K'_{a2}K'_{a3}} \quad (6.11)$$

$$\alpha_{HA^{2-}} = \frac{K'_{a1}K'_{a2}[H_3O^+]}{[H_3O^+]^3 + K'_{a1}[H_3O^+]^2 + K'_{a1}K'_{a2}[H_3O^+] + K'_{a1}K'_{a2}K'_{a3}} \quad (6.12)$$

$$\alpha_{A^{3-}} = \frac{K'_{a1}K'_{a2}K'_{a3}}{[H_3O^+]^3 + K'_{a1}[H_3O^+]^2 + K'_{a1}K'_{a2}[H_3O^+] + K'_{a1}K'_{a2}K'_{a3}} \quad (6.13)$$

Before groaning at the length of these equations, note that the denominators are the same in all of them and that the numerator is simply one of the terms from the sum in the denominator. Although a bit tedious, they are fairly simple to type into a spreadsheet. The equations for a diprotic acid are of the same form (compare these equations for a triprotic acid to the monoprotic case given previously).

As noted for the monoprotic acid case, when the $p[H_3O^+]$ is equal to one of the pK'_a values, the acid and conjugate base concentrations are equal. Phosphoric acid is an ideal case in that the K'_a values are spaced far apart. At any given $p[H_3O^+]$, only one or two predominant forms of the phosphoric acid are present.

Why might we want to know the fraction of one particular species as a function of $p[H_3O^+]$? Let us consider the example reaction of adenosine diphosphate (ADP) → adenosine triphosphate (ATP)

$$ADP^{3-}(aq) + HPO_4^{2-}(aq) + H^+(aq) \rightarrow ATP^{4-}(aq) + H_2O$$

This reaction has a positive ΔG of ≈ 30 kJ/mol. Producing ATP is the body's way of storing energy in cells, for example, doughnuts → ATP. When your body needs energy for important activities such as thinking, ATP reverts back to ADP to supply the energy. Intracellular $p[H_3O^+]$ varies from 6.1 in muscle cells to approximately 7 in most other cells. Looking at the HPO_4^{2-} alpha plot, you can see that α varies from 0.07 to 0.4 over this range. For the body to store energy it needs a reliable supply of HPO_4^{2-}, which is one of the many reasons why buffer systems are so important in biology.[4] Large changes in $p[H_3O^+]$ would convert HPO_4^{2-} to $H_2PO_4^-$ or PO_4^{3-}, and you would not be able to convert the chemical energy of doughnuts to a form that could be stored and used in your body cells.

You-Try-It 6.B
The polyprotic-acid worksheet in you-try-it-06.xlsx lists several polyprotic acid solutions. Predict $p[H_3O^+]$ for each example.

6.4.3 Polyprotic Buffer Solutions

We will work through an example of preparing a buffer solution of a polyprotic acid in some detail. Example 6.9 calculates the amount of 1.0 M KOH to add to 500.0 ml of 0.200 M phosphoric acid to prepare 1.0 l of a pH = 7.0 phosphate-based buffer. The total phosphate concentration should remain 0.1 M. The possible equilibria were given above. Looking at the alpha plots for the phosphoric acid system, to be near a pH of 7, we will need to have $H_2PO_4^-$ and HPO_4^{2-} present. The pK_{a2} value is closest to the desired pH of 7, so we want to use the K_{a2} equilibrium:

$$H_2PO_4^-(aq) + H_2O(aq) \rightleftharpoons HPO_4^{2-}(aq) + H_3O^+(aq) \quad K_{a2} = 6.32 \times 10^{-8}$$

The alpha plots show the approximate relative amounts of $H_2PO_4^-$ and HPO_4^{2-} to reach a pH of 7.0. We will need slightly more $H_2PO_4^-$ than HPO_4^{2-}, roughly a 60:40 ratio. The alpha plot was generated using thermodynamic K_a

---

[4]This chemical issue is just one of many disruptions if cellular pH changed significantly. The movie *Apollo 13* has a dramatization of the effects of acidosis due to above-normal CO_2 partial pressure.

values, and we will have a solution with an ionic strength greater than 0.1 M. To calculate the amounts of $H_2PO_4^-$ and HPO_4^{2-}, which we will need to reach a pH of 7.0, we will find the activity coefficients to correct the K_{a2}. Let us assume that we will be near a 60:40 mixture to find the ionic strength, that is 0.0600 mol of $H_2PO_4^-$ and 0.0400 mol of HPO_4^{2-}.

Example 6.9 Phosphoric Acid/Phosphate Buffer. Prepare 1.0 l of a phosphate-based buffer of pH = 7.0 starting with 500.0 ml of 0.200 M phosphoric acid and a 1.0 M KOH solution.

We want to add KOH to neutralize enough H_3PO_4 to get 0.0600 mol of $H_2PO_4^-$ and 0.0400 mol of HPO_4^{2-}. We can think about doing so stepwise:

$H_3PO_4(aq)$	+	$OH^-(aq)$	$\rightarrow$	$H_2PO_4^-(aq)$	+	H_2O
0.100 mol		0.100 mol				
≈ 0 mol		≈ 0 mol		0.0100 mol		0.0100 mol

$H_2PO_4^-(aq)$	+	$OH^-(aq)$	$\rightarrow$	$HPO_4^{2-}(aq)$	+	H_2O
0.100 mol		0.040 mol				
0.060 mol		≈ 0 mol		0.040 mol		0.010 mol

Expressing this reaction in words: 0.100 mol OH^- neutralizes 0.100 mol H_3PO_4. We now have 0.100 mol of $H_2PO_4^-$. Adding 0.0400 more moles of OH^- neutralizes 0.0400 mol of $H_2PO_4^-$ to produce 0.0400 mol of HPO_4^{2-}, leaving 0.0600 mol of $H_2PO_4^-$. To achieve the desired ratio, we added a total of 0.140 mol of KOH. After diluting the mixture to 1.00 l, we have 0.140 M K^+, 0.0600 M $H_2PO_4^-$, and 0.0400 M HPO_4^{2-}. Now we can calculate the ionic strength:

$$I_c = \frac{1}{2}\{(+1)^2(0.140\text{ M}) + (-2)^2(0.040\text{ M}) + (-1)^2(0.060\text{ M})\} = 0.180\text{ M}$$

Next calculate the activity coefficients from the Debye–Hückel equation (refer to Table 5.9 and Table 5.10) and use these coefficients to find K_a':

$$\gamma_{H_2PO_4^-} = 0.727 \quad \gamma_{HPO_4^{2-}} = 0.279 \quad \gamma_{H_3O^+} = 0.802$$

$$K_{a2} = 6.32 \times 10^{-8} = \frac{(a_{HPO_4^{2-}})(a_{H_3O^+})}{a_{H_2PO_4^-}}$$

$$= \frac{(\gamma_{HPO_4^{2-}})[HPO_4^{2-}])(\gamma_{H_3O^+})[H_3O^+]}{(\gamma_{H_3PO_4^-})[H_2PO_4^-]}$$

Rearranging

$$K'_{a2} = \frac{\gamma_{H_2PO_4^-}}{(\gamma_{HPO_4^{2-}})(\gamma_{H_3O^+})} K_{a2}$$

$$K'_{a2} = \frac{(0.727)}{(0.279)(0.802)} 6.32 \times 10^{-8} = \frac{[HPO_4^{2-}][H_3O^+]}{[H_2PO_4^-]}$$

$$K'_{a2} = 2.05 \times 10^{-7} = \frac{[HPO_4^{2-}](1.0 \times 10^{-7})}{[H_2PO_4^-]}$$

where $[H_3O^+] = 1.0 \times 10^{-7}$ for a pH = 7.0 buffer. The ratio of $[H_2PO_4^-]$ to $[HPO_4^{2-}]$ is thus 2.05. As the total concentration of all phosphate species is 0.1 M, the relative concentrations that fit this ratio are 0.067 M $H_2PO_4^-$ and 0.033 M HPO_4^{2-}. Preparing this buffer requires 0.133 mol of OH^-. To answer the original question, add 133 ml of the 1.0 M KOH to the phosphoric acid solution and dilute to 1.0 l.

After correcting for ionic strength, Example 6.9 gives us our best prediction of the amounts of the phosphoric acid species needed to create a pH = 7.0 buffer. The concentrations are slightly different from our initial estimate from the alpha plots, but not enough to bother recalculating the ionic strength.

You-Try-It 6.C
The polyprotic-buffer worksheet in you-try-it-06.xlsx lists several applications for polyprotic buffers. Calculate how to prepare each example from the listed starting materials.

6.4.3 Carbonic Acid Examples

We will look at one more polyprotic example, which is of great importance in aquatic systems and biology. Rainwater has a pH of approximately 5.6 (at sea level) rather than 7.0 as expected for pure water.[5] The lower pH of rain is due to the solubility of atmospheric CO_2, which forms equilibrium amounts of carbonic acid, H_2CO_3, in the rainwater (see Example 6.10):

---

[5]Many geographic areas have rain of lower pH due to sulfur and nitrogen oxide emissions from human combustion sources. Compare a recent National Atmospheric Deposition Program "Lab pH" map to one from 20 years ago: https://nadp.slh.wisc.edu/maps-data/ntn-gradient-maps; accessed August 2022.

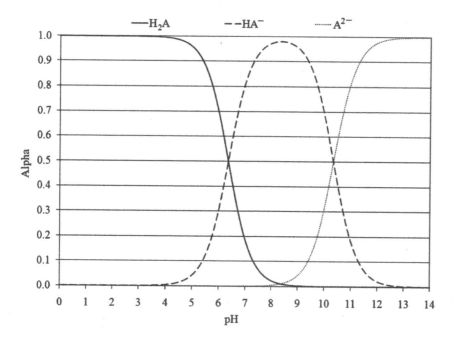

Figure 6.7 Alpha plots for the three forms of carbonic acid.

$$CO_2(g) \rightleftharpoons CO_2(aq)$$

$$CO_2(aq) + H_2O \rightleftharpoons H_2CO_3(aq)$$

$$H_2CO_3(aq) + H_2O \rightleftharpoons HCO_3^-(aq) + H_3O^+(aq)$$

$$HCO_3^-(aq) + H_2O \rightleftharpoons CO_3^{2-}(aq) + H_3O^+(aq)$$

The carbonic acid equilibria have K_a values of $K_{a1} = 4.45 \times 10^{-7}$, $pK_{a1} = 6.35$, and $K_{a2} = 4.7 \times 10^{-11}$, $pK_{a2} = 10.33$, which produces the alpha plots in Figure 6.7.

Example 6.10 Carbonic Acid. When pH = 5.6, what is the concentration of carbonic acid in water that is in equilibrium with atmospheric CO_2?

The pH of natural water of 5.6 leads to a H_3O^+ concentration of

$$[H_3O^+] = 10^{-5.6} = 2.5 \times 10^{-6} \text{ M}$$

When the only significant source of H_3O^+ and HCO_3^- is from the dissociation of the carbonic acid,

$$[HCO_3^-] = [H_3O^+] = 2.5 \times 10^{-6} \text{ M}$$

Now inserting these values in the K_a expression

$$4.45 \times 10^{-7} = \frac{(2.5 \times 10^{-6} \text{ M})(2.5 \times 10^{-6} \text{ M})}{[H_2CO_3]}$$

and

$$[H_2CO_3] = 1.4 \times 10^{-5} \text{ M}$$

As a final example, human blood has a pH of 7.4. Looking at the alpha plots in Figure 6.7, we see that this pH requires a ratio of bicarbonate ion to carbonic acid of approximately 11. Carbonate ion, CO_3^{2-}, is present at this pH, but at a relatively low concentration. Comparing the blood pH of 7.4 to the natural water pH of 5.6 shows that blood is obviously not in equilibrium with the atmospheric concentration of CO_2. The body is able to maintain the blood pH at 7.4 because the kidneys excrete excess acid and pump bicarbonate ion back into the blood. The carbonic acid/bicarbonate equilibrium is the primary pH buffer in blood, and the kidneys maintain the concentrations to fix the pH at 7.4.

You-Try-It 6.D
The carbonic-acid worksheet in you-try-it-06.xlsx contains several examples of carbonate-containing systems. Predict $p[H_3O^+]$ for each example.

Chapter 6. What Was the Point? The purpose of this chapter was to describe polyprotic acids and develop an understanding of acid–base buffer systems. There are three cases for predicting $p[H_3O^+]$ with polyprotic acids:

- end cases: only a weak acid or only a weak base;
- a buffer system: mixture of a weak acid and its conjugate base;
- amphiprotic species.

The three cases each have an appropriate approach for predicting $p[H_3O^+]$. We also introduced alpha plots, which provide a graphical means of visualizing the different forms of polyprotic acids as a function of pH.

PRACTICE EXERCISES

1. Use the pK_a values in Table 5.8 to identify suitable monoprotic acids to prepare buffer solutions at
   (a) pH = 5.0

(b) pH = 7.0

(c) pH = 9.0

2. Use the pK_a values in Table 5.8 to identify suitable polyprotic acids to prepare buffer solutions at
   (a) pH = 5.0
   (b) pH = 7.0
   (c) pH = 9.0

3. At any given pH, what is the sum of all alpha fractions for a given acid–base system? How will this value depend on the ionic strength of the solution?

4. How many moles of NaOH must be added to a solution containing 0.600 mol of phosphoric acid, H_3PO_4, to make a buffer solution containing equal amounts of HPO_4^{2-} and PO_4^{3-}?

5. How many moles of HCl must be added to 1.0 l of 0.050 M trisodium phosphate, Na_3PO_4, to obtain 0.050 M dihydrogen phosphate ion, $H_2PO_4^-$? (You many assume that the addition of the HCl produces an insignificant increase in the solution volume.)

6. What is the I_c of the solution in the previous problem before and after adding HCl?

7. For each of the following solutions, determine if the solution is (i) neutral with no buffer capacity, (ii) acidic with no buffer capacity, (iii) basic with no buffer capacity, or (iv) a buffer solution:
   (a) Adding 0.070 mol of NaOH to a solution containing 0.040 mol of a monoprotic weak acid.
   (b) Adding 0.040 mol of NaOH to a solution containing 0.070 mol of a monoprotic weak acid.
   (c) Adding 0.040 mol of NaOH to a solution containing 0.070 mol of a potassium hydrogen phthalate.
   (d) Adding 0.040 mol of HCl to a solution containing 0.070 mol of a potassium hydrogen phthalate.

8. What $p[H_3O^+]$ do you predict for the following solutions?
   (a) 0.010 M NaH_2PO_4
   (b) 0.010 M $NaHCO_3$
   (c) 1.0×10^{-8} M $NaHCO_3$

9. Approximately how many moles of HCl must be added to a solution containing 0.500 mol of trisodium phosphate to reach a $p[H_3O^+]$ of 3.0?

10. Predict the $p[H_3O^+]$ at the following points in the titration of 100 ml of a 0.10 M solution of o-phthalic acid using 0.10 M NaOH titrant. (You may neglect activity effects.)
    (a) Halfway point (for the first acidic proton)
    (b) The first equivalence point
    (c) The second equivalence point

11. On the basis of the results for the halfway point in the previous question, estimate the error that neglecting activity effects might introduce.

12. Predict the dominant form of phosphate in solutions that have the pH adjusted with strong acid or strong base to values of
    (a) pH = 4
    (b) pH = 6
    (c) pH = 8
    (d) pH = 10

13. A saturated solution of nitrilotriacetic acid (NTA) (0.13 g NTA/100 g solution) produces a pH of 2.3. Given that NTA has pK_a's of 1.7, 2.9, and 10.3, what is the predominant form of NTA when it is dissolved in water? You may assume a solution density of 1.0 g/ml in any calculation.

14. Predict $p[H_3O^+]$ of the solution that is prepared by mixing 0.0350 mol of HCl with 0.0250 mol of disodium phthalate, $C_6H_4(COONa)_2$, and diluting to 1.00 l. You may neglect ionic strength effects.

15. What is the alpha fraction of HPO_4^{2-} in a solution that is a 50:50 mixture of H_3PO_4 and $H_2PO_4^-$?

16. If the total concentration of all phosphate in the previous question is 0.0500 M, what is the concentration of HPO_4^{2-} in the solution?

17. Use the pH-calculation.xlsx (or your own spreadsheet) to determine $p[H_3O^+]$ in a carbonate solution for HCO_3^-/CO_3^{2-} ratios of 0.05–0.95 in increments of 0.05 pH units. You may neglect activity effects.

18. Example 6.1 treated carbonic acid as a monoprotic weak acid. Verify that this assumption is valid.

19. A recipe to make phosphate-buffered saline is to dissolve the following reagents in 800 ml of distilled water:
    (a) 80 g NaCl
    (b) 2.0 g KCl
    (c) 14.4 g Na_2HPO_4
    (d) 2.4 g KH_2PO_4.
    Adjust pH to 7.4. Adjust volume to 1 l with additional distilled H_2O. Sterilize by autoclaving. What is the ionic strength of this solution?

CHAPTER 7

METAL–LIGAND COMPLEXATION

Learning Outcomes

- Identify and describe inorganic coordination complexes.
- Write equilibrium constant expressions for metal–ligand complexes.
- Calculate equilibrium concentrations using stepwise and cumulative equilibrium constants.
- Identify competing equilibria and predict the pH dependence on metal–ligand equilibria.
- Choose appropriate ions or ligands to use as titrants or masking reagents.

We encountered metal–ligand complexes in titrations in Chapter 3 and as light–absorbing entities in Chapter 4. In a complexometric titration, the strong 1:1 reaction between a ligand, such as EDTA, and a metal ion provides the basis for the titration. In UV/Vis absorption spectroscopy (Chapter 4), metal ions are converted to a strong absorber by forming a metal–ligand complex. In these analytical applications, the reaction must go to completion. The metal–ligand complexes used in these applications have large formation constants so that the complexation equilibrium lies far to the right. Published methods give appropriate ligands and experimental conditions to meet this criterion and to minimize interferences. Using complexes in spectroscopy has an additional

Basics of Analytical Chemistry and Chemical Equilibria: A Quantitative Approach, Second Edition. Brian M. Tissue.
© 2023 John Wiley & Sons, Inc. Published 2023 by John Wiley & Sons, Inc.
Companion Website: www.wiley.com/go/tissue/analyticalchemistry2e

criterion that only one specific metal–ligand complex can be present. A test portion that contains a mixture of different metal–ligand complexes cannot be related to a standard solution because the different complexes will have different molar absorption coefficients at a given wavelength.

In this chapter, we review metal–ligand terminology and discuss equilibrium conventions. We then use the equilibrium constants to calculate quantitative predictions. Many analytical protocols add complexing agents or pH buffers to remove interferences or prevent competing equilibria. Comparing the formation constants of different metal–ligand combinations allows an analyst to choose an appropriate reagent based on its relative strength in forming complexes. Section 7.3 introduces equilibria with multiple competing pathways. These examples illustrate the pH dependence of complexation equilibria, which can determine appropriate experimental conditions for an analytical method. Recognizing and predicting the formation of complexes and competing equilibria is important to understand the chemistry of aqueous solutions in environmental, industrial, and biochemical systems.

7.1   COMPLEX TERMINOLOGY

A metal–ligand or coordination complex is an association in solution between a metal ion and one or more surrounding ligands. A ligand has at least one pair of unbonded electrons that can form a coordinate covalent bond by overlapping the empty electronic orbitals of a metal ion. The process of complexation is the same as the neutralization of a weak base by adding an H^+. In the following example, I write the chemical formulas in a nonstandard way to explicitly show the lone pair of electrons on the ammonia that allow the proton from water to bond to form the ammonium ion. The H^+:NH_3 or NH_4^+ is analogous to a metal and a ligand.

$$:NH_3(aq) + H:O:H \rightleftharpoons H^+:NH_3(aq) +^- :O:H(aq)$$

Before discussing metal–ligand complexes further, let me clarify two types of reactions that appear similar to complexation, but are inherently different. Complexation is an equilibrium association in solution at a given temperature. Displacement and other types of reactions break covalent bonds and create new ones. These reactions may occur when the reactants are mixed or the reactants might need to be heated for some length of time to make the reaction proceed. The following nucleophilic substitution or S_N2 reaction is common in organic synthesis:

$$X + CH_3Z \rightarrow X \cdot\cdot CH_3 \cdot\cdot Z \rightarrow CH_3X + Z$$

In this reaction, the middle "product" is an intermediate or transition state. The dots indicate the breaking of the C–Z bond and creation of the C–X bond. This type of reaction proceeds by raising the temperature, letting the reaction run for some amount of time, and then cooling the reaction mixture to isolate the products. Although these types of reactions can involve equilibria between different forms of the reactants, we do not describe the final product mixture as an equilibrium. The conversion of reactants to products depends on intrinsic reactivity and the reaction conditions.

Redox (reduction and oxidation) reactions can also appear similar to complexation. Chapter 9 will discuss redox chemistry in detail, but here, we must be clear on the difference between oxidation state and charge and between metal–ligand complexes and metal oxyanions. Metal ions with an oxidation state of +1, +2, or +3 can exist as complexes in aqueous solution when suitable ligands are present. (Ti^{4+} and Zr^{4+} can form complexes with some ligands, but they are uncommon.) Many transition metals can exist in oxidation states greater than +3. However, the charge density on the metal is so high that metals in high oxidation states make strong covalent bonds with oxygen to form oxyanions. Some examples are listed in Table 7.1. These oxyanions exist as stable entities in aqueous solution and solids. Unlike metal–ligand complexes, the oxygen atoms are not displaced by other ligands. A chemical reaction that changes an oxyanion is a redox reaction that requires the transfer of electrons to change the oxidation state of the metal.

We specify the oxidation state of a metal using the Roman numeral with the chemical symbol as shown in Table 7.1. The charge of an oxyanion is indicated by the superscript on the oxyanion. The oxidation state of the metal is not explicit in the chemical formula, but we can determine it given that oxygen has an oxidation state of −2. Note that the oxidation state of the metal and the charge of a complex or oxyanion can be different, and we refer to them separately. For metal ions in solution, such as Ag^+ or Cu^{2+}, the oxidation state and the charge are the same.

The terminology to describe complexes is as follows. Some terms introduced in Chapter 3 are repeated here for completeness.

Adduct: A general term for a new chemical species formed from two others: $A + B \rightarrow A\text{-}B$. The original structures of A and B are maintained in the adduct.

Coordination complex: An adduct between a central metal ion and the surrounding ligands.

TABLE 7.1 Examples of Oxyanions

Oxyanion	Name	Metal Oxidation State	Charge of Oxyanion
VO^{2+}	Vanadyl ion	V(IV) or +4	2+
CrO_4^{2-}	Chromate ion	Cr(VI) or +6	2−
$Cr_2O_7^{2-}$	Dichromate ion	Cr(VI) or +6	2−
MnO_4^-	Permanganate ion	Mn(VII) or +7	1−
MoO_4^{2-}	Molybdate ion	Mo(VI) or +6	2−

Ligand: A neutral or an anionic species with at least one pair of unbonded electrons that can form a coordinate covalent bond with a metal ion.

Coordination number: The number of bonds formed with a central metal ion, usually 2, 4, or 6.

Chelate: A specific type of complex in which at least one ligand contains more than one atom with lone pairs of electrons so that it can form multiple bonds with the central metal ion.

-dentate: A suffix that refers to the number of bonds that a ligand can form with a metal, for example, ethylenediamine is bidentate because it has two nitrogen atoms with lone pairs of electrons to form two coordinate covalent bonds with a metal ion.

Bridging ligand: A polydentate ligand that can bond with more than one metal ion.

Labile: An adjective that indicates that a given ligand is easily displaced.

Figure 7.1 shows the structures of hexaamminecobalt(II) and silver cyanide complexes. In Figure 7.1, the ammonia ligands, $:NH_3$, are neutral and the complex has the same overall charge as the cobalt ion. We usually do not show the electron lone pairs, but we are familiar with most common ligands because they are also weak bases. In Figure 7.1, the overall charge of -1 arises from the sum of the $+1$ charge of the silver ion and the negative charges from the two cyanide ions, CN^-. Example 7.1 gives another example. Figure 7.1 shows the convention of writing the overall charge of a complex outside the square brackets that enclose the complex. I usually omit the square brackets when writing reactions to avoid confusion with the square brackets that represent equilibrium concentrations.

Example 7.1 Complex Conventions. Cu^{2+} can form a complex with four OH^- ions. What is the charge of the complex and what is the oxidation state of the metal in such a complex?

The convention for complexation equilibria is to write the reaction as the complex forming:

$$Cu^{2+}(aq) + 4OH^-(aq) \rightleftharpoons Cu(OH)_4^{2-}(aq)$$

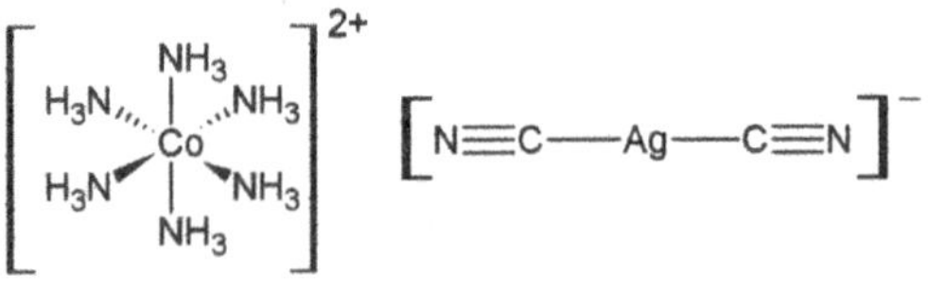

Figure 7.1 Examples of ligand–metal complexes: (left) hexaamminecobalt(II) complex and (right) silver cyanide complex.

Looking at the right side of the equilibrium, the complex has an overall charge of -2. As the four OH^- ions each have a charge of -1, the charge on the Cu^{2+} must remain $+2$. The charge and oxidation state are the same for a simple metal ion, so the oxidation state of copper does not change. It is $+2$ in the complex on the right side of the equilibrium and $+2$ on the left side of the equilibrium. (If it did change, this reaction would involve a redox reaction and the oxidation state of another species would also need to change.)

As a check, we can verify that the total charge on each side of the equilibrium is equal. The charge on the left side is

$$(+2) + (4 \times -1) = -2$$

which equals the charge of the complex on the right side.

Note that Cu^{2+} can also react with OH^- to form complexes with fewer OH^- or $Cu(OH)_2(s)$. The actual distribution of Cu^{2+} and $Cu–OH$ complexes will depend on the Cu^{2+} and OH^- concentrations. At low OH^- concentrations, $Cu(OH)^+$ might predominate. At some combinations of Cu^{2+} and OH^- concentrations, $Cu(OH)_2(s)$ will appear. At high OH^- concentration but low Cu^{2+} concentration, the equilibrium in Example 7.1 will predominate.

In the schematics of the aqueous complexes in Figure 7.1, we use a line analogous to any other covalent bond to indicate formation of a bond between the ligand and the metal ion. Unlike covalent bonds in hydrocarbons or oxyanions, the ligands can be *labile*, meaning they can be displaced by other ligands. In the following titration reaction, the ammonia ligands are more labile than EDTA and the EDTA displaces them to form $Cu(edta)^{2-}$.

$$Cu(NH_3)_4^{2+}(aq) + EDTA^{4-}(aq) \rightleftharpoons Cu(edta)^{2-}(aq) + 4NH_3(aq)$$

Note that we use EDTA as a common abbreviation, but ligand abbreviations in formulas are all lowercase. The relative lability of ligands is important in many applications. Oxygen, O_2, is transported from lungs or gills to other tissues by associating with a Fe^{2+} in the heme group of hemoglobin in blood. Carbon monoxide, CO, is poisonous because it binds more strongly to this Fe^{2+} than does oxygen. Carbon monoxide occupies a portion of the oxygen-binding sites and also causes an increase in the oxygen-binding strength at the other heme groups in hemoglobin.[1] Both of these effects reduce the transport of oxygen to cells leading to the poisoning. The consequences of using incorrect reagents or conditions are usually not as severe as poisoning, but failing to understand the competition between different ligands can lead to inaccurate measurements.

---

[1]Hemoglobin is a tetramer with four iron-containing heme cofactors.

7.2 COMPLEX EQUILIBRIA

7.2.1 Complexation Convention

When we prepare a solution, we label the bottle with the formal concentration of the solute, for example, 1 mM $Cu(NO_3)_2$. The actual equilibrium concentrations of neutral and ionic solutes in a solution will depend on the nature of the substance. Even unreactive metal ions will exist in solution as water or aqua complexes. Dissolving 1 mmol of $Cu(NO_3)_2$, a soluble blue solid, in 1 l of water produces the formal concentrations of 1 mM Cu^{2+} and 2 mM NO_3^-. The NO_3^-, being the anion of a strong acid, does not react with water and we treat it here as a spectator ion. Although we refer to the copper ion concentration, $c_{Cu^{2+}} = 1mM$, the Cu^{2+} exists in solution in different forms. The oxygen atom of water has two lone pairs of electrons that can form complexes with metal ions, in this case forming the hexaaqua complex, $Cu(H_2O)_6^{2+}$. Aqua complexes are fairly weak and other ligands can displace the water ligands to form a wide range of different complexes for any given metal ion. We will see in Section 7.3 that some fraction of the Cu^{2+} will react with water to form $Cu(OH)^+$. We should always think of metal ion complexes in aqueous solution as being in competitive equilibria with whatever ligands are present. Having said that, we still label our bottle as 1 mM $Cu(NO_3)_2$, knowing that the full amount of Cu^{2+} is available to act as a reagent.

Adding NH_3 to a copper nitrate solution results in a $Cu(NH_3)_4^{2+}$ complex because the NH_3 forms a complex with Cu^{2+} more strongly than does water. Assuming that the ammonia is present in a sufficient amount, the most stable product is the tetraamminediaquacopper(II) ion:

$$Cu(H_2O)_6^{2+}(aq) + 4NH_3(aq) \rightleftharpoons Cu(NH_3)_4(H_2O)_2^{2+}(aq) + 4H_2O(aq)$$

In practice, we simplify writing complexation equilibria by ignoring the water ligands, similar to the "net reactions" for acids and bases:

$$Cu^{2+}(aq) + 4NH_3(aq) \rightleftharpoons Cu(NH_3)_4^{2+}(aq)$$

Exceptions to this shorthand reaction are in writing complexes with different ligands when we want to show the water or coordination number explicitly, for example, $[Ru(bpy)_2(H_2O)_2](BF_4)_2$ or $[Ru(edta)H_2O]^-$. These complexes are less common in analytical chemistry but very common in catalysis and other applications or when working in a nonaqueous solution. In the first example, the square brackets enclose a Ru^{2+} complex with bipyridine (bpy) and water ligands. The tetrafluoroborate ion, BF_4^-, is the counteranion. It is not in the complex and it is shown outside of the brackets. In the second example of the ruthenium–edta complex, the water is shown explicitly to indicate that the Ru^{3+} has one opening in its coordination sphere to act as a site for catalysis.

Having just completed two chapters on weak acids and weak bases, do you notice anything about the copper–ammonia equilibrium that might be pH dependent? I hope that you said to yourself that ammonia is a base. What will happen to this equilibrium in an acidic solution? Adding acid to this solution will supply protons, H^+, which can compete with the Cu^{2+} for the ammonia. As written, the equilibrium reaction assumes that there is no competing equilibrium. We will look at such competitive or simultaneous equilibria in Section 7.3 and again in Chapter 8 for precipitates. The take-home message from this discussion is to always ask yourself, what else might be going on when you prepare or mix solutions?

Complexation equilibria are always written in the direction of the complex forming. The rules for writing equilibrium constants give an expression with the complex concentration over the metal and ligand concentrations, all raised to their stoichiometric coefficients. For the aforementioned reaction, the equilibrium constant expression is

$$\beta_4' = \frac{\left[Cu(NH_3)_4^{2+}\right]}{\left[Cu^{2+}\right][NH_3]^4}$$

The subscript "4" indicates that the equilibrium constant is a formation constant for a 4-coordinate complex. In earth sciences and other fields, the term *stability constant* is more common than *formation constant*. I mostly use the term formation constant, but the two terms are synonyms. Recall from Chapter 3 that we use the symbol β_n' for a cumulative complexation equilibrium and K_n' for stepwise equilibria. Cumulative and stepwise constants are the same for 1:1 complexes: $\beta_1' = K_1'$.

Table 7.2 lists cumulative formation constants for a selection of metal–ligand complexes.[2] The examples in the table are chosen for their analytical usefulness or because they are common in sample solutions, for example, Cl^- and OH^-. More complete tables are available in the reference sources.

7.2.2 Masking and Protecting Reagents

Many analytical methods add complexing agents to sample test portions to control aspects of the solution chemistry. The purpose is to ensure that sample matrix components do not affect the accuracy of measurements. Two common approaches are masking and protecting reagents:

Masking reagent: A reagent added to a test portion to prevent sample components from acting as interferences in an analytical method.
Protecting reagent: A reagent added to a test portion to prevent the analyte(s) from being lost.

[2]Adapted with permission from Speight, J. G., Ed. *Lange's Handbook of Chemistry*, 16th ed.; McGraw-Hill: New York, 2005.

TABLE 7.2 β **Values for Complexes**

Metal	Ligand	Formula	β_n
Aluminum	EDTA	$Al(edta)^-$	1.3×10^{16}
	Hydroxide	$Al(OH)^{2+}$	1.9×10^{9}
	Hydroxide	$Al(OH)_4^-$	1.1×10^{33}
	Oxalate	$Al(C_2O_4)_3^{3-}$	2×10^{16}
Cadmium	Ammine	$Cd(NH_3)_4^{2+}$	1.3×10^{7}
	Ethylenediamine	$Cd(en)_4^{2+}$	1.2×10^{12}
	Cyanide	$Cd(CN)_4^{2-}$	6.0×10^{18}
Calcium	EDTA	$Ca(edta)^{2-}$	1×10^{11}
Chromium(II)	EDTA	$Cr(edta)^{2-}$	4×10^{13}
Chromium(III)	EDTA	$Cr(edta)^-$	1×10^{23}
	Hydroxide	$Cr(OH)_4^-$	8×10^{29}
Cobalt(II)	Ethylenediamine	$Co(en)_3^{2+}$	8.7×10^{13}
	EDTA	$Co(edta)^{2-}$	2.0×10^{16}
	Thiocyanate	$Co(SCN)_4^{2-}$	1×10^{3}
Cobalt(III)	Ammonia	$Co(NH_3)_6^{3+}$	2×10^{5}
	Ethylenediamine	$Co(en)_3^{3+}$	4.9×10^{48}
	EDTA	$Co(edta)^-$	2.0×10^{16}
	Oxalate	$Co(C_2O_4)_3^{3-}$	5×10^{9}
Copper(I)	Chloride	$Cu(Cl)_2^-$	3×10^{5}
	Cyanide	$Cu(CN)_2^-$	1×10^{24}
	Cyanide	$Cu(CN)_4^{3-}$	2.0×10^{30}
Copper(II)	Ethylenediamine	$Cu(en)_2^{2+}$	1×10^{20}
	EDTA	$Cu(edta)^{2-}$	5×10^{18}
	Ammine	$Cu(NH_3)_4^{2+}$	2.1×10^{13}
	Oxalate	$Cu(C_2O_4)_2^{2-}$	3×10^{8}
Gold(I)	Cyanide	$Au(CN)_2^-$	2×10^{38}
Iron(II)	Cyanide	$Fe(CN)_6^{4-}$	1×10^{35}
	Ethylenediamine	$Fe(en)_3^{2+}$	5.0×10^{9}
	EDTA	$Fe(edta)^{2-}$	2.1×10^{14}
	Oxalate	$Fe(C_2O_4)_3^{4-}$	1.7×10^{5}
	Phenanthroline	$Fe(phen)_3^{2+}$	2×10^{21}
Iron(III)	Cyanide	$Fe(CN)_6^{3-}$	1×10^{42}
	EDTA	$Fe(edta)^-$	1.7×10^{24}
	Oxalate	$Fe(C_2O_4)_3^{3-}$	2×10^{20}
	Phenanthroline	$Fe(phen)_3^{3+}$	3×10^{23}
	Thiocyanate	$Fe(SCN)_2^+$	2.3×10^{3}
Lead	EDTA	$Pb(edta)^{2-}$	2×10^{18}
	Hydroxide	$Pb(OH)_3^-$	3.8×10^{14}
	Iodide	PbI_4^{2-}	3.0×10^{4}
	Oxalate	$Pb(C_2O_4)_2^{2-}$	3.5×10^{6}
Magnesium	EDTA	$Mg(edta)^{2-}$	4.4×10^{8}
Manganese	EDTA	$Mn(edta)^{2-}$	6×10^{13}
Mercury(II)	Chloride	$Hg(Cl)_4^{2-}$	1.2×10^{15}
	Cyanide	$Hg(CN)_4^{2-}$	3×10^{41}
	Ethylenediamine	$Hg(en)_2^{2+}$	2×10^{23}
	EDTA	$Hg(edta)^{2-}$	6.3×10^{21}

(*Continued*)

TABLE 7.2 (*Continued*)

Metal	Ligand	Formula	β_n
	Iodide	HgI_4^{2-}	6.8×10^{29}
	Oxalate	$Hg(C_2O_4)_2^{2-}$	9.5×10^6
Nickel	Ammonia	$Ni(NH_3)_6^{2+}$	5.5×10^8
	Cyanide	$Ni(CN)_4^{2-}$	2×10^{31}
	Ethylenediamine	$Ni(en)_3^{2+}$	2.1×10^{18}
	EDTA	$Ni(edta)^{2-}$	3.6×10^{18}
	Oxalate	$Ni(C_2O_4)_3^{4-}$	3×10^8
Platinum	Ammonia	$Pt(NH_3)_6^{2+}$	2×10^{35}
	Chloride	$PtCl_4^{2-}$	1×10^{16}
Silver	Ammonia	$Ag(NH_3)_2^{+}$	1.1×10^7
	Chloride	$AgCl_2^{-}$	1×10^2
	Cyanide	$Ag(CN)_2^{-}$	1.3×10^{21}
	Ethylenediamine	$Ag(en)_2^{+}$	5.0×10^7
	EDTA	$Ag(edta)^{3-}$	2.1×10^7
	Thiocyanate	$Ag(SCN)_4^{3-}$	1.2×10^{10}
Strontium	EDTA	$Sr(edta)^{2-}$	6.3×10^8
Zinc	Ammonia	$Zn(NH_3)_4^{2+}$	2.9×10^9
	Cyanide	$Zn(CN)_4^{2-}$	5×10^{16}
	Ethylenediamine	$Zn(en)_3^{2+}$	1.3×10^{14}
	EDTA	$Zn(edta)^{2-}$	3×10^{16}
	Hydroxide	$Zn(OH)_4^{2-}$	4.6×10^{17}
	Oxalate	$Zn(C_2O_4)_3^{4-}$	1.4×10^8

These auxiliary reagents might affect the analyte or target an interference, and a given procedure might use both masking and protecting reagents. Although an H^+ can be considered an auxiliary reagent, controlling pH is so common that buffering the test portion to an optimum pH is usually treated separately. The use of auxiliary reagents is best illustrated with a few examples.

An ion-selective electrode (ISE) generates an electrical signal in response to a specific ion in solution. If that ion is in a different form, the ISE does not sense it. The F^--selective electrode responds to free F^-, and fluoride ions that are protonated or bound in a complex are not sensed. Fluoride ion forms complexes with metal ions such as Al^{3+} and Fe^{3+}, e.g., AlF_2^+ and FeF_2^+. The F^--selective electrode does not respond to these complexes and the ISE signal will be lower than it should be if these metal ions are present in the sample solution. To prevent this problem, CDTA (cyclohexanediaminetetraacetic acid) is added to the test solution to complex the metal ions. We say that these interfering metal ions are masked by the CDTA complexing agent. In practice, the chelating CDTA is included in total ionic strength adjustment buffer (TISAB) that is added in an equal ratio to both standard and test solutions when using an F^--selective electrode.[3]

---

[3]The TISAB includes salt to equalize ionic strength in all solutions and a pH buffer so the fluoride ion is not protonated.

An electrochemical measurement of gold will be affected by the presence of silver. Example 7.2 provides a quantitative calculation that illustrates an example of "masking" the interfering silver from the gold analysis. Because of the large formation constant, ammonia serves as a common masking agent for silver ion.[4] As the calculation in the next example will show, masking and protecting reagents are usually added in excess to drive the equilibrium in the desired direction.

Example 7.2 Complex Stoichiometry. Find the fraction of Ag^+ that remains "free" in solution when a stoichiometric amount of ammonia is added to 5.0×10^{-4} M Ag^+ to form the 1:2 metal-to-ligand complex:

$$Ag^+(aq) + 2NH_3(aq) \rightleftharpoons Ag(NH_3)_2^+(aq) \quad \beta_2 = 1.1 \times 10^7$$

The formal concentrations are $c_{Ag} = 5.0 \times 10^{-4}$ M, $c_{NH_3} = 1.0 \times 10^{-3}$ M, and the equilibrium expression is

$$\beta_2' = \frac{\left[Ag(NH_3)_2^+\right]}{\left[Ag^+\right][NH_3]^2} = 1.1 \times 10^7$$

In this equilibrium expression, there is one ion in the numerator and one ion in the denominator. The activity coefficients will get canceled to some extent and it is not unreasonable to assume that $\beta_2' = \beta_2$.

We want to solve for the ratio of $[Ag^+]$ to c_{Ag}. To find $[Ag^+]$ from the β_2' expression, we must replace $[NH_3]$ and $[Ag(NH_3)_2^+]$ with quantities that we know, that is, with the formal concentrations of silver and ammonia, respectively. As the total amount of a substance must equal the sum of all of the forms that are present, we can always write mass-balance relationships to substitute for equilibrium concentrations in an equilibrium expression. For this problem, we have the following mass balance for the metal and the ligand, respectively:

$$c_{Ag} = \left[Ag^+\right] + \left[Ag(NH_3)_2^+\right]$$

$$c_{NH_3} = [NH_3] + 2\left[Ag(NH_3)_2^+\right]$$

Note that there is a factor of 2 in front of the $[Ag(NH_3)_2^+]$ because each complex contains two NH_3 ligands. Rearranging the first expression gives

$$\left[Ag(NH_3)_2^+\right] = c_{Ag} - \left[Ag^+\right]$$

---

[4]Cheng, K. L. "Increasing selectivity of analytical reactions by masking," *Anal. Chem.* **1961**, *33*, 783–790.

Rearranging the second expression and substituting in the first mass-balance expression gives

$$[NH_3] = c_{NH_3} - 2\left[Ag(NH_3)_2^+\right]$$

$$[NH_3] = c_{NH_3} - 2\left(c_{Ag} - \left[Ag^+\right]\right)$$

$$[NH_3] = c_{NH_3} - 2c_{Ag} + 2\left[Ag^+\right]$$

Substituting into the equilibrium expression gives

$$\beta_2' = \frac{c_{Ag} - [Ag^+]}{[Ag^+](c_{NH_3} - 2c_{Ag} + 2[Ag^+])^2} = 1.1 \times 10^7$$

Now we have an equation with only one unknown, but rearranging will produce a cubic expression. Cubic expressions are solvable, but let us think if we can simplify this equation. For this particular situation of adding a stoichiometric amount of ammonia to silver ion, we can see that $c_{NH_3} - 2c_{Ag}$ will be zero. Then the denominator is $[Ag^+](2[Ag^+])^2$.

$$\beta_2' = \frac{c_{Ag} - [Ag^+]}{[Ag^+](2[Ag^+])^2} = 1.1 \times 10^7$$

Now inserting the formal concentrations that we know gives

$$1.1 \times 10^7 = \frac{(5.0 \times 10^{-4} \text{ M} - [Ag^+])}{4[Ag^+]^3}$$

To solve this expression, I neglect the $[Ag^+]$ in the numerator and solve by successive approximations. The result is $[Ag^+] = 1.9 \times 10^{-4}$ M. Answering the original question about the fraction of free metal,

$$\alpha_{Ag} = \frac{1.9 \times 10^{-4} \text{ M}}{5.0 \times 10^{-4} \text{ M}} = 0.38 \quad (\text{or } 38\%)$$

Adding the stoichiometric amount of ammonia does not mask all of the silver ions. Adding an excess of ammonia drives the reaction in the forward direction and greatly reduces the concentration of free Ag^+ so that it does not interfere in a measurement of Au^+.

An example of a protecting reagent is a weak complexing agent that prevents metal ions from precipitating as insoluble hydroxides at high pH. A common case is to use an ammonia buffer when determining metal ions by EDTA titration. Recall from Chapter 3 that EDTA titrations are performed at high pH so the ETDA is not protonated. Some metal ions can form solid precipitates at higher pH. The ammonia buffer raises the pH and forms weak complexes with the metal ion, preventing precipitation. The ammonia ligands are easily displaced by the much stronger EDTA chelator as the titration proceeds.

In all of these cases, auxiliary reagents are chosen based on the relative strength of formation constants to provide discrimination to get an accurate measurement of the target analyte. Masking reagents must mask interferences without binding strongly to the analyte. Likewise, protecting reagents must form a complex with analyte ions that is weak enough to not interfere in the measurement. A similar situation to the relative strength of formation constants is in the speed or kinetics of reactions. A protocol that uses UV/Vis absorption of a complex to measure an analyte often specifies waiting some amount of time after adding reagents to the test solution for the color to fully develop. In a titration based on a metal–EDTA complex, waiting after each titrant addition is not feasible. If the metal–EDTA complex does not form rapidly, attempting a titration will not produce a clear end point. In cases when the titration reaction is slow or a competing equilibrium is present, we used the back-titration approach to obtain accurate results. Excess complexing agent is added to a test solution and sufficient time is allowed for all analyte to be totally complexed. The excess complexing agent is then back-titrated with a titrant that gives a clear endpoint.

You-Try-It 7.A
The free-metal worksheet in you-try-it-07.xlsx sets up the silver-ammonia equilibrium with different amounts of ammonia. Repeat the calculation for [Ag$^+$], i.e., "free silver," using Example 7.2 as a guide.

7.3 COMPETING EQUILIBRIA

For salts that contain both a strong cation and strong anion, we think of the ions as being "free" in solution and simply hydrated with water molecules. Ions that are not strong electrolytes will participate in competing or simultaneous equilibria in aqueous solution. Metal ions will react with water to form hydroxide complexes, making the resulting solution acidic. Similarly, ligands can accept a proton and act as a weak base. Adding a soluble salt containing weak electrolytes to distilled water can alter the solution pH. Likewise, the p[H$_3$O$^+$] of a solution will affect complexation equilibria by affecting the availability of the "free" metal ion and the availability of the "free" ligand. We have already discussed this competition concept with ion-exchange separations in Chapter 2. In this sample cleanup process, a class of analyte ions are bound to an immobilizing stationary phase and other species are rinsed away. Mobile phase conditions are then changed, usually by changing pH, to elute and collect the target analyte ions.

7.3.1 Metal Hydrolysis Reaction

Most metal ions in soluble salts will react with H_2O when dissolved in pure water and alter the pH. The strong-electrolyte metal ions that do not react with water to any appreciable extent are those that are the alkali metal ions in the first column of the periodic table and Sr^{2+} and Ba^{2+}. Ca^{2+} is a borderline case in that it reacts to some extent, but it will not affect the solution pH unless the concentration is relatively high. In the following discussion I assume that the anion of the soluble salt is a strong electrolyte and thus a spectator ion. If the anion is a weak base, the relative strength of the anion versus the cation will determine the effect on pH.

The hydrolysis reaction removes a hydroxide ion from a water molecule to form the metal–hydroxide complex and makes the solution acidic:

$$M^{m+}(aq) + 2H_2O \rightleftharpoons M(OH)^{(m-1)+}(aq) + H_3O^+(aq)$$

The extent to which a metal ion reacts with water depends on the metal charge and size. Metals with a higher positive charge will have a greater attraction to an electronegative oxygen. For metals of equal charge, the metal with the smaller ionic radius will be able to make a closer approach to other ions and thus have a stronger interaction. These rules are somewhat simplified, and there are other factors that cause exceptions in the trends.

As shown in Table 7.3, the metal ions that are common in strong electrolytes (left column) are very weak acids and do not affect solution pH when dissolved in water. Divalent, trivalent, and tetravalent transition metals have a correspondingly stronger interaction. They will react with water to a greater extent and thus have a larger K_a value. The data in Table 7.3 provides a quantitative measure to classify the acidity of metal cations.[5] In Table 7.3, there is a lack of metal ions with oxidation states greater than +4 because these metals exist as polyatomic oxyanions.

Examples 7.3 and 7.4 show how to determine the pH change and ion concentrations due to a metal hydrolysis reaction. Note that the metal hydrolysis reaction is a complexation equilibrium involving a metal cation and it is not

TABLE 7.3 Acidity due to Metal Ion Hydrolysis

Weakly Acidic	pK_a	Acidic	pK_a	Very Acidic	pK_a
Li^+	13.8	Co^{2+}	9.7	Ti^{3+}	2.55
Ba^{2+}	13.4	Cu^{2+}	7.5	Zr^{4+}	-0.32
Sr^{2+}	13.2	Co^{3+}	6.6		
Ca^{2+}	12.6	Cr^{2+}	5.5		
Mg^{2+}	11.4	Al^{3+}	5.0		

[5]Data for 25°C, adapted with permission from Speight, J. G., Ed. *Lange's Handbook of Chemistry*, 16th ed.; McGraw-Hill: New York, 2005.

the same process as adding a reactive metal to water. In this type of reaction, a redox reaction occurs with a transfer of electrons, for example:

$$Mg(s) + 2H_2O \rightarrow Mg^{2+}(aq) + H_2(g) + 2OH^-(aq)$$

There is a change in solution pH caused by this redox reaction, but this type of reaction is fundamentally different from the following metal hydrolysis equilibrium.

Example 7.3 Metal Hydrolysis Reaction. Predict the solution pH when 0.0010 mol of $Cu(NO_3)_2$ is dissolved in 1.0 l of water. You may assume that the water has been purged of CO_2 and has a starting pH of 7.0.

The solution contains formal concentrations of 0.0010 M Cu^{2+} and 0.0020 M NO_3^-. The NO_3^- ion does not react with water and the metal hydrolysis between Cu^{2+} and water is the only reaction affecting pH:

$$Cu^{2+}(aq) + 2H_2O \rightleftharpoons Cu(OH)^+(aq) + H_3O^+(aq)$$

The pK_a in Table 7.3 gives $K_a = 3.2 \times 10^{-8}$, so the equilibrium constant expression is

$$3.2 \times 10^{-8} = \frac{[Cu(OH)^+][H_3O^+]}{[Cu^{2+}]}$$

As in other weak acid problems, $[Cu(OH)^+] = [H_3O^+]$ and we will neglect $[H_3O^+]$ compared to $c_{Cu^{2+}}$. Our expression then simplifies to

$$3.2 \times 10^{-8} = \frac{[H_3O^+]^2}{0.0010\,M}$$

Solving and taking the negative log gives

$$[H_3O^+] = 5.6 \times 10^{-6}\,M$$
$$p[H_3O^+] = 5.2$$

Even for this moderate metal ion concentration, using it as a reagent in an analytical procedure might require a buffer to control the solution pH.

Example 7.4 Quantitative Metal Hydrolysis Reaction. What are the predicted equilibrium concentrations of the copper species in Example 7.3?

As the only source of $Cu(OH)^+$ was the reaction of Cu^{2+} and water,

$$[Cu(OH)^+] = [H_3O^+] = 5.6 \times 10^{-6}\,M$$

Assuming that there are no other copper complexes,

$$[Cu^{2+}] = 0.001\,M - 5.6 \times 10^{-6}\,M = 0.00099\,M$$

Although the hydrolysis reaction does affect the solution pH, only a small fraction of the Cu^{2+} actually reacts.

You-Try-It 7.B

The metal–hydrolysis worksheet in you-try-it-07.xlsx lists several different metal ion solutions. Predict the pH for each solution.

7.3.2 Ligand Protonation

Because ligands have functional groups with unbonded pairs of electrons, for example, carboxylate ($-COO^-$), amine ($-NH_2$), and sulfide ($-S^-$), they can also accept protons. The acid–base properties of ligands affect complexation equilibria through the following types of reaction:

$$L + H_2O \rightleftharpoons LH^+ + OH^-$$

$$L^- + H_2O \rightleftharpoons LH + OH^-$$

where the L and L^- ligands are generic examples. Ligands can have other charges, the only requirement is that the charge should be balanced in the overall reaction. A good example of this chemistry is EDTA, which is an important reagent in analytical chemistry and other applications. It is added to many consumer products to complex metal ions. As examples, it forms complexes with Ca^{2+} and Mg^{2+} to prevent soap scum formation when using shampoo and detergents and with Fe^{2+} and Zn^{2+} to prevent spoilage and bad tastes in canned foods.[6] EDTA forms six bonds with metal ions and forms 1:1 complexes with large formation constants (see values in Table 3.6 or Table 7.2). These properties make it very useful as a titrant or masking agent in complexometric titrations. Figure 7.2 shows the alpha plots for EDTA, abbreviated as H_4Y, H_3Y^-, etc. Note that the charges are not shown on the labels in the figure. EDTA titrations are usually done in solutions buffered to alkaline pH to control the protonation of the EDTA titrant. At neutral pH, EDTA exists in protonated forms, which means that the metal ion being titrated must compete with protons to complex with EDTA.

The end point of a complexometric titration can be determined with a visual indicator or by an electrode that is sensitive to metal concentration. Figure 7.3, repeated from Figure 3.5, shows the structure of the calmagite indicator. This molecule produces a red color in solution when bound with Ca^{2+} or Mg^{2+} and blue when unbound.[7] Adding a small amount of indicator to a test portion turns the solution a red color. The solution remains red as EDTA titrant is

---

[6]Soap scum is a precipitate of metal ions and long-chain fatty acids used in soaps.
[7]Refer to Chapter 4 to think about how the energy levels change to produce this color change.

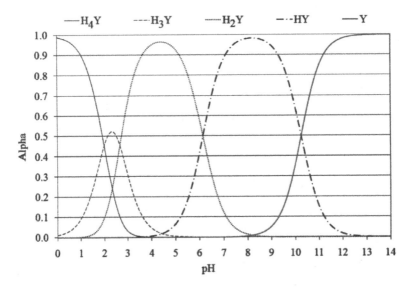

Figure 7.2 Alpha plots for the different forms of EDTA.

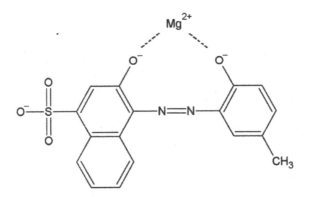

Figure 7.3 Calmagite indicator bound to Mg^{2+}.

added, that is, until one equivalent of EDTA has been added. At this point, all of the metal ions form a complex with EDTA, leaving the indicator unbound, causing the color change to indicate the end point. A requirement for any complexometric indicator is that it should have a formation constant for the metal that is smaller than the formation constant of the EDTA–metal complex. Otherwise, the EDTA cannot displace the indicator.

7.3.3 pH Dependence of Complexes

Owing to the possibility of metal hydrolysis and ligand protonation reactions, formation of a metal–ligand complex will be pH dependent. Overall, we can write competing equilibria in the following manner:

$$M^{m+}(aq) \;+\; L(aq) \;\rightleftharpoons\; ML^{m+}(aq)$$
$$+ \qquad\qquad\quad +$$
$$OH^-(aq) \qquad H^+(aq)$$
$$\Updownarrow \qquad\qquad\quad \Updownarrow$$
$$M(OH)^{(m-1)+} \quad LH^+(aq)$$

where the $\Updownarrow$ symbol has the same meaning as the $\rightleftharpoons$ symbol. This example uses a generic neutral ligand, but the ligand can be negatively charged with the charge of the complexes adjusted appropriately.

How pH will affect complex formation will depend on the pK_a of the metal and the pK_b of the ligand. As an example, the availability of metals, such as Cd^{2+}, in water and soil is greatly affected by pH. Natural organic matter contains carboxylic acids that act as ligands to bind and immobilize metals to soil. As water in contact with the soil becomes more acidic, a proton will displace the metal ion from the organic matter to which it is bound, releasing the metal into solution. At higher pH, that is, higher [OH⁻] concentration, the formation of hydroxide complexes can increase the total amount of soluble forms of a metal in solution. As we will see in Chapter 8, alkaline conditions can produce metal hydroxide precipitates, making the pH dependence of metal species in solution difficult to predict. Overall, any given metal will have a complicated bioavailability in groundwater due to the details of the competing equilibria. As a practical example, soils contaminated with heavy metals (Cd, Cu, Ni, Pb, and Zn are common industrial pollutants) can be remediated by growing and removing plants that concentrate the metal. This process is called *phytoremediation* and is of interest because it has a lower cost compared to landfilling contaminated soil. Scientists at the US Department of Agriculture (USDA) and the University of Maryland found a greater uptake of cadmium by alpine pennycress (*Thlaspi caerulescens*) when the soil pH was decreased from 7 to 6.[8] The lower pH increases the bioavailability of the cadmium, which in this case improves the phytoremediation of the site. In the next chapter we'll see that lowering pH is usually problematic by releasing heavy metals into drinking water supplies.

Discrete calculations of competing equilibria can become quite complicated. Rather than developing all of the simultaneous equations for such cases, we will make use of alpha plots to predict pH effects. We can combine alpha plots for metal hydrolysis and ligand protonation to determine the optimal pH for formation of metal–ligand complexes. We will do so for the example of metal–EDTA complexes, which will explain why test portions in EDTA titrations are buffered to different pH values when measuring different metal ions.

[8]See USDA press release: https://www.ars.usda.gov/news-events/news/research-news/2005/acidifying-soil-helps-plant-remove-cadmium-zinc-metals; accessed August 2022.

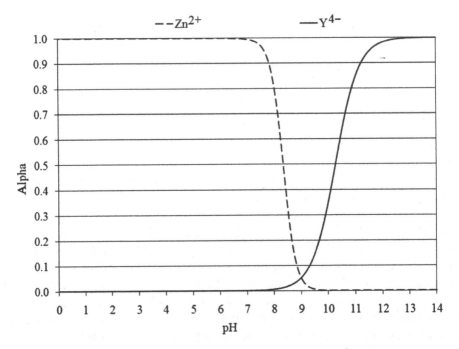

Figure 7.4   Alpha plots to find optimum pH for $Zn(edta)^{2-}$ titration.

TABLE 7.4   **Predicted Optimal pH for Metal–EDTA Titrations**

Metal Ion	Optimal pH
Pb^{2+}	8.3
Cu^{2+}	9.0
Zn^{2+}	9.0
Fe^{2+}	9.5
Ni^{2+}	9.5
Mg^{2+}	11
Ca^{2+}	12

Figure 7.4 shows the alpha plot for unprotonated EDTA overlaid with the alpha plot for the free Zn^{2+} ion as a function of pH. The point where the two curves cross is the optimal pH for forming $Zn(edta)^{2-}$ complexes. When titrating Zn^{2+} with EDTA, we predict the sample solution should be buffered to a pH of approximately 9. At lower pH, protons compete for EDTA. At higher pH, hydroxide ion complexes compete for Zn^{2+}. Table 7.4 lists the predicted optimal pH for EDTA titrations of different metal ions.

In practice, metal–EDTA titrations are not always done at the optimal pH. The alpha plots do not show the problem of precipitation of metal ions at a high pH, which depends on the metal ion concentration in addition to $[OH^-]$. It is common to perform EDTA titrations of transition metal ions, which have very large metal–EDTA β_1 values, at pH values lower than those in Table 7.4.

For example, a validated procedure titrates Cd^{2+}, Pb^{2+}, and Zn^{2+} at a solution pH of 5.5–6.[9] Adjusting the pH can also provide some selectivity for a specific metal ion, with titrations at lower pH being less affected by interference by Ca^{2+} or Mg^{2+}. As discussed previously, appropriate masking agents can also be added to discriminate against interference when using EDTA to titrate a specific metal analyte.

To obtain some predictive ability in competitive equilibria, we can calculate a conditional or effective formation constant, β'_{eff}. This constant depends on experimental conditions, which must be stated with the constant for it to have meaning. At low pH, the metal exists as the free metal ion and precipitation will not occur. However, most of the EDTA will have protons attached to one or two of the carboxylic acid groups. These metal ions titrate well with EDTA at lower than optimal pH because they have large β_1 values. Even though the EDTA will be partly protonated, the metal ion can displace the protons because of the tight binding of the metal–EDTA complex. You can rationalize this process by Le Chatelier's principle. As the added metal ion forms complexes with the very small amount of free EDTA, protonated EDTA molecules will release their proton to maintain equilibrium.

In calculating a conditional formation constant, β'_{eff}, I drop the other subscripts for simplicity. When equilibria get this complicated, the constants are useful to an order of magnitude at best. For a 1:1 complex, β'_{eff} is the β'_1 constant times the alpha fraction of each species:

$$\beta'_{\text{eff}} = \alpha_M \alpha_L \beta_1$$

For the Zn–EDTA titration example at a pH of 6, the alpha value for EDTA is 2×10^{-5} and the alpha value for Zn^{2+} is 1. The Zn–EDTA β_1 is 3×10^{16}. The effective formation constant, β'_{eff}, is

$$\beta'_{\text{eff}} = (2 \times 10^{-5})(1.0)(3 \times 10^{16}) = 6 \times 10^{11}$$

Even for the diminished fraction of unprotonated EDTA at a pH of 6, the effective formation constant for the Zn–EDTA complex is still large enough to allow complete reaction during the titration.

7.4 STEPWISE COMPLEXATION

In most of the previous examples, we looked at cases in which there was one dominant complex. Some metals that bind multiple ligands will do so stepwise. The following equations show an example for a metal that forms mono-, bi-, tri-, and tetradentate complexes. For clarity in these equilibria, I have used a neutral ligand and I have left off the charge of the metal ion. For a specific

[9]Suzuki, T.; Tiwari, D.; Hioki, A. "Precise chelatometric titrations of zinc, cadmium, and lead with molecular spectroscopy," *Anal. Sci.* **2007**, *23*, 1215–1220.

example, the overall charge of each complex will be the sum of the metal charge and the charge on the ligands. K_n values are given numerical subscripts to indicate the equilibrium to which they refer:

$$M + L \rightleftharpoons ML \quad K_1$$

$$ML + L \rightleftharpoons ML_2 \quad K_2$$

$$ML_2 + L \rightleftharpoons ML_3 \quad K_3$$

$$ML_3 + L \rightleftharpoons ML_4 \quad K_4$$

The relationships between stepwise, K_n, and cumulative, β_n, formation constants are

$$\beta_1 = K_1$$
$$\beta_2 = K_1 K_2$$
$$\beta_3 = K_1 K_2 K_3$$

and so on (see Example 7.5). These relationships are verified easily by multiplying the equilibrium constant expressions and simplifying:

$$K_1' K_2' = \frac{[ML]}{[M][L]} \frac{[ML_2]}{[ML][L]} = \frac{[ML_2]}{[M][L]^2} = \beta_2'$$

Example 7.5 Stepwise K_n Constants.   Formation or stability constants are often tabulated as $\log \beta_n$ values. Determine the stepwise K_n values for zinc oxalate complexes given the following listings: $\log \beta_1 = 4.89$, $\log \beta_2 = 7.60$, and $\log \beta_3 = 8.15$.

As there are three β_n values, we know that zinc and oxalate can make 1:1, 1:2, and 1:3 complexes. $K_1 = \beta_1$ so we merely need to take the inverse log to find K_n:

$$K_1 = \beta_1 = 10^{4.89} = 7.8 \times 10^4$$

The simplest way to find the other constants is to work with the log values and then take the inverse log:

$$\log K_2 = \log \beta_2 - \log \beta_1 = 7.60 - 4.89 = 2.71$$
$$K_2 = 10^{2.71} = 5.1 \times 10^2$$

K_3 is found in the same way:

$$\log K_3 = \log \beta_3 - \log \beta_2 = 8.15 - 7.60 = 0.55$$
$$K_2 = 10^{0.55} = 3.5$$

Whether a given metal ion and ligand will form stepwise is not predictable. To determine what might happen in any specific case, simply refer to a table of K_n values. The complexes that form stepwise will have separate K_n values for each possible metal–ligand combination that are well separated from each other.

The complex or complexes that will be present in appreciable equilibrium concentrations will depend on the ligand concentration, and we can generate alpha fraction plots to visualize how the dominant complex(es) change as the ligand concentration changes. The alpha value equations for a generic tetradentate complex are

$$\alpha_M = \frac{1}{1 + K_1[L] + K_1K_2[L]^2 + K_1K_2K_3[L]^3 + K_1K_2K_3K_4[L]^4}$$

$$\alpha_{ML} = \frac{K_1[L]}{1 + K_1[L]^2 + K_1K_2[L]^2 + K_1K_2K_3[L]^3 + K_1K_2K_3K_4[L]^4}$$

$$\alpha_{ML_2} = \frac{K_1K_2[L]^2}{1 + K_1[L] + K_1K_2[L]^2 + K_1K_2K_3[L]^3 + K_1K_2K_3K_4[L]^4} \qquad (7.1)$$

$$\alpha_{ML_3} = \frac{K_1K_2K_3[L]^3}{1 + K_1[L] + K_1K_2[L]^2 + K_1K_2K_3[L]^3 + K_1K_2K_3K_4[L]^4}$$

$$\alpha_{ML_4} = \frac{K_1K_2K_3K_4[L]^4}{1 + K_1[L] + K_1K_2[L]^2 + K_1K_2K_3[L]^3 + K_1K_2K_3K_4[L]^4}$$

Analogous to the polyprotic acids, the alpha fraction equations have a common denominator and only the numerator changes for each form of the metal ion. The equations for metal ions with fewer or greater than four ligands follow the same pattern. Table 7.5 lists stepwise formation constants for several ligands that can be used to calculate the metal–ligand alpha plots.[10]

TABLE 7.5 Selected Stepwise K_n Values for Metal–Ligand Complexes

Complex	K_1	K_2	K_3	K_4
Ammonia				
Ag^+	2.0×10^3	8.3×10^3		
Cu^{2+}	9.8×10^3	2.2×10^3	5.4×10^2	9.3×10^1
Zn^{2+}	1.5×10^2	1.8×10^2	2.0×10^2	9.1×10^1
Chloride				
Ag^+	5.0×10^3	8.3×10^1	6.0	
Hydroxide				
Fe^{2+}	3.2×10^4	7.9×10^2	4.0×10^2	
Fe^{3+}	7.41×10^{11}	2.00×10^9	3.16×10^8	
Pb^{2+}	2.5×10^6	3.2×10^4	1×10^3	
Iodide				
Ag^+	3.8×10^6	1.4×10^5	8.7×10^1	

[10]Adapted with permission from Speight, J. G., Ed. *Lange's Handbook of Chemistry*, 16th ed.; McGraw-Hill: New York, 2005.

Let us illustrate the use of these alpha plots with an example. Consider the dependence of iron hydroxide complexes as a function of pH. The presence of iron is a common issue in many surface and underground water systems. From Table 7.5, the K_n values are

$$K_1 = 7.41 \times 10^{11} \quad K_2 = 2.00 \times 10^9 \quad K_3 = 3.16 \times 10^8$$

Alpha plots on linear and log scales are shown in Figures 7.5 and 7.6, respectively. Note that the plot labels use generic ML notation, but the equilibrium

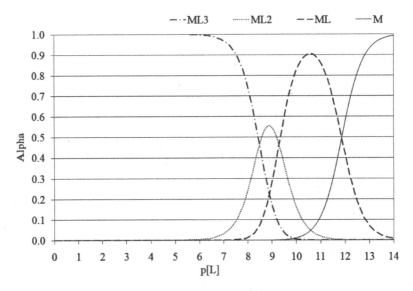

Figure 7.5   Alpha plots for iron hydroxide complexes.

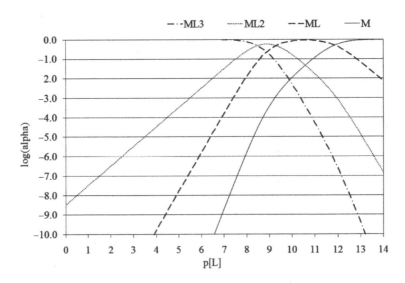

Figure 7.6   Log(alpha) plots for iron hydroxide complexes.

species are $Fe(OH)_3$, $Fe(OH)_2^+$, $FeOH^{2+}$, and Fe^{3+} and the x-axis label is pOH. Because it is easy to confuse pOH and pH values, you might want to routinely write pH values or hydroxide concentrations below the pOH values on the x-axis.

The plots show that at very low hydroxide concentrations, that is, pOH >12 or pH <2, the iron does not complex with hydroxide strongly. Very few natural water systems have pH this low unless they are polluted from mining waste. At higher hydroxide concentrations, $FeOH^{2+}$ forms (the dashed line), and it is the dominant form at pOH values between 12 and 9.5. As hydroxide concentration increases further, the complexes containing two and three hydroxide ions predominate. (Due to the closeness of the K_2 and K_3 values, it would be difficult to isolate the $Fe(OH)_2^+$ species.) The plots show that iron will exist in aqueous solutions near neutral pH as $Fe(OH)_2^+$ and $Fe(OH)_3$ complexes. Reality is always more complicated. Even neglecting interactions with organic matter, iron in natural water systems will form iron oxyhydroxide colloids, which are solid particles that are small enough to stay suspended in solution. Our plots are a starting point for simple systems, but natural water systems can get quite complicated depending on the details of the local soils and geology.

Examples 7.6 and 7.7 illustrate the usefulness of alpha and log(alpha) plots to give us a graphical means to quickly determine the form of complexes in aqueous solution. The linear alpha plots show us predominant equilibrium species; however, they are not very useful for fractions less than 0.01. The log(alpha) plots are useful to show the very small values of alpha, that is, less than $\alpha = 0.01$ and less than $\log \alpha = -2$, and to visualize order-of-magnitude changes. You can also multiply the α fraction by a given formal concentration of Fe^{3+} to get a plot of $[Fe^{3+}]$ or $\log[Fe^{3+}]$ versus pOH.

Example 7.6 Stepwise Complexation.   Use Figures 7.5 and 7.6 to predict the dominant Fe(III) species at a pH of 7.

A pH of 7 corresponds to a pOH of 7. From Figure 7.5, we can see that there are two significant forms present at this pH, $Fe(OH)_3$ and $Fe(OH)_2^+$; $\approx 96\%$ of the iron will be present as $Fe(OH)_3$ and $\approx 4\%$ of the iron will be present as $Fe(OH)_2^+$. Using the log(alpha) plot, we can determine the fraction of the species in the next highest concentration, $FeOH^{2+}$, to be present at less than 0.1%.

Example 7.7 Stepwise Complexation.   Use Figures 7.5 and 7.6 to predict the equilibrium concentrations of all Fe species at a pH of 7 if $c_{Fe} = 0.1$ mM.

The equilibrium concentration of each Fe species will be the total iron concentration times the alpha value of each species at a pH of 7. Three alpha values were found in Example 7.6, and using the log(alpha) plot, we can determine the fraction Fe^{3+} to be 1×10^{-9}.

The equilibrium concentrations are thus

$$[Fe(OH)_3] = (0.96)(1\times10^{-4}\,M) = 9.6\times10^{-5}\,M$$
$$[Fe(OH)_2{}^+] = (0.04)(1\times10^{-4}\,M) = 4\times10^{-6}\,M$$
$$[FeOH^{2+}] = (0.001)(1\times10^{-4}\,M) = 1\times10^{-7}\,M$$
$$[Fe^{3+}] = (1\times10^{-9})(1\times10^{-4}\,M) = 1\times10^{-13}\,M$$

You-Try-It 7.C
The stepwise complexation worksheet in you-try-it-07.xlsx lists a series of metal and ligand combinations. Use β_n values from Table 7.2 to find stepwise K values and generate alpha plots.

7.5 IMMUNOASSAYS

Biological macromolecules (biomolecules) can form strong complexes with very specific targets, both other biomolecules and small molecules. This "complexation" is often referred to as a *lock-and-key model*. The site of interaction on the biomolecule is called the receptor site for the target. This model is used for enzyme-substrate, antigen-antibody, and hormone-receptor interactions.[11] The specificity arises from the large size and 3D structure of the biomolecule. The binding site will have an opening that exactly matches the shape of the specific target and it will usually make multiple interactions with a target. The binding can include electrostatic, hydrogen bonding, and van der Waals or hydrophobic attractions.

The high specificity of this type of complexation can be applied to many analytical applications. I broadly refer to the many variations that use the lock-and-key type of interactions as *immunoassays*. There are numerous approaches, but they all take advantage of the strong and specific association between a biomolecule and the target analyte. Immunoassays might be single-use or incorporated into flow cell instrumentation. The receptor is usually an antibody, so I will refer to the target analyte as an antigen. The other necessary component is a *reporter* that provides the signal to quantitate the amount of antigen. In some cases, it is as simple as a color change to indicate the presence or absence of the antigen.

Figure 7.7 shows the steps to perform a sandwich immunoassay. An inert surface, such as a glass slide or the plastic well in a microtiter plate, is functionalized

---

[11]Note that in biology, the term "ligand" can refer to any molecule that binds reversibly to a receptor site. This chapter has used it in the narrow chemical sense of a species with unbonded electrons that form a coordinate covalent bond with a metal ion.

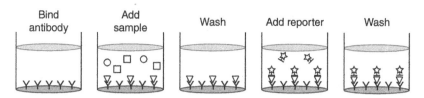

Figure 7.7 Steps in an immunoassay measurement.

with a capture protein or antibody **Y**. Unreacted **Y** is rinsed away and any open surface is blocked with another protein to prevent nonspecific binding to the glass or plastic (not shown in Figure 7.7). On adding the sample solution, only the analyte that is specific to the antibody will bind to **Y** and be immobilized. The analyte, or antigen, is represented in the figure by the inverted triangle symbol. All other sample components are removed by washing. The reporter, a labeled antibody that will bind to the antigen, is added next. The label, indicated by the star symbol in the figure, provides the analytical signal. The most common labels are molecules that absorb or fluoresce at specific wavelengths of light, but other detection mechanisms are also available. Steps 2 and 4 in Figure 7.7, where binding occurs, are often incubated with gentle shaking for some set amount of time. Excess reporter is removed by washing. If no analyte was present in the sample, no reporter will be present and there is no analytical signal. If an analyte was present, as shown in Figure 7.7, some amount of signal is recorded. The signal is proportional to the amount of reporter that was retained, which is proportional to the amount of antigen in the test solution.

An enzyme-linked immunoassay (ELISA) is a variation of a sandwich immunoassay. The overall process is the same as shown in Figure 7.7, but the reporter is created by an enzyme. Excess substrate is added after the last wash step and the enzyme converts substrate into a new product. Detection of the enzyme product is similar to detection in a regular immunoassays, with absorbance or fluorescence being the most common. The advantage of this approach is that the enzyme amplifies the analytical signal. One enzyme molecule might produce hundreds or thousands of new reporter molecules. The lock-and-key mechanism provides specific complexation of the analyte and the protein provides amplification to make the assay very sensitive.

Immunoassay methods are sold commercially in kit form for many common analytes. The majority of kits are for biological targets and are used in clinical settings. Instrument manufacturers have developed microtiter plate readers to automate the analysis of immunoassay results. Portable kits have also been developed for on-the-spot drug and pesticide screening. Another approach is to attach the **Y** antibody to a transducer surface in a flow cell. The transducer signal changes when antigen binds to the immobilized antibody. This is an active research area because it has the advantage of reusing the antibody-coated surface for multiple measurements. However, there are numerous challenges, with a key one being repeatability. After each test measurement, the transducer surface must be rinsed with a pH buffer or other solution to remove all antigen that is bound to the antibody.

As one final example, many of us became familiar with home tests for the SARS-CoV-2 coronavirus which causes the COVID-19 disease. At-home COVID and pregnancy tests are examples of immunoassay-based rapid diagnostic tests (RDTs).[12] RDTs are also used in clinical settings for malaria, influenza A and B, and other diseases.[13] Figure 7.8 shows a schematic of how these tests work. For SARS-CoV-2, mucous sample is collected with a swab from the nose or throat and transferred to a buffer solution that contains detergent to disperse the viral protein. This solution is transferred to the sample pad shown in the figure. The solution moves laterally to the right by capillary action through the reagent pad. If the virus antigen is present, it binds with the SARS reporter reagent. The virus antigen and reporter will then be immobilized and visible at the test line (middle diagram). If there no virus antigen, the SARS reporter flows to the end of the device and nothing is visible at the test line (bottom diagram). The control reporter is visible in both negative and positive cases. It is necessary to verify that the device operates correctly. A limitation of these devices is that they are directly dependent on the amount of virus antigen transferred to the device.

Chapter 7. What Was the Point?

The main point of this chapter was to review complexation and to develop the formalism for making quantitative predictions. This chapter also introduced

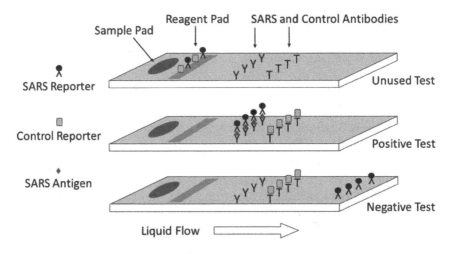

Figure 7.8   Schematic of a lateral flow assay.

[12]These tests are known by many other names such as lateral flow devices, rapid tests, quick tests, test strip, or dipsticks.
[13]See for example: https://www.cdc.gov/flu/professionals/diagnosis/rapidlab.htm; accessed August 2022.

the concept of simultaneous or competitive equilibria. These competing equilibria are sufficiently complicated to make graphical analysis using plots of alpha fractions more useful than direct calculations. As the later examples illustrated, we have now built up our knowledge of aqueous solution chemistry to include multiple competing equilibria involving weak acids, bases, and complexes. The next chapter will extend this concept further to include a solid precipitate in contact with an aqueous solution.

PRACTICE EXERCISES

1. Determine the metal oxidation state and the charge of any ions and describe the bonding in the following manganese compounds or complexes. Indicate if any of the listed complexes are unlikely to exist as written.
   (a) $KMnO_4(s)$
   (b) $MnCl_6^{3-}$ (aq)
   (c) $Mn(H_2O)_6^{2+}$ (aq)
   (d) $Mn(H_2O)_6^{7+}$ (aq)
   (e) $MnO_2(s)$

2. Explain how ligands form complexes with metal ions in solution, that is, describe the bonding. Compare complexation between a metal ion and a ligand to protonation of a weak base such as acetate ion or ammonia. How are these two types of reactions similar and different?

3. In a back-titration procedure to determine Ca^{2+}, a known amount of EDTA is added to the test portion to complex all of the analyte ions. The remaining amount of "free" EDTA is then determined by titrating with a metal ion such as Pb^{2+}. The end point is determined by the color change of a complexometric indicator such as Eriochrome Black T. What will be the effect on the end point detection if some of the Pb^{2+} titrant displaces Ca^{2+} from the $Ca(edta)^{2-}$ complex? What might be the effect if you were detecting the end point using an ion-selective electrode that was sensitive to Pb^{2+} and insensitive to any other ions in solution?

4. Use the pK_a values in Table 7.2 to calculate the solution pH for 1.0 mM solutions of LiCl, $CaCl_2$, and $CuCl_2$. You may assume that the water has been purged of carbon dioxide before adding each of these soluble salts.

5. Calculate the concentrations of $CaCl_2$ and $CuCl_2$ that will change the pH of pure water by 0.5 pH units. You may assume that the water has been purged of carbon dioxide before adding each of these soluble salts and that the starting pH is 7.0.

6. Explain how changing the solution pH will affect the equilibrium concentration of the following complexes:
   (a) $Mn(edta)^{2-}$ (aq)
   (b) $Mn(H_2O)_6^{2+}$ (aq)
   (c) $Cu(NH_3)_4^{2+}$ (aq)

7. Potassium cyanide, KCN, can serve as a masking reagent for water hardness analysis using EDTA titration.[14] The cyanide ion does not complex with the analytes such as Ca^{2+} and Mg^{2+}. Use the values from Table 7.2 to explain how the cyanide ion, CN^-, can mask transition metal ions that might otherwise interfere in an EDTA titration.

8. Use the alpha plots in Figures 7.5 and 7.6 to answer the following questions:
   (a) What are the predominant soluble iron species (alpha >0.01) at hydroxide concentrations of 10^{-9} and 10^{-13} M?
   (b) What is the $[Fe^{3+}]$ concentration in the solutions in (a) if the total iron concentration is $c_{total} = 1.0$ mM?
   (c) What is the ionic strength of a solution that is prepared by adding 0.005 mol of $FeCl_3$ to 1.0 l of a pH = 9.2 buffer consisting of 0.10 M ammonia and 0.10 M NH_4Cl? (Assume no volume change on adding the $FeCl_3$.)

9. The log cumulative formation constants for Ag^+ and Br^- are $\log \beta_1 = 4.38$, $\log \beta_2 = 7.33$, $\log \beta_3 = 8.00$, and $\log \beta_4 = 8.73$.
   (a) What are the stepwise formation constants for Ag^+ and Br^-?
   (b) 1,10-Phenanthroline has log cumulative formation constants with Ag^+ of $\log \beta_1 = 5.02$ and $\log \beta_1 = 12.07$. Will 1,10-phenanthroline form a complex with Ag^+ in the presence of Br^-?

10. Explain how binding a metal affects a complexometric indicator causing it to change color. Answer in terms of both chemical equilibria and spectroscopic and electronic changes (refer to Chapter 4).

11. A number of transition metal ions, for example, Cu^{2+}, Al^{3+}, and Fe^{3+}, will form oxalate complexes that will crystallize with counterions and precipitate from solution. Several examples are

    $K_2[Cu(C_2O_4)_2(H_2O)_2]$

    $K_3[Al(C_2O_4)_3] \cdot 3H_2O$

    $K_3[Fe(C_2O_4)_3] \cdot 3H_2O.$

    Use the alpha-plot-ML-complexes.xls spreadsheet to predict if any of these complexes might be formed with less than the maximum number of oxalate ligands.

12. Determine the effective formation constant, β'_{eff}, for EDTA complexes with Cd^{2+}, Pb^{2+}, and Zn^{2+} at pH values of 5 and 6.

13. As a refresher, write the equilibria reactions and K' expressions for a diprotic weak acid, a weak base, an ML_4 complex, and an M_2X_3 precipitate. Use

---

[14]This particular masking reagent can only be used in an alkaline solution, $pK_a = 9.2$, to prevent formation of toxic HCN fumes.

generic symbols such as H_2A, M, L, and X. What similarities do you notice and what details are different in these different types of equilibria?

14. Review Figure 7.8. What is the purpose of including the control line? What problems could cause a lateral flow test to fail?

15. Use your preferred Internet search function to find examples of lateral flow tests. Specify the sampling method and identify the antigen if possible. Note that searching the primary literature will provide more detailed information than a general search.

CHAPTER 8

PRECIPITATION EQUILIBRIA

Learning Outcomes

- Predict the relative solubility of insoluble salts, given their K_{sp} constants.
- Write equilibrium constant expressions and calculate the intrinsic solubility of insoluble salts.
- Determine common-ion and ionic strength effects on a precipitation equilibrium.
- Identify competing equilibria and predict the pH dependence on precipitation equilibria.
- Describe factors that can affect the concentration of toxic species in drinking water.

Recognizing and predicting the formation of complexes, precipitates, and competing equilibria is important to understand the chemistry of aqueous solutions in environmental, industrial, and biochemical systems. In this chapter, we describe precipitation equilibria in more detail. This type of equilibrium describes the ions remaining in solution when an insoluble salt is in contact with the solution. These systems are also called saturated solutions, and the equilibrium constant is called the solubility constant, K_{sp}. As a preview of the complexity of this type of equilibrium, the main forms of

Basics of Analytical Chemistry and Chemical Equilibria: A Quantitative Approach, Second Edition. Brian M. Tissue.
© 2023 John Wiley & Sons, Inc. Published 2023 by John Wiley & Sons, Inc.
Companion Website: www.wiley.com/go/tissue/analyticalchemistry2e

magnesium in aqueous solution are Mg^{2+} and $Mg(OH)^+$ due to the following equilibria:[1]

$$Mg(OH)_2(s) \rightleftharpoons Mg^{2+}(aq) + 2\,OH^-(aq) \quad K_{sp} = \left[Mg^{2+}\right]\left[OH^-\right]^2 \gamma_\pm^3$$

$$Mg^{2+}(aq) + OH^-(aq) \rightleftharpoons Mg(OH)^+(aq) \quad K_1 = \left[Mg(OH)^+\right]/\left[Mg^{2+}\right]\left[OH^-\right]\gamma_\pm$$

I use the thermodynamic equilibrium constants and include $\gamma_\pm$ as an average activity coefficient for each ion. It is immediately apparent that these equilibria will depend on the pH of the solution. As we know, solution pH is affected by the equilibrium with atmospheric CO_2. The net result is that the total amount of magnesium in solution will depend on pH, ionic strength, and complexation equilibria.

Chapter 3 described how differences in solubility can be used to separate an analyte ion from other sample components. The ideal gravimetric approach of precipitating the analyte ion selectively from the test portion is not always possible. Sample preparation procedures might precipitate a general class of material, such as all proteins by adding ethanol, or precipitate a class of interferences for removal. The separation is completed by filtering or centrifuging the precipitate from the test solution.

An implicit assumption in gravimetric analysis is that the precipitation reaction goes to completion. The reality is that equilibrium concentrations of ions remain in solution when a solid precipitates from solution. Precipitating agents are selected with small K_{sp} values so that the concentration of analyte that remains in solution is very low. These reagents are tabulated in analytical handbooks and interferences are known. These validated methods will often specify experimental conditions and a minimum sample amount so that the amount of analyte that is not precipitated and collected will be insignificant.

8.1 PRECIPITATE EQUILIBRIUM

This section discusses the details of precipitation equilibria and how to calculate the equilibrium concentration of ions that remain in solution. Understanding precipitation equilibria is necessary to implement gravimetric methods correctly and to understand ionic equilibria in aqueous systems that are in contact with minerals or other solids. The limitation of the predictive capacity of these calculations is that a system must be in a true chemical equilibrium. Many systems maintain steady-state concentrations by regulating inputs and outputs. A biological example is homeostasis, where regulatory processes will keep concentrations of important ions, such as calcium and

[1]Lambert, I.; Clever, H. L. *Alkaline Earth Hydroxides in Water and Aqueous Solutions* (1992), IUPAC Solubility Data Series, vol. 52; https://iupac.org/what-we-do/databases/solubility-data-series; accessed August 2022.

inorganic phosphate, nearly constant in blood and cellular fluid. In the simplest description, dietary Ca^{2+} is absorbed through the small intestines to the blood, where approximately half of the blood calcium is bound to protein. Bone serves as a mineral reservoir and the kidneys reabsorb Ca^{2+} to prevent loss through excretion. Using a single equilibrium constant to predict the Ca^{2+} concentration in blood would be a very crude approximation.

Our convention for precipitation equilibria is to write the reaction in the direction of the insoluble salt dissolving:

$$M_aX_b(s) \rightleftharpoons aM^{m+}(aq) + bX^{x-}(aq)$$

where M^{m+} is the cation, X^{x-} is the anion, and a and b are their respective stoichiometric coefficients. Precipitation is not unlike a metal and a ligand forming a complex, and mechanistically a complex of cations and anions might be the first step in precipitation. The convention for complexation equilibria is to write the reaction in the direction of the ions forming the complex. However, the convention for an insoluble salt is the opposite, with the precipitate dissolving to produce ions. Once the separate phase of the solid precipitate is present, it is more intuitive to think about the ion concentrations that remain dissolved in solution.

The equilibrium constant expression is written following the general rules given in Chapter 3:

$$K'_{sp} = [M^{m+}]^a[X^{x-}]^b$$

The "sp" subscript specifies that this equilibrium constant is a "solubility product" for an insoluble salt. Table 8.1 lists K_{sp} values for insoluble salts, ordered alphabetically by cation.[2] Note that there is a wide range of K_{sp} values; some materials are extremely insoluble in water and other salts might be considered slightly soluble.

Let us start our solubility calculations with a qualitative example in Example 8.1. The following is a common question that arises in analytical, industrial, and environmental applications: Will a precipitate form under a certain set of experimental conditions? The general procedure is to calculate the reaction quotient, Q, and compare it to K'_{sp}. Recall that the reaction quotient, Q, has the same mathematical form as the equilibrium constant expression, but it is calculated using nonequilibrium concentrations. It provides a measure of how far a system is from equilibrium.

Example 8.1 Precipitation. Will a precipitate form when 0.1 mM Fe^{3+} in sulfuric acid, H_2SO_4, is neutralized to a pH of 7?

---

[2]Adapted with permission from Speight, J. G., Ed. *Lange's Handbook of Chemistry*, 16th ed.; McGraw-Hill: New York, 2005.

TABLE 8.1 K_{sp} **Values of Insoluble Salts**

Metal	Salt	Formula	fw (g/mol)	K_{sp}
Aluminum	Hydroxide	$Al(OH)_3$	78.004	1.3×10^{-33}
	Phosphate	$AlPO_4$	121.953	9.84×10^{-21}
	8-Quinolinolate	$Al(C_9H_6NO)_3$	459.432	1.00×10^{-29}
Barium	Carbonate	$BaCO_3$	197.336	2.58×10^{-9}
	Chromate	$Ba(CrO_4)_2$	253.321	1.17×10^{-10}
	Fluoride	BaF_2	175.324	1.84×10^{-7}
	Hydroxide	$Ba(OH)_2 \cdot 8H_2O$	315.464	2.55×10^{-4}
	Oxalate	BaC_2O_4	225.346	1.6×10^{-7}
	Phosphate	$Ba_3(PO_4)_2$	601.924	3.4×10^{-23}
	Sulfate	$BaSO_4$	233.390	1.08×10^{-10}
Bismuth	Hydroxide	$Bi(OH)_3$	260.002	6.0×10^{-31}
	Phosphate	$BiPO_4$	303.952	1.3×10^{-23}
	Sulfide	Bi_2S_3	514.156	1×10^{-97}
Cadmium	Carbonate	$CdCO_3$	172.420	1.0×10^{-12}
	Oxalate	$CdC_2O_4 \cdot 3H_2O$	254.476	1.42×10^{-8}
	Phosphate	$Cd_3(PO_4)_2$	527.176	2.53×10^{-33}
	Sulfide	CdS	144.476	8.0×10^{-27}
Calcium	Carbonate	$CaCO_3$ (calcite)	100.087	3.36×10^{-9}
	Oxalate	$CaC_2O_4 \cdot H_2O$	146.112	2.32×10^{-9}
	Chromate	$CaCrO_4$	156.072	7.1×10^{-4}
	Fluoride	CaF_2	78.075	5.3×10^{-9}
	Hydroxide	$Ca(OH)_2$	74.093	5.5×10^{-6}
	Phosphate	$Ca_3(PO_4)_2$	310.177	2.07×10^{-29}
	Sulfate	$CaSO_4$	136.141	4.93×10^{-5}
Cerium(III)	Hydroxide	$Ce(OH)_3$	191.138	1.6×10^{-20}
	Iodate	$Ce(IO_3)_3$	664.824	3.2×10^{-10}
	Oxalate	$Ce_2(C_2O_4)_3 \cdot 9H_2O$	706.427	3.2×10^{-26}
	Phosphate	$CePO_4$	235.087	1×10^{-23}
Chromium(III)	Fluoride	CrF_3	108.991	6.6×10^{-11}
	Hydroxide	$Cr(OH)_3$	103.018	6.3×10^{-31}
Cobalt(II)	Carbonate	$CoCO_3$	118.942	1.4×10^{-13}
	Iodate	$Co(IO_3)_2$	408.739	1.0×10^{-4}
	Phosphate	$Co_3(PO_4)_2$	366.742	2.05×10^{-35}
	8-Quinolinolate	$Co(C_9H_6NO)_2$	347.233	1.6×10^{-25}
Cobalt(III)	Hydroxide	$Co(OH)_3$	109.955	1.6×10^{-44}
Copper(I)	Bromide	$CuBr$	143.450	6.27×10^{-9}
	Chloride	$CuCl$	98.999	1.72×10^{-7}
	Cyanide	$CuCN$	89.563	3.47×10^{-20}
	Iodide	CuI	190.450	1.27×10^{-12}
	Sulfide	Cu_2S	159.157	2.5×10^{-48}
Copper(II)	Carbonate	$CuCO_3$	123.555	1.4×10^{-10}
	Chromate	$CuCrO_4$	179.540	3.6×10^{-6}
	Iodate	$Cu(IO_3)_2$	413.351	6.94×10^{-8}
	Hydroxide	$Cu(OH)_2$	97.561	2.2×10^{-20}
	Oxalate	CuC_2O_4	151.565	4.43×10^{-10}
	Phosphate	$Cu_3(PO_4)_2$	380.581	1.40×10^{-37}
	8-Quinolinolate	$Cu(C_9H_6NO)_2$	351.846	2.0×10^{-30}
	Sulfide	CuS	95.611	6.3×10^{-36}

(Continued)

TABLE 8.1   (*Continued*)

Metal	Salt	Formula	fw (g/mol)	K_{sp}
Iron(II)	Carbonate	$FeCO_3$	115.854	3.13×10^{-11}
	Hydroxide	$Fe(OH)_2$	89.860	4.87×10^{-17}
	Oxalate	$FeC_2O_4 \cdot 2H_2O$	179.895	3.2×10^{-7}
	Sulfide	FeS	87.910	6.3×10^{-18}
Iron(III)	Hydroxide	$Fe(OH)_3$	106.867	2.79×10^{-39}
	Phosphate	$FePO_4 \cdot 2H_2O$	186.847	9.91×10^{-16}
Lanthanum	Fluoride	LaF_3	195.901	7×10^{-17}
	Hydroxide	$La(OH)_3$	189.928	2.0×10^{-19}
	Iodate	$La(IO_3)_3$	663.614	7.50×10^{-12}
	Oxalate	$La_2(C_2O_4)_3 \cdot 9H_2O$	704.005	2.5×10^{-27}
	Phosphate	$LaPO_4$	233.877	3.7×10^{-23}
	Sulfide	La_2S_3	374.006	2.0×10^{-13}
Lead	Bromide	$PbBr_2$	367.008	6.60×10^{-6}
	Carbonate	$PbCO_3$	267.209	7.4×10^{-14}
	Chloride	$PbCl_2$	278.106	1.70×10^{-5}
	Chromate	$PbCrO_4$	323.194	2.8×10^{-13}
	Fluoride	PbF_2	245.197	3.3×10^{-8}
	Hydroxide	$Pb(OH)_2$	241.215	1.43×10^{-15}
	Iodide	PbI_2	461.009	9.8×10^{-9}
	Iodate	$Pb(IO_3)_2$	557.005	3.69×10^{-13}
	Oxalate	PbC_2O_4	295.219	4.8×10^{-10}
	Phosphate	$Pb_3(PO_4)_2$	811.543	8.0×10^{-43}
	Sulfate	$PbSO_4$	303.263	2.53×10^{-8}
	Sulfide	PbS	239.265	8.0×10^{-28}
	Thiocyanate	$Pb(SCN)_2$	323.365	2.0×10^{-5}
Magnesium	Ammonium phosphate	$MgNH_4PO_4$	137.315	2.5×10^{-13}
	Carbonate	$MgCO_3$	84.314	6.82×10^{-6}
	Carbonate	$MgCO_3 \cdot 3H_2O$	138.360	2.38×10^{-6}
	Fluoride	MgF_2	62.302	5.16×10^{-11}
	Hydroxide	$Mg(OH)_2$	58.320	5.61×10^{-12}
	Iodate	$Mg(IO_3)_2 \cdot 4H_2O$	446.171	3.2×10^{-3}
Manganese	Carbonate	$MnCO_3$	114.947	2.34×10^{-11}
	Hydroxide	$Mn(OH)_2$	88.953	1.9×10^{-13}
	Oxalate	$MnC_2O_4 \cdot 2H_2O$	178.988	1.70×10^{-7}
	8-Quinolinolate	$Mn(C_9H_6NO)_2$	343.238	2.0×10^{-22}
Mercury(I)	Bromide	Hg_2Br_2	560.988	6.40×10^{-23}
	Chloride (calomel)	Hg_2Cl_2	472.086	1.43×10^{-18}
	Hydroxide	$Hg_2(OH)_2$	435.195	2.0×10^{-24}
	Iodide	Hg_2I_2	654.989	5.2×10^{-29}
Mercury(II)	Bromide	$HgBr_2$	360.398	6.2×10^{-20}
	Hydroxide	$Hg(OH)_2$	234.605	3.2×10^{-26}
	Iodide	HgI_2	454.399	2.9×10^{-29}
Nickel	Carbonate	$NiCO_3$	118.702	1.42×10^{-7}
	Hydroxide	$Ni(OH)_2$	92.708	5.48×10^{-16}
	Iodate	$Ni(IO_3)_2$	408.499	4.71×10^{-5}
	Oxalate	$NiCrO_4$	146.712	4×10^{-10}
	Phosphate	$Ni_3(PO_4)_2$	366.023	4.74×10^{-32}
	8-Quinolinolate	$Ni(C_9H_6NO)_2$	346.993	8×10^{-27}

(*Continued*)

TABLE 8.1 (*Continued*)

Metal	Salt	Formula	fw (g/mol)	K_{sp}
Potassium	Hexafluorosilicate	K_2SiF_6	220.273	8.7×10^{-7}
	Iodate	KIO_3	214.001	3.74×10^{-4}
Silver	Bromide	$AgBr$	187.772	5.35×10^{-13}
	Chloride	$AgCl$	143.321	1.77×10^{-10}
	Chromate	Ag_2CrO_4	331.730	1.12×10^{-12}
	Iodide	AgI	234.773	8.52×10^{-17}
	Iodate	$AgIO_3$	282.771	3.17×10^{-8}
	Oxalate	$Ag_2C_2O_4$	303.755	5.40×10^{-12}
	Thiocyanate	$AgSCN$	165.951	1.03×10^{-12}
Sodium	Hexafluoroaluminate	Na_3AlF_6	209.941	4.0×10^{-10}
Strontium	Carbonate	$SrCO_3$	147.629	5.60×10^{-10}
	Chromate	$SrCrO_4$	203.614	2.2×10^{-5}
	Fluoride	SrF_2	125.617	4.33×10^{-9}
	Iodate	$Sr(IO_3)_2$	437.425	1.14×10^{-7}
	Iodate	$Sr(IO_3)_2 \cdot H_2O$	455.441	3.77×10^{-7}
	Oxalate	$SrC_2O_4 \cdot H_2O$	193.654	1.6×10^{-7}
	Phosphate	$Sr_3(PO_4)_2$	452.803	4.0×10^{-28}
	8-Quinolinolate	$Sr(C_9H_6NO)_2$	375.920	5×10^{-10}
	Sulfate	$SrSO_4$	183.683	3.44×10^{-7}
Thallium(I)	Bromide	$TlBr$	284.287	3.71×10^{-6}
	Chloride	$TlCl$	239.836	1.86×10^{-4}
	Chromate	Tl_2CrO_4	524.760	8.67×10^{-13}
	Iodate	$TlIO_3$	379.286	3.12×10^{-6}
	Oxalate	$Tl_2C_2O_4$	496.786	2×10^{-4}
	Sulfide	Tl_2S	440.832	5.0×10^{-21}
	Thiocyanate	$TlSCN$	262.466	1.57×10^{-4}
Thallium(III)	Hydroxide	$Tl(OH)_3$	255.405	1.68×10^{-44}
	8-Quinolinolate	$Tl(C_9H_6NO)_3$	636.833	4.0×10^{-33}
Thorium	Hydroxide	$Th(OH)_4$	300.067	4.0×10^{-45}
	Iodate	$Th(IO_3)_4$	931.649	2.5×10^{-15}
	Oxalate	$Th(C_2O_4)_2$	408.076	1×10^{-22}
	Phosphate	$Th_3(PO_4)_4$	1076.000	2.5×10^{-79}
Tin(II)	Hydroxide	$Sn(OH)_2$	152.725	5.45×10^{-28}
	Sulfide	SnS	150.775	1.0×10^{-25}
Tin(IV)	Hydroxide	$Sn(OH)_4$	186.739	1×10^{-56}
Titanium(III)	Hydroxide	$Ti(OH)_3$	98.889	1×10^{-40}
Titanium(IV)	Oxide hydroxide	$TiO(OH)_2$	97.881	1×10^{-29}
Yttrium	Fluoride	YF_3	145.901	8.62×10^{-21}
	Hydroxide	$Y(OH)_3$	139.928	1.00×10^{-22}
	Oxalate	$Y_2(C_2O_4)_3$	441.869	5.3×10^{-29}
Zinc	Carbonate	$ZnCO_3$	125.418	1.46×10^{-10}
	Hydroxide	$Zn(OH)_2$	99.424	3×10^{-17}
	Iodate	$Zn(IO_3)_2 \cdot 2H_2O$	451.245	4.1×10^{-6}
	Oxalate	$ZnC_2O_4 \cdot 2H_2O$	277.478	1.38×10^{-9}
	Phosphate	$Zn_3(PO_4)_2$	386.170	9.0×10^{-33}
	Sulfide	ZnS (alpha form)	97.474	1.6×10^{-24}
Zirconium	Oxide hydroxide	$ZrO(OH)_2$	141.238	6.3×10^{-49}
	Phosphate	$Zr_3(PO_4)_4$	653.557	1×10^{-132}

Iron(III) sulfate is a slightly soluble salt and the likely precipitate is iron(III) hydroxide. The solubility equilibrium is

$$Fe(OH)_3(s) \rightleftharpoons Fe^{3+}(aq) + 3OH^-(aq)$$

and the equilibrium constant expression, neglecting ionic strength and using $K'_{sp} \approx K_{sp}$, is

$$K'_{sp} = 2 \times 10^{-39} = [Fe^{3+}][OH^-]^3$$

To determine if a precipitate will form, find the reaction quotient, $Q = (c_{Fe^{3+}})(c_{OH^-})^3$, and compare it to K'_{sp}:

- If $Q > K'_{sp}$, the ion concentrations are greater than their equilibrium concentrations and a precipitate will form.
- If $Q < K'_{sp}$, the ion concentrations are less than their equilibrium concentrations and no precipitate will form.

For $[Fe^{3+}] = 1 \times 10^{-4}$ M and $[OH^-] = 1 \times 10^{-7}$ M:

$$Q = (c_{Fe^{3+}})(c_{OH^-})^3 = (1 \times 10^{-4} M)(1 \times 10^{-7} M)^3 = 1 \times 10^{-25}$$

$1 \times 10^{-25} \gg 2 \times 10^{-39}$, so we predict that a precipitate will form at a pH of 7. As Q is orders of magnitude larger than our approximate K'_{sp}, correcting K'_{sp} for ionic strength is not necessary. Gaining intuition for when such corrections are necessary or not necessary takes time, but doing so is a sign that you are developing a thorough understanding of these chemical concepts.

An exact calculation of the pH at which a given concentration of Fe^{3+} will precipitate will depend on ionic strength and competing equilibria, so we will stop with this qualitative answer. In real water systems, the total amount of iron in solution can be much higher than that predicted by a simple K'_{sp} calculation. The iron can exist in a variety of forms such as iron complexes and soluble iron hydroxide colloids that do not settle out of solution.

Example 8.1 was not solely an academic exercise. Iron and sulfuric acid are present in mine drainage because of the oxidation of pyrite, FeS_2. One reaction that occurs is

$$FeS_2(s) + 14Fe^{3+}(aq) + 8H_2O \rightleftharpoons 15Fe^{2+}(aq) + 2SO_4^{2-}(aq) + 16H^+(aq)$$

Recent research has identified a biological aspect to acid mine drainage.[3] A microorganism oxidizes Fe^{2+} to Fe^{3+}, which keeps the pyrite oxidation reaction

---

[3]Edwards, K. J.; Bond, P. L.; Gihring, T. M.; Banfield, J. F. "An archaeal iron-oxidizing extreme acidophile important in acid mine drainage," *Science* **2000**, *287*, 1796–1799.

cycling, producing more acidic runoff than expected. When the mine drainage is diluted sufficiently to raise the pH, the iron precipitates as iron hydroxide and creates an orange sediment in streams and rivers.

Similar to the above example, we often ask the question of what ions will precipitate and what ions will remain in solution in an analytical procedure. This issue can arise when trying to isolate an analyte ion or to remove interferences by precipitation. Examples 8.2 and 8.3 illustrate the use of the K'_{sp} expression to predict the concentrations at which different species will precipitate. We will again neglect ionic strength and competing equilibria in these calculations. Keep in mind that experimental conditions can change these simple predictions intentionally or unintentionally.

Example 8.2 Precipitation Order. What precipitates first as SO_4^{2-} is added to a solution containing Ba^{2+}, Ca^{2+}, and Pb^{2+}? The metal ions are present in equal concentrations. You may assume that ionic strength affects each equilibrium equally and that the pH is adjusted to avoid competing equilibria.

The metal ions all form 1:1 ionic solids with SO_4^{2-}. The equilibrium reaction and equilibrium constant expression are each of the form:

$$MSO_4(s) \rightleftharpoons M^{2+}(aq) + SO_4^{2-}(aq)$$
$$K'_{sp} = [M^{2+}][SO_4^{2-}]$$

Because the metal ions are present in equal concentrations, the metal sulfate with the lowest K'_{sp} will be the one that begins precipitating first as the SO_4^{2-} concentration increases. The way to analyze this type of problem is to assume the SO_4^{2-} concentration is at 0.0 M initially and then increases as SO_4^{2-} is added. At some value of $[SO_4^{2-}]$, the most insoluble metal sulfate will begin precipitating. It will continue to precipitate as more SO_4^{2-} is added until that metal reaches an equilibrium concentration. Only after the most insoluble metal sulfate is precipitated from solution completely will the next metal begin to precipitate. Predicting the amount of SO_4^{2-} to reach each stage is basically a limiting reagent calculation.

Checking the K_{sp} values in Table 8.1, we find that $BaSO_4$ has the lowest K_{sp} of the three metal sulfates. It is the least soluble and we predict that the Ba^{2+} will be the first metal ion to be removed from this solution by precipitation. In practice, these predictions require experimental verification, as some metal ions can coprecipitate together.

Example 8.3 Precipitation Order. What precipitates first when an acidic solution containing 1.0 mM each of Cu^{2+}, Al^{3+}, and Th^{4+} is neutralized with a strong base? You may again neglect ionic strength effects and competing equilibria.

As OH^- is added to the solution, a precipitate will form when the solubility of one of the metal hydroxides is exceeded. As the stoichiometries of these metal hydroxides are all different, this problem is a little more involved than the previous one. It is not easy to see which one is least soluble from the K'_{sp} values, and we must calculate the [OH⁻] at which each metal ion will precipitate. The equilibrium constant expressions are

$$2.2\times10^{-20} = [Cu^{2+}][OH^-]^2$$
$$1.3\times10^{-33} = [Al^{3+}][OH^-]^3$$
$$4.0\times10^{-45} = [Th^{4+}][OH^-]^4$$

We solve for the OH⁻ concentration at which each metal hydroxide precipitates by entering 1.0 mM for the metal ion concentration:

$$Cu^{2+}: [OH^-] = 4.7\times10^{-9}$$
$$Al^{3+}: [OH^-] = 1.1\times10^{-10}$$
$$Th^{4+}: [OH^-] = 4.5\times10^{-11}$$

From these results, we predict that $Th(OH)_4$ will precipitate first from the solution. After all the Th^{4+} is removed from the solution, $Al(OH)_3$ will precipitate. Once all Al^{3+} is removed, $Cu(OH)_2$ will begin to precipitate.

You-Try-It 8.A
The precipitation order worksheet in you-try-it-08.xlsx lists a series of combinations of cations and anions. Generate a table of concentrations at which precipitates are predicted to form.

8.2 MOLAR SOLUBILITY

As Table 2.12 showed, it is common to find solubility data as g/100 ml, g/l, mg/l, etc. In analytical chemistry, it is more useful to work in molar units. We define the *molar solubility*, *s*, as moles of the solid that dissolves per liter of solution. As defined, the total amount of precipitate in contact with solution does not affect molar solubility. Once a precipitate has formed, whether 1 mg or 1 kg, the concentrations of the ions in solution are determined by the equilibrium. In some applications, the amount of precipitate or the amount of solution might be important. If you are adding acid to dissolve $CaCO_3$, the amount of the precipitate will determine how much acid must be added. Similarly, for a given equilibrium concentration, the total amount of the dissolved species depends on the concentrations *and* the volume of solution.

Figure 8.1 shows a representation of solid CaF_2 in the bottom of a beaker and the equilibrium concentrations of the ions that remain in solution. The arrows in the figure indicate that equilibrium is a dynamic process with ions

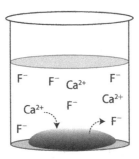

Figure 8.1 Equilibrium between a solid CaF_2 precipitate and ions in solution.

being exchanged constantly between the solid and the solution. One way to describe equilibrium is the condition when the rate of the forward reaction (ions dissolving into solution) is equal to the rate of the reverse reaction (ions in solution "sticking" to the solid). This dynamic process is observed experimentally. Allowing freshly precipitated solids to sit in solution for some amount of time is known as *aging*. The aging allows rearrangement at the atomic level to form larger crystalline grains or eliminate trapped water or impurities.

In the simple schematic in Figure 8.1, three Ca^{2+} ions and six F^- ions are shown in solution. In the absence of other equilibria or other sources of Ca^{2+} or F^-, the precipitation equilibrium produces the following equalities for the solution:

$$2[Ca^{2+}] = [F^-] \quad \text{or} \quad [Ca] = 0.5[F^-] \tag{7.2}$$

If you enter 3 and 6 from the simple picture in Figure 8.1 for $[Ca^{2+}]$ and $[F^-]$, respectively, you will see that these two relationships are correct. Entering simple, hypothetical values based on stoichiometry can help you catch simple mistakes when doing solubility calculations.

I am neglecting additional equilibria in the representation of the CaF_2 equilibrium. F^- is a weak base and Ca^{2+} is a weakly acidic cation. In this case, the fluoride has a stronger effect on the pH of the solution than does calcium ion. Since ion concentrations due to insoluble salts are usually small, the change in pH due to these ions is also usually small. This situation is different from the metal-hydrolysis and ligand-protonation examples for soluble salts that were discussed in the previous chapter. It is important to remember that the following calculation for *intrinsic molar solubility*, s, will usually underestimate the solubility of a precipitate. We will address the issue of competing equilibria later in the chapter, but any additional equilibrium pathway that converts an ion to another form will increase the solubility of a precipitate.[4] In the CaF_2 case, a small amount of the F^- will be protonated to form HF, slightly increasing the amount of CaF_2 that dissolves.

[4]Le Chatelier's principle strikes again.

To calculate the intrinsic molar solubility using the K'_{sp} expression, we must replace the ion concentrations in the expression with the one variable s. We make the substitutions based on the stoichiometry of the insoluble salt. Using $CaF_2(s)$ as an example, we can relate s to the solution concentrations of $[Ca^{2+}]$ and $[F^-]$. The precipitate equilibrium is

$$CaF_2(s) \rightleftharpoons Ca^{2+}(aq) + 2F^-(aq)$$

The equilibrium shows that each formula unit of CaF_2 that dissolves will produce one calcium ion in solution. The molar solubility of CaF_2 is therefore equivalent to the concentration of Ca^{2+}:

$$s = [Ca^{2+}]$$

Each CaF_2 formula unit that dissolves also produces two fluoride ions in solution. Thus, $[F^-]$ is twice as large as $[Ca^{2+}]$ (Eq. 7.2). We express the molar solubility in terms of F^- as

$$s = \frac{1}{2}[F^-] \quad \text{or} \quad [F^-] = 2s$$

Now writing the K'_{sp} expression and substituting s for ion concentrations,

$$K'_{sp} = [Ca^{2+}][F^-]^2 = (s)(2s)^2 = 4s^3$$

Rearranging gives us

$$s = \sqrt[3]{\frac{K'_{sp}}{4}}$$

Using the CaF_2 K_{sp} value from Table 8.1, $s = (5.3 \times 10^{-9}/4)^{0.333} = 1.1 \times 10^{-3}$ M. Given the stoichiometric relationships, $[Ca^{2+}] = 1.1 \times 10^{-3}$ M and $[F^-] = 2.2 \times 10^{-3}$ M.

You-Try-It 8.B
The intrinsic solubility worksheet in you-try-it-08.xlsx lists a number of insoluble salts. Determine the intrinsic solubility of each salt and the resulting concentrations of their constituent ions.

The solubility of an insoluble salt, that is, the equilibrium ion concentrations in solution, depends to a great extent on the details of the solution. So far, we have assumed on $K'_{sp} = K_{sp}$ and that there are no significant competing equilibria. Real aqueous samples such as surface water, well water, body fluids, and industrial waste will rarely be so simple. The factors that we must consider include

- ionic strength effects
- common-ion effects
- competing equilibria for one or both ions.

As we saw for the equilibria of metal–ligand complexes, pH will often be a factor in solubility. Example 8.4 illustrates another "simple" solubility calculation where there are no competing equilibria and we merely correct K'_{sp} for the effect of ionic strength.

Example 8.4 Molar Solubility. What is the molar solubility of barium iodate, $Ba(IO_3)_2$, in pure water at 25°C?
  The solubility equilibrium is

$$Ba(IO_3)_2(s) \rightleftharpoons Ba^{2+}(aq) + 2IO_3^-(aq) \quad K_{sp} = 1.5 \times 10^{-9}$$

We must get an approximate ionic strength to calculate K'_{sp}, so let us first do the calculation assuming $K'_{sp} = K_{sp}$.

$$K'_{sp} \approx 1.5 \times 10^{-9} = [Ba^{2+}][IO_3^-]^2$$

Given our definition of s, we know that $[Ba^{2+}] = s$ and $[IO_3^-] = 2s$. Substituting these equalities into the K'_{sp} expression, let us solve for s:

$$K'_{sp} = (s)(2s)^2 = 1.5 \times 10^{-9}$$
$$4s^3 = 1.5 \times 10^{-9}$$
$$s = 7.2 \times 10^{-4} M$$

From this first estimate of s, we know that the approximate concentrations of $[Ba^{2+}] = 7.2 \times 10^{-4}$ M and $[IO_3^-] = 1.4 \times 10^{-3}$ M. Then the ionic strength is

$$I_c = 0.5\{(+2)^2(7.2 \times 10^{-4}) + (-1)^2(1.4 \times 10^{-3})\} = 2.1 \times 10^{-3} M$$

With this value of I_c, we can use the Debye–Hückel equation to find the activity coefficients:

$$\gamma_{Ba} = 0.80 \quad \text{and} \quad \gamma_{IO_3} = 0.95$$

Now use these activity coefficients to find K'_{sp}:

$$K_{sp} = (a_{Ba})(a_{IO_3})^2$$

$$K_{sp} = \gamma_{Ba}[Ba^{2+}]\gamma_{IO_3}^2[IO_3^-]^2$$

$$K_{sp} = (\gamma_{Ba})(\gamma_{IO_3}^2)K_{sp}'$$

$$K_{sp}' = \frac{K_{sp}}{(\gamma_{Ba})(\gamma_{IO_3}^2)}$$

$$K_{sp}' = \frac{1.5 \times 10^{-9}}{(0.80)(0.95)^2} = 2.1 \times 10^{-9}$$

Now redoing the calculation with this better value for K_{sp}':

$$K_{sp}' = [Ba^{2+}][IO_3^-]^2 = 2.1 \times 10^{-9}$$

$$4s^3 = 2.1 \times 10^{-9}$$

$$s = 8.1 \times 10^{-4} \, M$$

Not correcting for ionic strength led to a calculated intrinsic solubility that was ≈10% low. We will often neglect this degree of error given that ion concentrations are low for insoluble salts. In cases where additional spectator ions are present, ionic strength effects can have a significant effect on precipitate equilibria. Such is the case in brackish or ocean water and biological solutions, where spectator ion concentrations are high. Depending on the stoichiometry, raising the activity coefficients to exponential powers amplifies the effect of ionic strength on K_{sp}'.

There is a reason that I picked a rather uncommon salt such as barium iodate for Example 8.4. Determining the solubility of $Ba(IO_3)_2$ is simple compared to determining the solubility of many other insoluble salts. Ba^{2+} is a very weak acidic ion, so we do not worry about it reacting with water. What about the iodate anion? It is a weak base and it can react with water.

$$IO_3^-(aq) + H_2O \rightleftharpoons HIO_3(aq) + OH^-(aq)$$

The extent to which this reaction proceeds is given by the value of K_b'. The K_a' of iodic acid is 0.17 (it is a rather strong weak acid), and the K_b' is

$$K_b' = \frac{K_w'}{K_a'} = 1 \times 10^{-14} / 0.17 = 5.9 \times 10^{-14}$$

This value of K_b' is so small that we can be confident that the hydrolysis of water by iodate does not affect the solubility of $Ba(IO_3)_2$. To verify this assumption, we can calculate the fraction of iodate that is protonated at a pH of 7 using the K_b' expression:

$$K_b' = 5.9 \times 10^{-14} = \frac{[HIO_3](1 \times 10^{-7})}{[IO_3^-]}$$

$$\frac{[HIO_3]}{[IO_3^-]} = \frac{5.9 \times 10^{-14}}{1 \times 10^{-7}}$$

$$\frac{[HIO_3]}{[IO_3^-]} = 5.9 \times 10^{-7}$$

The amount of IO_3^- that is protonated, $\approx 0.00006\%$, is an insignificant amount, and we do not expect the competing equilibrium to affect the solubility of $Ba(IO_3)_2$.

It's very common to need to express solubility in various units. Knowing the limit when trying to make a concentrated solution is easier knowing solubility in grams per some volume. For environmental applications, knowing the amount of a toxic metal is often more important than the counterion. Example 8.5 shows how to do a couple of these conversions.

Example 8.5 Solubility Units. Convert the molar solubility, s, found in the previous example to units of g $Ba(IO_3)_2$/100 ml and g Ba/100 ml.

We can perform this calculation using the formula weight and converting the denominator from liter to 100 ml:

$$s = \frac{8.1 \times 10^{-4} \, mol}{1} \left(\frac{487.132 \, g}{mol} \right) \frac{0.1 \, l}{100 \, ml} = 0.039 \, g \, Ba(IO_3)_2 / 100 \, ml$$

Repeating the same calculation using the formula weight of Ba:

$$s = \frac{8.1 \times 10^{-4} \, mol}{1} \left(\frac{137.327 \, g}{mol} \right) \frac{0.1 \, l}{100 \, ml} = 0.011 \, g \, Ba / 100 \, ml$$

In the first approximate calculation of the solubility of $Ba(IO_3)_2$, using K_{sp} in place of K_{sp}' resulted in a small error in the calculation (8×10^{-4} vs 7×10^{-4} M). Table 8.2 shows calculational results for $BaSO_4$ with the presence of KCl spectator ions. Besides its use in gravimetric analysis, $BaSO_4$ and related materials cause clogging problems when brine or seawater are injected into the ground for oil recovery. At moderate ionic strength, the solubility increases approximately 50%. Experimental measurements actually show an increase of about twofold at an ionic strength of 0.1–0.2 M.[5] Given our understanding of activity effects, do you always expect solubility to increase with ionic strength (in the absence of other effects)?

---

[5]Adapted with permission from Collins, A. G.; Davis, J. W. "Solubility of barium and strontium sulfates in strong electrolyte solutions," *Environ. Sci. Technol.* **1971**, 5, 1039–1043. Copyright 1971 American Chemical Society.

TABLE 8.2 Calculated Values of BaSO₄ Solubility Versus Ionic Strength

I_c (M)	K_{sp}'	S_{BaSO_4} (M)	S_{BaSO_4} (mg/l)
0.0	1.1×10^{-10}	1.1×10^{-5}	2.4
0.01	1.3×10^{-10}	1.2×10^{-5}	2.7
0.02	1.4×10^{-10}	1.2×10^{-5}	2.8
0.05	1.7×10^{-10}	1.3×10^{-5}	3.0
0.10	1.8×10^{-10}	1.3×10^{-5}	3.1
0.20	2.1×10^{-10}	1.4×10^{-5}	3.4

You-Try-It 8.C
The ionic-strength worksheet in you-try-it-08.xlsx lists several insoluble salts in different solutions. Determine the K_{sp} as a function of I_c and calculate solubility as grams of the salt per 100 ml of solution for each solution.

8.3 COMMON-ION EFFECT

The previous discussion assumed that spectator ions affected K_{sp}' but did not act on the equilibrium directly. For the results in Table 8.2, neither K⁺ nor Cl⁻ is involved directly in the BaSO₄ equilibrium. As we mentioned in the first example of this chapter, the solubility of Mg(OH)₂ will depend on the pH of the solution. We should clarify that this is the case when discussing a static equilibrium situation. If solid Mg(OH)₂ is added to an acidic solution, a limiting reagent calculation is necessary to determine if the amount of solid Mg(OH)₂ or the amount of acid in solution is the limiting reagent. After determining the limiting reagent, then it is appropriate to consider a precipitate equilibrium if solid remains. We'll look at a case involving OH⁻ after working through a "simple" case. Example 8.6 illustrates the effect on solubility when one of the ions that is *common* to the precipitate equilibrium is already present in the solution. We again use the insoluble salt Ba(IO₃)₂ because we know that we don't have to worry about acid–base equilibria affecting the solubility.

Example 8.6 Solubility with a Common Ion. What is the molar solubility of barium iodate, Ba(IO₃)₂, at 25°C in a solution that contains 0.010 M barium nitrate, Ba(NO₃)₂?
The problem is set up the same as before to calculate K_{sp}':

$$K_{sp} = (a_{Ba})(a_{IO_3})^2$$
$$K_{sp} = \gamma_{Ba}[Ba^{2+}]\gamma_{IO_3}^2[IO_3^-]^2$$
$$K_{sp} = (\gamma_{Ba})(\gamma_{IO_3}^2)K_{sp}'$$
$$K_{sp}' = \frac{K_{sp}}{(\gamma_{Ba})(\gamma_{IO_3}^2)}$$

The ionic strength, neglecting the relatively small concentration of Ba^{2+} and IO_3^- ions from the insoluble barium iodate, is

$$I_c = 0.5\{(+2)^2[Ba^{2+}] + (-1)^2[NO_3^-]\}$$

$$I_c = 0.5\{(+2)^2(0.010\,M) + (-1)^2(0.020\,M)\} = 0.030\,M$$

With the value of I_c, we can use the Debye–Hückel equation to find the activity coefficients:

$$\gamma_{Ba} = 0.53 \quad \text{and} \quad \gamma_{IO_3} = 0.85$$

$$K'_{sp} = \frac{1.5 \times 10^{-9}}{(0.53)(0.85)^2} = 3.9 \times 10^{-9}$$

Now we can set up the equilibrium problem:

$$Ba(IO_3)_2(s) \rightleftharpoons Ba^{2+}(aq) + 2IO_3^-(aq)$$

$$K'_{sp} = [Ba^{2+}][IO_3^-]^2 = 3.9 \times 10^{-9}$$

The major difference from the simple solubility calculation is that there are two sources of Ba^{2+} in this case: Ba^{2+} from the 0.010 M $Ba(NO_3)_2$ and Ba^{2+} from the dissolution of $Ba(IO_3)_2$. The equilibrium concentration is the total that comes from both sources:

$$[Ba^{2+}] = 0.010\,M + s$$

Thus

$$K'_{sp} = (0.010\,M + s)(2s)^2 = 3.9 \times 10^{-9}$$

We can solve a cubic equation or we can first try neglecting s compared to 0.010 M:

$$K'_{sp} = (0.010\,M)(2s)^2 = 3.9 \times 10^{-9}$$

$$s^2 = 9.81 \times 10^{-8}$$

$$s = 3.1 \times 10^{-4}\,M$$

The main thing to verify when doing these types of calculations is that s is small compared to the common-ion concentration. Checking our assumption: $s = 0.00031$ M $\ll 0.010$ M, so the approximation in the calculation was reasonable.

At the same ionic strength, but without the presence of Ba^{2+}, the solubility is $s = 1.0 \times 10^{-3}$ M. With the additional 0.01 M Ba^{2+} present, the solubility of barium iodate is reduced by a factor of approximately 3. The common-ion effect is a general effect that can be understood qualitatively by Le Chatelier's principle. The additional Ba^{2+} from the $Ba(NO_3)_2$ shifts the equilibrium toward the solid precipitate, reducing the solubility. The common-ion effect will always suppress solubility in the absence of other equilibria.

The common-ion effect involving hydroxide ion is a little more complicated. Taking the case of $Mg(OH)_2$, first look at the solubility of a saturated solution assuming the solid was added to pH = 7 water. The tabulated value for the solubility of $Mg(OH)_2$ is 0.00064 g/100 ml, which is 0.00011 M. Given the low ionic strength in the absence of spectator ions, I will use $K'_{sp} = K_{sp} = 5.61 \times 10^{-12}$. The equilibrium is

$$Mg(OH)_2(s) \rightleftharpoons Mg^{2+}(aq) + 2\,OH^-(aq)$$

and the equilibrium expression is

$$K'_{sp} = 5.61 \times 10^{-12} = \left[Mg^{2+}\right]\left[OH^-\right]^2 = (s)(2s)^2 = 4s^3$$

Solving this expression gives $s = 0.00056$ M. Since $[OH^-] = 2s$, $[OH^-] = 0.00011$ M or a value of $p[OH^-] = 4$ and $p[H_3O^+] = 10$. In this case, the common-ion effect will not suppress the solubility of $Mg(OH)_2(s)$ unless the solution pH is higher than ≈ 10.

As noted at the beginning of this section, adding a solid like $Mg(OH)_2$ to water of pH < 7 will neutralize acid. A limiting reagent calculation is necessary to determine if the $Mg(OH)_2$ or acid is consumed first. If the acid is the limiting reagent, the $[Mg^{2+}]$ in solution is now a common ion and the solubility and $p[OH^-]$ can be calculated assuming $s \ll [Mg^{2+}]$.

You-Try-It 8.D
The common-ion worksheet in you-try-it-08.xlsx lists insoluble salts of different stoichiometry. Determine the solubility dependence on common-ion concentration for each case.

8.4 PRECIPITATION AND COMPETING EQUILIBRIA

I've hinted at these issues already, but now we get to the complicated equilibria. Consider the precipitate equilibrium of silver chloride, AgCl:

$$AgCl(s) \rightleftharpoons Ag^+(aq) + Cl^-(aq)$$
$$K'_{sp} = [Ag^+][Cl^-]$$

How will s_{AgCl} change if KCl is added to the solution that is in contact with the solid AgCl? On the basis of the common-ion effect, we expect s_{AgCl} to decrease. Measurements of s_{AgCl}, determined from the total concentration of silver in solution, agree with common-ion calculations below approximately 10^{-4} M KCl. As the chloride concentration increases above 10^{-4} M, the concentration of silver, c_{Ag}, in solution also increases. Note that I use the c_{Ag} variable to indicate total silver concentration in any form. Table 8.3 compares the calculated AgCl solubility, s_{AgCl}, to the measured solubility, c_{Ag}, as a function of KCl concentration.[6] K'_{sp} is corrected for ionic strength at each KCl concentration.

The increase in c_{Ag} as Cl⁻ is added above 0.001 M is the opposite of what we expect based on the common-ion effect. What is happening? Formation of the following silver chloride complexes can explain these results:

$$Ag^+(aq) + Cl^-(aq) \rightleftharpoons AgCl(aq) \qquad K'_1 = 1.1 \times 10^3$$
$$AgCl(aq) + Cl^-(aq) \rightleftharpoons AgCl_2^-(aq) \qquad K'_2 = 1 \times 10^2$$
$$AgCl_2^-(aq) + Cl^-(aq) \rightleftharpoons AgCl_3^{2-}(aq)$$
$$AgCl_3^{2-}(aq) + Cl^-(aq) \rightleftharpoons AgCl_4^{3-}(aq) \quad K'_4 = 1 \times 10^2$$

As the [Cl⁻] increases, formation of the complexes increases the solubility of AgCl. You can again think in terms of Le Chatelier's principle. Adding chloride takes Ag^+ away from the AgCl(s) equilibrium, causing more solid AgCl to dissolve to counteract the effect:

$$AgCl(s) \rightleftharpoons Ag^+(aq) + Cl^-(aq)$$
$$+$$
$$Cl^-(aq)$$
$$\Updownarrow$$
$$AgCl(aq)$$
$$+$$
$$Cl^-(aq)$$
$$\Updownarrow$$
$$AgCl_2^-(aq)$$

TABLE 8.3 Calculated and Measured Solubility of AgCl

c_{KCl}, M	K'_{sp}	Predicted s_{AgCl}, M	Measured c_{Ag}, M
0.0	1.8×10^{-10}	1.3×10^{-5}	1.2×10^{-5}
0.0001	1.8×10^{-10}	1.8×10^{-6}	6.0×10^{-7}
0.0005	1.9×10^{-10}	3.8×10^{-7}	1.9×10^{-7}
0.0010	1.9×10^{-10}	1.9×10^{-7}	1.6×10^{-7}
0.0050	2.1×10^{-10}	4.2×10^{-8}	2.2×10^{-7}
0.0100	2.2×10^{-10}	2.2×10^{-8}	3.5×10^{-7}

[6]Adapted with permission from Barney, J. E.; Argersinger, W. J.; Reynolds, C. A. "A study of some complex chlorides and oxalates by solubility measurements," *J. Am. Chem. Soc.* **1951**, *73*, 3785–3788. Copyright 1951 American Chemical Society.

where I have shown only the first two silver chloride complexes that can form, AgCl(aq) and $AgCl_2^-$ (aq). Depending on the total Cl^- concentration, one or more of the species listed in the stepwise complexation reactions might dominate as the soluble form of silver in solution. The KCl concentration at which the solubility reaches a minimum is a consequence of the balance between the common-ion effect suppressing the solubility and the formation of the soluble silver chloride complexes increasing the solubility. Writing the first complex equilibrium constant expression,

$$K_1' = \frac{[AgCl(aq)]}{[Ag^+][Cl^-]}$$

we can substitute K_{sp}' for the denominator and rearrange the expression to obtain

$$[AgCl(aq)] = K_1' K_{sp}' = (1\times10^3)(2\times10^{-10}) = 2\times10^{-7}\,M$$

On the basis of this calculation, we can predict that the solubility of silver chloride will be approximately 2×10^{-7} M or greater at any concentration of Cl^-. The discrepancy between this prediction and the experimental minimum, 1.6×10^{-7} M, is not unusual given the typical precision in K_{sp} and K_n values. Predicting s_{AgCl} for higher c_{Cl} is done in the same way using the equilibria for K_2', etc. It is also possible to use alpha plots to find the dominant form of a complex as a function of ligand concentration and apply it to solubility equilibria. This process is illustrated in Example 8.7.

Example 8.7 Solubility.   Predict the AgCl solubility at 0.01 M Cl^-. The alpha fraction for Ag^+ at this chloride ion concentration is 0.044.

We have the K_{sp}' value from Table 7.7:

$$K_{sp}' = 2.2\times10^{-10} = [Ag^+][Cl^-]$$

At this relatively high concentration of the common ion, we know that

$$[Cl^-] = c_{Cl}$$

From our definition of alpha fractions, we know that

$$[Ag^+] = (\alpha_{Ag^+})(c_{Ag})$$

Now substituting for the equilibrium concentrations in the K_{sp}' expression:

$$2.2\times10^{-10} = (\alpha_{Ag^+})(c_{Ag})(c_{Cl})$$
$$2.2\times10^{-10} = (0.044)(c_{Ag})(0.01M)$$

Because the solubility of AgCl is equal to the total concentration of all silver species,

$$s_{AgCl} = c_{Ag} = 5 \times 10^{-7}\,M$$

This prediction is not far from the experimental value of $3.5 \times 10^{-7}\,M$.

In real water systems, the total amount of metal ions will often be determined by the presence of complex-forming species, acid–base chemistry, and the formation of colloids. These effects are especially important in predicting the bioavailability of trace metal nutrients and harmful metal species. Example 8.8 shows another quantitative calculation that includes competitive equilibria.

Example 8.8 Solubility. What is the solubility of $CaCO_3$ in pH = 7 water at 25°C?

The dissolution equilibrium is

$$CaCO_3(s) \rightleftharpoons Ca^{2+}(aq) + CO_3^{2-}(aq) \qquad K'_{sp} = 5 \times 10^{-9}$$

Before writing any K expressions, think about what competing equilibria might be possible. Ca^{2+} could form complexes or precipitates with hydroxide. The K'_n for $Ca(OH)^+$ is 20. That is a pretty small K'_n, so we probably do not need to worry about that complex at a pH of 7. As a check find the ratio of $Ca(OH)^+$ to Ca^{2+} when $[OH^-] = 1 \times 10^{-7}\,M$:

$$K'_n = \frac{[Ca(OH)^+]}{[Ca^{2+}][OH^-]} = 20$$

$$\frac{[Ca(OH)^+]}{[Ca^{2+}]} = 20(1 \times 10^{-7})$$

$$\frac{[Ca(OH)^+]}{[Ca^{2+}]} = 2 \times 10^{-6}$$

This calculation shows that $[Ca(OH)^+]$ will be insignificant at a pH of 7. K'_{sp} for $Ca(OH)_2$ is 6×10^{-6}, and doing a similar calculation for $Ca(OH)_2(s)$ also shows that no $Ca(OH)_2$ precipitate will form at a pH of 7. Because we usually treat Ca^{2+} as a strong electrolyte, it should not be surprising that the K'_n and K'_{sp} values show that Ca^{2+} will not have any competing equilibria at a pH of 7. What about the CO_3^{2-}? Carbonate takes part in the following competing acid–base equilibria:

$$CO_3^{2-}(aq) + H_2O \rightleftharpoons HCO_3^-(aq) + OH^-(aq)$$

$$HCO_3^-(aq) + H_2O \rightleftharpoons H_2CO_3(aq) + OH^-(aq)$$

To determine solubility, s, we will use the following K'_{sp} expression and we will find equalities for $[Ca^{2+}]$ and $[CO_3^{2-}]$ to get the expression in terms of solubility:

$$K'_{sp} = [Ca^{2+}][CO_3^{2-}] = 5 \times 10^{-9}$$

The $[Ca^{2+}]$ is equal to the solubility, s, but because of the competing equilibria, $[CO_3^{2-}]$ is not equal to s. The solubility is equal to the total carbonate concentration.

$$s = c_{carbonat}$$

where

$$c_{carbonate} = [H_2CO_3] + [HCO_3^-] + [CO_3^{2-}]$$

From the definition of an alpha fraction, we know that

$$\alpha_{CO_3^{2-}} = \frac{[CO_3^{2-}]}{c_{carbonate}}$$

or, on rearranging

$$[CO_3^{2-}] = (\alpha_{CO_3^{2-}})(c_{carbonate}) = (\alpha_{CO_3^{2-}})(s)$$

Now we can substitute back into the K'_{sp} expression and solve for s. $\alpha_{CO_3^{2-}}$ can be calculated from equations of the form discussed with Figure 6.7.

$$K'_{sp} = (s)(\alpha_{CO_3^{2-}})(s)$$

$$s = \sqrt{\frac{K'_{sp}}{\alpha_{CO_3^{2-}}}} = \sqrt{\frac{5 \times 10^{-9}}{4 \times 10^{-4}}}$$

$$s = \sqrt{\frac{5 \times 10^{-9}}{4 \times 10^{-4}}}$$

$$s = 4 \times 10^{-3} \, M$$

This result shows that the solubility is approximately two orders of magnitude greater than it would be in the absence of the competing equilibrium. (In the absence of competing equilibrium, $s = \sqrt{5 \times 10^{-9}} = 7 \times 10^{-5} \, M$.)

Examples 8.7 and 8.8 showed the various conditions that will affect the solubility of an insoluble precipitate. Predicting solubility will usually involve determining any common-ion effects and the dominant competing equilibria. Doing so requires knowledge of the species in the sample solution. Depending on the desired precision of a calculation, it is not uncommon to perform a rough calculation and then to repeat the calculation with a better value for ionic strength, etc.

8.5 DRINKING WATER

This section describes issues with drinking water due to contact with solid material. We won't try to calculate solubility for these cases, but you can see how remediation strategies depend on solubility, redox chemistry, and other factors. Examples of arsenic and lead are discussed here, but these are only two species that can dissolve in water to dangerous levels. The US EPA has set maximum contaminant levels (MCLs) for other toxic metals, coliform bacteria, fluoride, nitrate, and numerous other organic compounds.

Arsenic occurs naturally as various minerals, with arsenopyrite (iron arsenic sulfide, FeAsS) being common. This mineral consists of Fe^{3+} and AsS^{3-} with arsenic in the $+3$ oxidation state. Fresh rainwater that filters through the ground can cause chemical weathering of the rocks and minerals in contact with the groundwater. As we saw for pyrite oxidation, microbial action can also contribute to release of metals from solids. Arsenic released by weathering exists in solution as oxyanions of As(III) or As(V). At the slightly acidic pH of rainwater, the predominant species in solution are H_3AsO_3 and $H_2AsO_4^-$.

In the United States, locations in the southwest and the upper Midwest are most likely to have geologic conditions that leach arsenic into well water. Municipal water authorities use precipitating agents and flocculant to remove arsenic and other metals. Water from private wells is usually treated with an adsorptive filter or ion-exchange system. For ion-exchange, the As(III) must be oxidized to As(V) in a pretreatment step so the arsenic form is charged and retained.

A major hotspot for arsenic globally is Bangladesh. A well-drilling campaign was initiated to reduce the incidence of waterborne diseases due to contaminated surface water. An unintended consequence was the incidence of skin lesions and other disease due to arsenic in ground water from some wells. After much work, inexpensive filters were developed to remediate the arsenic contamination. Water is poured through bucket filters that contain sand, iron metal, and other media. The arsenic is removed by reaction and adsorption with the zero valent Fe metal.

These next two examples are legacy issues with lead service lines that were installed in US water systems prior to the 1950s.[7] Estimates are that approximately 9 million lead service lines remain in use in the US. The US EPA lead and copper rule sets action levels of 15 ppb for lead and 1.3 ppm for copper.[8] This rule requires action, such as remediation, public education, or more

---

[7]A service line is the pipe that connects a water main to a residence or other building.
[8]United States Environmental Protection Agency; https://www.epa.gov. Lead and copper rule; https://www.epa.gov/dwreginfo/lead-and-copper-rule; accessed August 2022.

frequent monitoring, if greater than 10% of households test above an action level. There is no accepted safe level of lead exposure for children. The US CDC lowered the blood lead reference value (BLRV) in 2021 from 5.0 to 3.5 µg/dl (µg lead per dl blood).[9] This value is meant to identify the top 2.5% of young children with the highest blood lead level for intervention.

First, a little background on disinfectants. Public water utilities add chlorine or chloramine (monochloramine, NH_2Cl) to drinking water to kill pathogens that cause waterborne diseases. Many utilities have switched from chlorine to less reactive chloramine to reduce formation of disinfection by-products such as chloroform, $CHCl_3$, and haloacetic acids. These compounds form when the disinfectant reacts with natural organic matter in the water. The chloramine also generally has longer persistence in the water compared to chlorine.

The Washington, D.C. water authority switched from chlorine to chloramine in 2000. Elevated lead levels began to be detected in home tap water in 2001. Subsequent monitoring of blood lead levels in children also showed an increase in high-risk neighborhoods.[10] Switching from chlorine to chloramine disinfectant affected the leaching of lead from lead service lines. These pipes are generally coated internally with a layer of precipitates and corrosion products. Disrupting this layer can expose fresh lead to the water. The Washington, D.C. water supplier began adding orthophosphate as a corrosion inhibitor to the water supply in 2004. Lead levels decreased gradually, and the D.C. area was below the EPA action level for lead in 2005.

At pH near 7, orthophosphate exists primarily as a combination of $H_2PO_4^-$ and HPO_4^{2-}. The orthophosphate does serve as a source for the low equilibrium concentration of PO_4^{3-}. The orthophosphate is added to water at ≈1 mg/l, which is an approximately 100-fold excess compared to 15 ppb lead. Lab studies and the water testing results show that the orthophosphate additive suppresses the leaching of lead into the water. We won't try to predict an equilibrium lead concentration. In addition to phosphate, the water contains carbonate, hydroxide, and other species. If the source of the soluble lead is corrosion of the metallic lead pipe, there is also a redox process to convert zero valent lead to Pb^{2+}. Suffice to say that even a relatively simple system such as drinking water can have complicated equilibrium chemistry.

A similar experience occurred in Flint, Michigan in 2014. The municipal water system switched the source of drinking water from Lake Huron, via the Detroit Water Authority, to the Flint River. The Flint Water System did not add corrosion inhibitors to the water supply and residents began to complain of discolored tap water. The Flint River water was more corrosive than the Lake Huron water source and began corroding steel and lead service lines.

[9]United States Centers for Disease Control and Prevention; https://www.cdc.gov; accessed August 2022.
[10]In the United States, blood lead levels in children have been dropping steadily since lead was removed from gasoline in 1996. Use of leaded gasoline ended worldwide in 2021.

Elevated levels of lead in children's blood was documented during 2015. A cascade of effects occurred, including detection of coliform bacteria in the water supply. Leaching of iron is thought to have reduce the activity of chlorine disinfectant. Increasing the chlorine dosing led to higher amounts of total trihalomethanes, TTHM, in the water.

In late 2015, the Flint water source was switched back to the Detroit supply and residents were advised to obtain filters or to drink bottled water. Lead levels in the municipal water supply returned to less than the EPA action level of 15 ppb by 2017. After legal action and emergency declarations over the years, a program is in place to replace the lead service lines in Flint, Michigan. Research of this incident will probably continue for years. It has raised questions about the method and frequency for sampling drinking water. The current US EPA lead and copper rule specifies collecting "first draw samples" from homes, i.e., 1 liter of water from a tap that has not been used for at least six hours.[11] This process captures water in contact with the home plumbing rather than being drawn in from the water main. It might or might not represent water that was in contact with a lead service line for some amount of time before use.

Chapter 8. What Was the Point?   The main point of this chapter was to review precipitation equilibria and to develop the formalism for making quantitative predictions. The simplest calculation of intrinsic solubility is really just a starting point. We must then determine if ionic-strength effects, common ions, or simultaneous or competitive equilibria are important. These competing equilibria are sufficiently complicated to make graphical analysis using plots of alpha fractions more useful than direct calculations. We have now built up our knowledge of aqueous solution chemistry to include multiple competing equilibria involving acids, bases, complexes, and precipitates.

PRACTICE EXERCISES

1. Write the equilibrium reaction(s) for each of the following solutions.
    (a) Adding 0.1 mol of NH_3 to 1.0 l of a saturated solution of $Cu_3(PO_4)_2$.
    (b) Mixing 0.1 mol of NH_4OH and 0.1 mol of $Cu(NO_3)_2$ in 1.0 l of water.

2. Predict the precipitation order for the following solutions.
    (a) Neutralizing an acidic solution that contains 0.005 M Cd^{2+}, Cu^{2+}, and Zn^{2+}.
    (b) Adding PO_4^{3-} to a solution that is 0.1 mmol each of Cu^{2+} and Al^{3+}.

[11]US EPA, *Suggested Directions for Homeowner Tap Sample Collection Procedures*, Revised Version: May 2019.

3. Express the solubility, s, in terms of the equilibrium concentration of each ion for the following insoluble salts. You may neglect competing equilibria.
   (a) $BaCO_3$
   (b) $Ba(OH)_2$
   (c) $Pb_3(PO_4)_2$
   (d) ZnF_2
   (e) $CaC_2O_4 \cdot H_2O$

4. For each one of the insoluble salts listed in the previous question, describe qualitatively how adjusting the solution pH will affect the solubility.

5. Will a CaF_2 precipitate form if enough F^- is added to water containing 0.10 mM Ca^{2+} to make the solution 0.10 mM F^-? You may neglect ionic strength effects and competing equilibria.

6. Explain how increasing the ionic strength of a solution will affect the solubility of a precipitate. What happens if one of the strong electrolyte ions that increases the ionic strength is also involved directly in the precipitation equilibrium, for example, adding Na_2SO_4 to a solution that is saturated with $BaSO_4$?

7. What is the molar solubility of $La(IO_3)_3$ in water? You may neglect ionic strength effects and competing equilibria.

8. What is the molar solubility of $La(IO_3)_3$ in a solution that is 0.010 M sodium iodate, $NaIO_3$? You may neglect competing equilibria.

9. Consider the following precipitates.
   - $Cd(OH)_2$ $K_{sp} = 4.5 \times 10^{-15}$
   - $Mg(OH)_2$ $K_{sp} = 7.1 \times 10^{-12}$
   - $Pb(OH)_2$ $K_{sp} = 1.2 \times 10^{-15}$
   - $Zn(OH)_2$ $K_{sp} = 1 \times 10^{-17}$
   (a) Which one of these metal hydroxides will precipitate first when an acidic solution containing equal concentrations of these metal ions is made more basic by adding NaOH? Explain.
   (b) Which one of these metal hydroxides will dissolve first when a basic solution containing all of these precipitates is made more acidic by slowly adding HCl? Explain.

10. For which one of the following insoluble salts will a competing equilibrium most affect the solubility? Following is the trend in K_b' of the anions: $CO_3^{2-} > SO_4^{2-} > IO_3^- > NO_3^-$.
    (a) $BaCO_3$
    (b) $BaSO_4$
    (c) $Ba(IO_3)_2$
    (d) $Ba(NO_3)_2$

11. Use an online search engine to find the results of drinking water analysis in the region where you live. These results are usually published by a local

water authority or government. Are any of the analytes over an EPA action level?

12. What classes of contaminants are listed in the EPA primary drinking water standards? What type of analytical methods are used for each of these types of contaminants?

13. What is the purpose of the EPA secondary drinking water standards? What type of analytical methods are used for each of these types of contaminants?

14. Use an online search engine to find maps or information about arsenic levels in groundwater in the region where you live. What types of arsenic remediation strategies are recommended to make groundwater safe to drink?

Part II. What Was the Point?

Part II of this text described the details of chemical equilibria. We started with the equilibria of weak acids and weak bases in Chapter 5 and added ionic strength effects to improve the predictive capability of equilibrium calculations. Chapters 6 and 7 on buffers and complexes, respectively, illustrated simultaneous equilibria involving multiple forms of an analyte. Alpha plots gave us a graphical method to plot the dominant forms versus concentration of a ligand (H^+ in the case of acids). Chapter 8 showed a more complex case of aqueous equilibria in contact with a second phase, that is, the dependence of precipitate solubility on other aqueous-phase species.

The effects of ionic strength and simultaneous equilibria must often be controlled to obtain accurate analytical measurements. These factors can apply to both the classical methods of Chapter 3 and to the instrumental methods discussed in Part III. The form of an analyte in solution can affect everything from completing sample preparation steps or wet-chemical reactions quantitatively to the accuracy of an instrumental measurement. Ionic strength impacts sample preparation and wet-chemical methods only indirectly by affecting equilibrium concentrations, but we'll see Chapter 9 that it can have direct effects on the response of electrochemical measurements to an analyte.

Part III. What Is Next?

The next part of this text describes the concepts and details of instrumental methods. These methods are more sensitive than wet-chemical methods for most types of analytes. I have made a somewhat arbitrary distinction between the automation of wet-chemical procedures, such as the autotitrators discussed briefly in Chapter 3, and instrumental methods that rely on spectroscopy or measurement of an electrical charge or current. This classification is based only on the difference in sensitivity between automated wet-chemical procedures and more complex instrumental methods. The primary driving force to develop the large number of instrumental methods that are now available is their greater sensitivity, and often selectivity, compared to wet-chemical methods. However, all instrumental methods rely on good standards for their accuracy. Unlike the direct relationship to primary standards that we have with wet-chemical methods, instrumental methods are more reliant on the proper use of QA and QC control samples to ensure accurate calibration.

Often, more than one type of instrumental method will provide accurate results for a given analyte or a type of sample. Selection of a method is often based on subtle differences between methods or on practical matters such as equipment availability, speed of analysis, or cost. To give you the knowledge to be able to select appropriate methods when presented with analytical challenges, the following chapters stress the underlying principles, differentiating characteristics, and the nuts and bolts of instrumentation. We will start with electrochemical instruments and then move to atomic spectrometry. Recall that we introduced molecular spectroscopy in Chapter 4 to provide context for the equilibrium topics in Part II. Chapter 11 describes instruments for determining the structure of a molecule. We will then finish with analytical separations, where some type of instrumental detector is coupled to a separations column to provide separation and quantitation in one process.

PART III

INSTRUMENTAL METHODS AND ANALYTICAL SEPARATIONS

CHAPTER 9

ELECTROANALYTICAL CHEMISTRY

Learning Outcomes

- Determine oxidation states and balance redox reactions.
- Use standard reduction potentials to predict if redox reactions will be spontaneous.
- Describe the major components of electrochemical cells.
- Calculate cell potentials given concentration information.
- Use potentiometric and voltammetric data to calculate analyte concentration.

9.1 INTRODUCTION

Redox reactions appear frequently in industrial, environmental, and biological processes, and Table 9.1 lists a small sampling of common processes. The last column of Table 9.1 lists applications in analytical chemistry where redox reactions can control oxidation state during sample preparation, form the basis of redox titrations, or make measurements with electrochemical-based instruments.

I begin this chapter with a review of the basic terminology and concepts of oxidation state and charge. We describe the degree of oxidation or reduction of each atom in a chemical species with the oxidation state. The *oxidation state* is the difference between the number of valence electrons assigned to an atom

Basics of Analytical Chemistry and Chemical Equilibria: A Quantitative Approach, Second Edition. Brian M. Tissue.
© 2023 John Wiley & Sons, Inc. Published 2023 by John Wiley & Sons, Inc.
Companion Website: www.wiley.com/go/tissue/analyticalchemistry2e

TABLE 9.1 Common Redox Processes

Industrial	Environmental	Biological	Analytical
Batteries	Elemental speciation	Photosynthesis	Redox buffers
Corrosion	Biological oxygen demand	Cellular respiration	Redox titrations
Electroplating	Nitrogen fixation	Disinfection	Coulometric titrations
Smelting	Contaminant remediation		Potentiometry
Combustion			Voltammetry

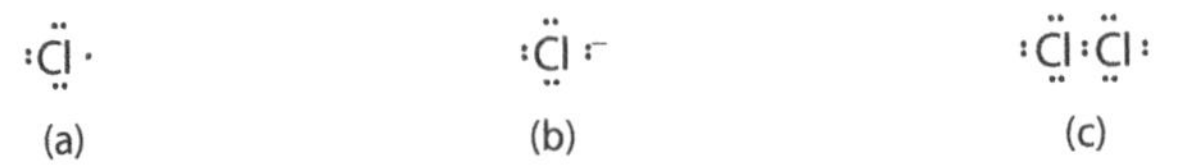

$$:\!\ddot{C}l\cdot \qquad :\!\ddot{C}l\,^- \qquad :\!\ddot{C}l:\!\ddot{C}l:$$

$$\text{(a)} \qquad\qquad \text{(b)} \qquad\qquad \text{(c)}$$

Figure 9.1 Chlorine oxidation states: (a) chlorine atom, oxidation state 0; (b) chloride ion, oxidation state -1; (c) chlorine molecule, oxidation state 0.

when it exists in a given species and the number of valence electrons that the atom has when it exists as a neutral atom. The calcium ion, Ca^{2+}, has zero valence electrons compared to two electrons in a Ca atom. Having two electrons less than Ca gives Ca^{2+} an oxidation state of $+2$. The chloride ion, Cl^-, has eight valence electrons compared to seven electrons for the Cl atom. Having one electron more than elemental Cl gives Cl^- an oxidation state of -1 (see Figure 9.1).

In polyatomic species the oxidation state of each atom is determined after "giving" the bonding electrons to the most electronegative atom in a bond. Recall that electronegativity increases as we go up or to the right in the periodic table (values are listed in Table 5.1). For bonds between atoms of the same element, *homonuclear bonds*, the bonding electrons are shared equally. Atoms retain any lone pairs of electrons that they have. For the Cl_2 molecule in Figure 9.1(c), each chlorine atom has six electrons in lone pairs and gets one electron from the bond between the two atoms. Each atom in Cl_2 is assigned seven electrons, the same as in its elemental atomic form (Figure 9.1(a)). The correct description is that each atom in Cl_2 has an oxidation state of zero. As another example, in carbon dioxide, CO_2 or $:\ddot{O}=C=\ddot{O}:$, the bonding electrons are given to the more electronegative oxygen atoms. The oxygen atoms are allocated eight electrons compared to six in the neutral atom. The carbon atom is allocated zero electrons compared to four in the neutral atom. The net result is that each oxygen atom has two extra electrons and an oxidation state of -2 and the carbon atom is missing four electrons and has an oxidation state of $+4$.

Oxidation state and charge are not equivalent. An oxidation state can be assigned to each atom. The *charge* of a chemical species is the sum of all oxidation states. In a monoatomic species, the charge is equal to the oxidation state. For example, the chloride ion, Cl^-, has an oxidation state of -1 and a charge of -1. In polyatomic species, the term charge refers to the overall molecule. The sum of oxidation states for carbon dioxide is $(-2)+(-2)+(+4)=0$, and we say that the molecule is neutral.

TABLE 9.2 Common Oxidation States in the Periodic Table

+1	+2	+3	+4							+1	+2	+3			−2	−1	0
1 H																	2 He
3 Li	4 Be												5 B	6 C	7 N	8 O	9 F	10 Ne
11 Na	12 Mg		More electronegative → ↑										13 Al	14 Si	15 P	16 S	17 Cl	18 Ar
19 K	20 Ca	21 Sc	22 Ti	23 V	24 Cr	25 Mn	26 Fe	27 Co	28 Ni	29 Cu	30 Zn	31 Ga	32 Ge	33 As	34 Se	35 Br	36 Kr
37 Rb	38 Sr	39 Y	40 Zr	41 Nb	42 Mo	43 Tc	44 Ru	45 Rh	46 Pd	47 Ag	48 Cd	49 In	50 Sn	51 Sb	52 Te	53 I	54 Xe
55 Cs	56 Ba	57 La	72 Hf	73 Ta	74 W	75 Re	76 Os	77 Ir	78 Pt	79 Au	80 Hg	81 Tl	82 Pb	83 Bi	84 Po	85 At	86 Rn

Assigning an oxidation state to all of the atoms in a polyatomic species is done by the process of elimination after assigning the atoms that have predictable oxidation states (see Example 9.1). Table 9.2 shows a partial periodic table with the common oxidation states listed at the top of each column. Where no value is listed, the oxidation state is variable and should be predicted after assigning the other atoms in the polyatomic species. Lanthanide and actinide ions are not shown in the table. Oxidation states of lanthanide ions tend to be +3 (except for cerium, which is also often +4) and those for the actinides are most commonly +3 or +4.

The common oxidation states in Table 9.2 should look familiar and match your chemical intuition. The proton, H^+, and +1 alkali metal ions, and −1 halide ions appear in many strong electrolytes. Likewise, the oxidation state of oxygen and sulfur is almost always −2 (two exceptions are noted below). If we approach the example of carbon dioxide in this way, the two oxygen atoms have oxidation states of $2 \times (-2)$. Since carbon dioxide is neutral, the oxidation state of carbon must be +4. As another example, the permanganate ion, MnO_4^-, has four oxygen atoms with oxidation states of $4 \times (-2)$. To achieve the overall charge of −1, the Mn must have an oxidation state of +7. After assigning oxidation states, it is always useful to check that the sum of the oxidation states does equal the overall charge. For MnO_4^- ion, that is $(+7) + 4(-2) = -1$.

A few common exceptions to the oxidation states listed in Table 9.2 are species that occur as dimers such as the peroxide ion, O_2^{2-}, disulfide ion, S_2^{2-}, and the mercury(I) ion, Hg_2^{2+}. These oxidation states are usually identified on the basis of the chemical name or on finding suitable oxidation states for other atoms in a compound. For example, pyrite has the chemical formula FeS_2. For the S atom to have an oxidation state of −2 would require that Fe have an oxidation state of +4. This is not a common oxidation state for iron, and signals that the sulfur is present as the disulfide ion S_2^{2-}. Another exception to the common cases in Table 9.2 is the formation of hydrides such as NaH or $NaBH_4$. The hydrogen has an oxidation state of -1 in these compounds. These species are common in organic synthesis, but they do not exist in aqueous solution.

Example 9.1 Assigning Oxidation States.   Assign the oxidation state of each atom in (a) NH_4^+, (b) $Na_2S_2O_3$, and (c) CH_3COOH (acetic acid).

(a) Nitrogen is more electronegative than hydrogen, so in NH_4^+ the H atoms have an oxidation state of $+1$. Since there are four H atoms, $4\times(+1)$, the N atom must have an oxidation state of -3 to achieve an overall charge of +. Checking: $4\times(+1) + 1\times(-3) = +1$.

(b) First, recognize that $Na_2S_2O_3$ is an ionic salt. Per Table 9.2, the sodium ion has an oxidation state of $+1$. Since $Na_2S_2O_3$ has two Na^+, the anion is $S_2O_3^{2-}$. The oxidation state of oxygen is almost always -2. Since there are three oxygen atoms, $3\times(-2)$, the sulfur atoms must each have an oxidation state of $+2$, $2\times(+2)$, to achieve an overall charge of -2 for $S_2O_3^{2-}$. Checking for $Na_2S_2O_3$: $2\times(+1) + 3\times(-2) + 2\times(+2) = 0$.

(c) For hydrocarbons, we expect hydrogen atoms to have an oxidation state of $+1$ and oxygen atoms to have an oxidation state of -2. The carbon atoms in most organic molecules will have varying oxidation states and you should assign each carbon separately.[1] We follow the same general rule and assign electrons in a bond to the most electronegative atom. Electrons in the C–H bond are assigned to the more electronegative carbon atom. Electrons in the C–O and C=O bonds are assigned to the more electronegative oxygen atoms. The oxygen atoms also have lone pairs of electrons that they retain. Electrons in the homonuclear C–C bond are assigned equally. First, consider the methyl carbon. Since carbon is more electronegative than hydrogen, the methyl carbon gets six electrons from the C–H bonds and one electron from the C–C bond. This total of seven electrons is three more than the valence electrons of a neutral carbon atom, and the methyl carbon has an oxidation state of -3. For the carboxyl carbon atom, the oxygen atoms are more electronegative than carbon and the carbon has only the one electron from the C–C bond. The carboxyl carbon has three less electrons than a neutral carbon atom and thus an oxidation state of $+3$. As a check, sum the oxidation states of all H, O, and C atoms: $4\times(+1) + 2\times(-2) + (-3) + (+3) = 0$.

9.2   STANDARD REDUCTION POTENTIALS

All of the different types of chemical equilibria that we discussed previously can be described as competitive processes. Acid–base reactions involve the exchange of a proton. Complexation involves the displacement of water (or

---

[1]Notable exceptions are symmetric molecules such as ethane, ethene, oxalic acid, and benzene.

other ligand) from the coordination sphere around a metal ion with a new ligand. Solubility involves the displacement of solvent molecules from the immediate surroundings of an ion in solution with counter ions, resulting in precipitation. Just as different acids have varying affinities to hold on to protons, different chemical species will have varying affinities to "hold on to" electrons. Reactions that involve the transfer of electrons are called *redox reactions*. When a chemical species loses one or more electron, we say that it is being oxidized. When a species gains one or more electron, we say that it is being reduced.

Oxidation is the process of losing one or more electrons.
Reduction is the process of gaining one or more electrons.

Just as a neutralization reaction does not occur without both an acid and a base, redox reactions must occur in pairs with one substance being oxidized and the other being reduced.[2] Before describing the relative affinities of electrons quantitatively, let us review some everyday experience for relative redox behavior. Some elements, such as gold and platinum, are difficult to oxidize—they "hold" on to their electrons tightly. Halogens also have a high affinity for electrons, and in nature they exist as the negative ions, for example, F^- and Cl^-. Other elements, such as Na and Ca, give up electrons to other elements easily. These elements exist in nature as the positive ions, for example, Na^+ and Ca^{2+}. We've discussed many examples of these anions and cations as they are so common in strong electrolyte solutions.

Just as we use equilibrium constants to quantitate the relative acidity of weak acids, we need a quantitative measure of the relative affinity that different chemical species have for electrons. To use weak acids as an analogy, we can generate a list of acids using K_a to describe their relative strength, including examples that have multiple acidic protons (Table 9.3). The larger the K_a, the more readily the acid gives up a proton in water (see Example 9.2). We can use the equilibrium constants to predict the effect that different acids will have on pH. Similarly, we will use standard reduction potentials to predict the redox reactions that will occur.

TABLE 9.3 Relative Strength of Different Weak Acids

Monoprotic Acids		Polyprotic Acids		
Iodic acid	1.57×10^{-1}	Phthalic acid	K_{a1}	1.12×10^{-3}
Chloroacetic acid	1.36×10^{-3}		K_{a2}	3.91×10^{-6}
Formic acid	1.77×10^{-4}			
Acetic acid	1.75×10^{-5}	Phosphoric acid	K_{a1}	7.11×10^{-3}
Hypochlorous acid	2.90×10^{-8}		K_{a2}	6.34×10^{-8}
Ammonium ion	5.68×10^{-10}		K_{a3}	4.8×10^{-13}

[2]An exception to this statement occurs when a photovoltaic or mechanical generator replaces one half of a redox reaction.

Example 9.2 Competitive Reactions. Use the K_a values in Table 9.3 to determine which acid (or acidic proton) will react first when a strong base is added to solutions containing (a) a mixture of chloroacetic acid or acetic acid, and (b) phthalic acid, that is, are the two protons neutralized simultaneously or sequentially?

(a) Looking at the K_a values, chloroacetic acid has a much larger K_a (1.36×10^{-3}) than does acetic acid (1.75×10^{-5}). The relative difference in these K_a values indicates that chloroacetic acid is more acidic and it loses its proton more easily than does acetic acid. On addition of a strong base to a mixture of these two acids, the base will first react with chloroacetic acid, and completely neutralize the stronger acid before reacting with acetic acid.

(b) By definition, K_{a1} for a polyprotic acid is the one that is most acidic. For phthalic acid with pK_{a1} (1.12×10^{-3}) > pK_{a2} (3.91×10^{-6}), the first proton will be neutralized by the added base completely before the second acidic proton begins to react. This competitive process is also visualized in the alpha plots in Figure 6.5.

The point of these examples is to illustrate the competitive nature of reactions of a mixture. In redox processes there are often multiple reactions possible, but the one that occurs first will be the one with the largest driving force.

We could tabulate equilibrium constants, K_{eq}, for redox reactions, but that would result in large tables of many possible redox reactions. It is more flexible and useful to work with voltages of half reactions. A *half reaction* is a reduction or oxidation reaction that shows the oxidized form, the reduced form, and the number of electrons transferred. We call these reactions "half reactions" because there must always be another reaction or electrical component for an overall redox reaction to occur. Tabulating redox reactions as half reactions gives us a flexible method to combine different half reactions to describe a variety of overall reactions.

Many of the main group nonmetals and transition metal elements exist in a large range of oxidation states (as oxyanions for their high oxidation states). Chromium, for example, exists in the common forms of Cr metal, Cr^{3+}, CrO_4^{2-}, and $Cr_2O_7^{2-}$, with oxidation states of 0, +3, +6, and +6, respectively. A given element will have multiple half reactions depending on the number of oxidation states in which it can exist. For tin, which exists as Sn metal, Sn^{2+}, and Sn^{4+}, the possible half reactions to convert between these different forms are

$$Sn^{4+}(aq) + 2e^- \rightleftharpoons Sn^{2+}(aq)$$
$$Sn^{2+}(aq) + 2e^- \rightleftharpoons Sn(s)$$

The relative affinity of a given species to accept or donate electrons in an electron transfer reaction is described by a reduction potential, E, or a standard reduction potential, $E°$. $E°$ is the value for a reaction at standard conditions: 1 M activity for solutes, 1 atm partial pressure for gases, and temperature = 298 K. We use $E°$ values analogous to K_a values to describe the relative "strength" of chemical species "to hold on to" electrons. Table 9.4 lists the $E°$ values of selected half reactions for acidic solutions.[3] Reactions are written with the oxidized forms on the left gaining one or more electrons to form the reduced species on the right.

Chemical species that have half reactions with high positive $E°$ values tend to occur in their reduced forms. Chemical species in half reactions with high negative $E°$ values tend to lose electrons easily and occur in their oxidized forms. In redox reactions, the species that is reduced is called an *oxidizing agent* and the species that is oxidized is called a *reducing agent*. The species in the upper left of Table 9.4 are strong oxidizing agents because they have a high affinity for electrons and readily remove electrons from other species. Species in the lower right are strong reducing agents because these metals lose electrons easily to be oxidized to their ionic form.

9.2.1 The H⁺ Half Reaction

The standard reduction potential, $E°$, for any half reaction is measured relative to

$$2H^+(aq) + 2e^- \rightleftharpoons H_2(g)$$

This half reaction is assigned the value $E° = 0.00$ V. A positive $E°$ means that a half reaction (at standard concentrations) will go in the direction indicated (reduction) when paired with the hydrogen half reaction, for example,

$$Cu^{2+}(aq) + 2e^- \rightarrow Cu(s) \quad E° = +0.337\,V$$
$$2H^+(aq) + 2e^- \rightarrow H_2(g) \quad E° = 0.00\,V$$

Comparing these two half reactions, we can predict that Cu^{2+} will become Cu metal and H_2 will be oxidized to $2H^+$ on mixing Cu^{2+} and H_2. The reaction mechanism and kinetics will depend on experimental conditions, but you can find online videos that demonstrate the following reaction:

$$CuO(s) + H_2(g) \rightarrow Cu(s) + H_2O(l)$$

---

[3] $E°$ values are for reactants and products at standard activity: 1.0 M (aq) or 1 atm partial pressure (g). Note that there are two different half reactions for bromine and iodine. Use the $Br_2(l)$ and $I_2(s)$ $E°$ values for saturated solutions and the $Br_2(aq)$ and $I_2(aq)$ $E°$ values when concentrations are below the solubility limit. Adapted with permission from *Lange's Handbook of Chemistry*, 16th ed., Speight, J. G., Ed.; McGraw-Hill: New York, 2005.

TABLE 9.4 $E°$ **Values for Half Reactions in Acidic Solution (25°C)**

Half Reaction				$E°$ (V)*
F_2 (aq)	+	$2e^-$	$\rightleftharpoons$ $2F^-$ (aq)	2.87
Au^+ (aq)	+	e^-	$\rightleftharpoons$ Au(s)	1.83
H_2O_2 (aq) + $2H^+$ (aq)	+	$2e^-$	$\rightleftharpoons$ $2H_2O$	1.763
Ce^{4+} (aq)	+	e^-	$\rightleftharpoons$ Ce^{3+} (aq)	1.70
MnO_4^- (aq) + $8H^+$ (aq)	+	$5e^-$	$\rightleftharpoons$ Mn^{2+} (aq) + $4H_2O$	1.51
PbO_2 (s) + $4H^+$ (aq)	+	$2e^-$	$\rightleftharpoons$ Pb^{2+} (aq) + $2H_2O$	1.46
Cl_2 (g)	+	$2e^-$	$\rightleftharpoons$ $2Cl^-$ (aq)	1.396
$Cr_2O_7^{2-}$ (aq) + $14H^+$ (aq)	+	$6e^-$	$\rightleftharpoons$ $2Cr^{3+}$ (aq) + $7H_2O$	1.36
O_2 (g) + $4H^+$ (aq)	+	$4e^-$	$\rightleftharpoons$ $2H_2O$	1.229
MnO_2(s) + $4H^+$ (aq)	+	$2e^-$	$\rightleftharpoons$ Mn^{2+} (aq) + $2H_2O$	1.224
Br_2 (aq)	+	$2e^-$	$\rightleftharpoons$ $2Br^-$ (aq)	1.087
Br_2 (l)	+	$2e^-$	$\rightleftharpoons$ $2Br^-$ (aq)	1.066
Ag^+ (aq)	+	e^-	$\rightleftharpoons$ Ag(s)	0.799
Fe^{3+} (aq)	+	e^-	$\rightleftharpoons$ Fe^{2+} (aq)	0.771
I_2 (aq)	+	$2e^-$	$\rightleftharpoons$ $2I^-$ (aq)	0.621
I_2 (s)	+	$2e^-$	$\rightleftharpoons$ $2I^-$ (aq)	0.5355
Cu^+ (aq)	+	e^-	$\rightleftharpoons$ Cu(s)	0.520
$Fe(CN)_6^{3-}$ (aq)	+	e^-	$\rightleftharpoons$ $Fe(CN)_6^{4-}$(aq)	0.361
Cu^{2+} (aq)	+	$2e^-$	$\rightleftharpoons$ Cu(s)	0.340
AgCl(s)	+	e^-	$\rightleftharpoons$ Ag(s) + Cl^- (aq)	0.222
Sn^{4+} (aq)	+	$2e^-$	$\rightleftharpoons$ Sn^{2+} (aq)	0.154
$2H^+$ (aq)	+	$2e^-$	$\rightleftharpoons$ H_2(g)	0.00
Pb^{2+} (aq)	+	$2e^-$	$\rightleftharpoons$ Pb(s)	−0.126
Sn^{2+} (aq)	+	$2e^-$	$\rightleftharpoons$ Sn(s)	−0.1375
Ni^{2+} (aq)	+	$2e^-$	$\rightleftharpoons$ Ni(s)	−0.257
Co^{2+} (aq)	+	$2e^-$	$\rightleftharpoons$ Co(s)	−0.277
$Ag(CN)_2^-$ (aq)	+	e^-	$\rightleftharpoons$ Ag(s) + $2CN^-$ (aq)	−0.31
$PbSO_4$ (s)	+	$2e^-$	$\rightleftharpoons$ Pb(s) + SO_4^{2-} (aq)	−0.356
Cd^{2+} (aq)	+	$2e^-$	$\rightleftharpoons$ Ca(s)	−0.403
Fe^{2+} (aq)	+	$2e^-$	$\rightleftharpoons$ Fe(s)	−0.44
$Cu(CN)_2^-$ (aq)	+	e^-	$\rightleftharpoons$ Cu(s) + $2CN^-$ (aq)	−0.44
$Au(CN)_2^-$ (aq)	+	e^-	$\rightleftharpoons$ Au(s) + $2CN^-$ (aq)	−0.596
Cr^{3+} (aq)	+	$3e^-$	$\rightleftharpoons$ Cr(s)	−0.74
Zn^{2+} (aq)	+	$2e^-$	$\rightleftharpoons$ Zn(s)	−0.7626
Mn^{2+} (aq)	+	$2e^-$	$\rightleftharpoons$ Mn(s)	−1.17
V^{2+} (aq)	+	$2e^-$	$\rightleftharpoons$ V(s)	−1.175
Al^{3+} (aq)	+	$3e^-$	$\rightleftharpoons$ Al(s)	−1.676
Ce^{3+} (aq)	+	$3e^-$	$\rightleftharpoons$ Ce(s)	−2.34
Mg^{2+} (aq)	+	$2e^-$	$\rightleftharpoons$ Mg(s)	−2.356
Na^+ (aq)	+	e^-	$\rightleftharpoons$ Na(s)	−2.713
Ca^{2+} (aq)	+	$2e^-$	$\rightleftharpoons$ Ca(s)	−2.84
Sr^{2+} (aq)	+	$2e^-$	$\rightleftharpoons$ Sr(s)	−2.899
Ba^{2+} (aq)	+	$2e^-$	$\rightleftharpoons$ Ba(s)	−2.92
Cs^+ (aq)	+	e^-	$\rightleftharpoons$ Cs(s)	−2.92
K^+ (aq)	+	e^-	$\rightleftharpoons$ K(s)	−2.924
Rb^+ (aq)	+	e^-	$\rightleftharpoons$ Rb(s)	−2.98
Li^+ (aq)	+	e^-	$\rightleftharpoons$ Li(s)	−3.040

Adapted with permission from Speight, J. G., Ed., *Lange's Handbook of Chemistry*, 16th ed.; McGraw-Hill: New York, 2005.

A negative $E°$ means that the half reaction (at standard concentrations) will go in the reverse direction (to the left) when paired with the hydrogen half reaction. As another example, compare

$$2H^+(aq) + 2e^- \rightarrow H_2(g) \quad E° = 0.00\,V$$
$$Zn^{2+}(aq) + 2e^- \rightarrow Zn(s) \quad E° = -0.763\,V$$

We can predict that mixing H^+ and Zn metal will produce H_2 and Zn^{2+}. The overall reaction is

$$Zn(s) + 2H^+(aq) \rightarrow Zn^{2+}(aq) + H_2(g)$$

Producing hydrogen gas by placing a piece of zinc metal in acidic solution is another demonstration of a redox reaction that you can view by searching for online videos.

The two overall reactions described here are examples of a *spontaneous reaction*. That is, the reactions occur on mixing the reactants. We make these predictions based on the $E°$ values, which are a measure of the relative affinity for electrons. Section 9.4 discusses these concepts in more detail and connects reduction potentials to Gibbs energy.

9.3 USING HALF REACTIONS

Since the $E°$ values are all relative, having a positive or negative sign has no special meaning except to indicate where a half reaction falls relative to the hydrogen half reaction. For any two half reactions (at standard concentrations), the half reaction with the more positive $E°$ will proceed as a reduction (as written) and the other will proceed as an oxidation (in reverse). For the case of concentrations not being at 1 M, we must calculate the E of each half reaction using the Nernst equation and then compare the values to determine if a spontaneous redox reaction will occur. (The Nernst equation is discussed in Section 9.5.) Consider the addition of a piece of Cu metal to a solution containing Ag^+

$$Cu(s) + Ag^+(aq) \xrightarrow{?} Cu^{2+}(aq) + Ag(s)$$

Without doing the experiment, can we predict if this reaction will be spontaneous as written?

Let us consider the reaction at standard concentrations, 1 M Ag^+ and 1 M Cu^{2+} in contact with Cu and Ag metal. First, write the half reactions as reductions with their $E°$ values

$$Ag^+(aq) + e^- \rightarrow Ag(s) \quad E° = +0.799\,V$$
$$Cu^{2+}(aq) + 2e^- \rightarrow Cu(s) \quad E° = +0.337\,V$$

The half reaction with the more positive $E°$ will occur as written (as a reduction). For the Cu^{2+} and Ag^+ half reactions, the Ag^+ $E°$ (+0.799 V) is more positive than the $Cu^{2+}E°$ (+0.337 V). We can predict that the Ag^+ will be reduced and the Cu metal will be oxidized.

This is another experiment that you can find online as a video demonstration, although it is generally not done at standard concentrations. On placing a shiny piece of copper wire in a clear solution of $AgNO_3$, the solution gradually turns bluish and a "fuzzy" solid forms on the copper wire. In less than 1 h, the solution is light blue and the wire is covered with white or grayish needles. What happened? As predicted,

- The Cu(s) loses electrons to be oxidized to Cu^{2+} (aq), creating the blue solution.
- The Ag^+(aq) gains electrons to be reduced to Ag(s), creating the silver needles.

Now, writing the two half reactions in the direction that they occur, we can balance the reaction by summing up the half reactions so that we eliminate electrons from the overall reaction

$$Cu(s) \rightarrow Cu^{2+}(aq) + 2e^-$$
$$\underline{2[Ag^+(aq) + e^- \rightarrow Ag(s)]}$$
$$Cu(s) + 2Ag^+(aq) \rightarrow Cu^{2+}(aq) + 2Ag(s)$$

To do so, we multiply the silver half reaction by two so that we eliminate the electrons from each side of the reaction when we combine the half reactions. We will balance more complicated redox reactions later. Given the $E°$ values for these two half reactions, what would you predict if you placed a piece of Ag metal in a solution of 1 M Cu^{2+}? Example 9.3 provides another predictive case.

Example 9.3 Predicting Redox Reactions. Does a reaction occur when bromine is added to a 1.0 M solution of tin(II) chloride?

1. The species present are Sn^{2+}(aq), Cl^-(aq), and Br_2(l)
2. Looking at Table 9.4, possible changes in oxidation state for these reactants are

$$Sn^{2+}(aq) \rightarrow Sn^{4+}(aq)$$
$$Sn^{2+}(aq) \rightarrow Sn(s)$$
$$Cl^-(aq) \rightarrow Cl_2(g)$$
$$Br_2(l) \rightarrow Br^-(aq)$$

3. Now write the half reactions (all as reductions) to compare $E°$ values

    A. $Cl_2(g) + 2e^- \rightarrow 2Cl^-(aq)$ $E° = 1.360\,V$

    B. $Br_2(l) + 2e^- \rightarrow 2Br^-(aq)$ $E° = 1.07\,V$

    C. $Sn^{4+}(aq) + 2e^- \rightarrow Sn^{2+}(aq)$ $E° = 0.154\,V$

    D. $Sn^{2+}(aq) + 2e^- \rightarrow Sn(s)$ $E° = -0.138\,V$

4. Looking at the half reactions, what can happen? (Hint: Start at either the top of the list and work down or start at the bottom of the list and work up.)

- Starting at the top: There is no half reaction with a more positive $E°$ value than half reaction A, so there is nothing that can be reduced by removing electrons from Cl^- to oxidize it to Cl_2.

- Now, consider half reaction B. Br_2 is present and it has a very high positive $E°$ value, that is, it has a high affinity for electrons. It will remove electrons from any species that is more easily oxidized, to be reduced to Br^-. Sn^{2+} can be oxidized to Sn^{4+}, providing a source of electrons for Br_2. Since half reaction B has a more positive $E°$ value than half reaction C, Br_2 will be reduced to Br^- and Sn^{2+} will be oxidized to Sn^{4+}.

- As stated above, Sn^{2+} will be oxidized to Sn^{4+} as Br_2 is reduced to Br^-.

- There is no half reaction with a less positive $E°$ value than half reaction D, so Sn^{2+} cannot be reduced to tin metal.

    So yes, at standard concentrations a spontaneous reaction can occur. The Cl^- is a spectator ion, and the overall reaction is

$$Sn^{2+}(aq) \rightarrow Sn^{4+}(aq) + 2e^-$$
$$\underline{Br_2(l) + 2e^- \rightarrow 2Br^-(aq)}$$
$$Sn^{2+}(aq) + Br_2(l) \rightarrow Sn^{4+}(aq) + 2Br^-(aq)$$

9.3.1 Balancing Redox Reactions

Before we discuss electrochemical cells, we must be able to balance redox reactions that are more complicated than the examples that we have considered so far. The general rules for balancing redox reactions in acidic solutions are the following:

1. Assign oxidation states to determine the species being oxidized and reduced and write the relevant half reactions.
2. Balance each half reaction.

    a. Balance the atoms being oxidized or reduced.
    b. Balance oxidation numbers by adding electrons.

   c. Balance other elements that are present, for example, S with SO_4^{2-} or Cl with Cl^-.
   d. Balance oxygen atoms by adding H_2O.
   e. Balance H atoms by adding H^+.
  3. Combine the half reactions to eliminate the electrons from the overall reaction.

This procedure assumes that the redox reactions occur in an acidic solution. If a reaction occurs in a basic solution, having H^+ in a reaction is clearly not realistic. A simple way to balance redox reactions that occur in basic solution is to follow the procedure above, and then eliminate the H^+ by adding equal numbers of OH^- to each side of the reaction. Often, after this step, some number of H_2O molecules can be eliminated from each side of the reaction. Examples 9.4 and 9.5 illustrate the process for acidic and basic conditions, respectively.

Example 9.4 Balancing Redox Reactions. Dichromate ion, $Cr_2O_7^{2-}$, is a strong oxidizing agent that is used in redox titrations. When we mix $Cr_2O_7^{2-}$ with Fe^{2+} we get Cr^{3+} and Fe^{3+}.

$$Fe^{2+}(aq) + Cr_2O_7^{2-}(aq) \rightarrow Fe^{3+}(aq) + Cr^{3+}(aq)$$

What is the balanced reaction?
  To balance this reaction, follow the general rules stated above:

  1. The oxidation states of the iron half reaction are easy; Fe(II) changes to Fe(III). The oxidation state of Cr in $Cr_2O_7^{2-}$ is +6 (to give an overall charge of −2) and it becomes Cr(III). The unbalanced half reactions are

$$Fe^{2+}(aq) \rightarrow Fe^{3+}(aq)$$
$$Cr_2O_7^{2-}(aq) \rightarrow Cr^{3+}(aq)$$

  2. Now, balance each half reaction. The iron half reaction simply needs addition of an electron to the right side of the half reaction.

$$Fe^{2+}(aq) \rightarrow Fe^{3+}(aq) + e^-$$

Since the $Cr_2O_7^{2-}$ contains two chromium atoms, we need two Cr^{3+} on the right side of the half reaction.

$$Cr_2O_7^{2-}(aq) \rightarrow 2Cr^{3+}(aq)$$

Since there are two chromium atoms that go from Cr(VI) to Cr(III), we need a total of six electrons in the Cr half reaction:

$$Cr_2O_7{}^{2-}(aq) + 6e^- \rightarrow 2Cr^{3+}(aq)$$

Now, we balance the oxygen atoms by adding water

$$Cr_2O_7{}^{2-}(aq) + 6e^- \rightarrow 2Cr^{3+}(aq) + 7H_2O$$

and we balance the hydrogen atoms by adding H^+

$$Cr_2O_7{}^{2-}(aq) + 14H^+(aq) + 6e^- \rightarrow 2Cr^{3+}(aq) + 7H_2O$$

3. The individual half reactions are now balanced. Now combine them to give the overall reaction. To eliminate the electrons from each side of the reaction the iron half reaction is multiplied by 6:

$$6Fe^{2+}(aq) \rightarrow 6Fe^{3+}(aq) + 6e^-$$
$$\underline{Cr_2O_7{}^{2-}(aq) + 14H^+(aq) + 6e^- \rightarrow 2Cr^{3+}(aq) + 7H_2O}$$
$$6Fe^{2+}(aq) + Cr_2O_7{}^{2-}(aq) + 14H^+(aq) \rightarrow 6Fe^{3+}(aq) + 2Cr^{3+}(aq) + 7H_2O$$

Example 9.5 Balancing Redox Reactions (Basic Conditions).   Silver and gold metal can be extracted from their ores with cyanide ion, CN^-. The extraction is done under basic conditions to prevent losing the ligand as HCN. For silver, the overall unbalanced reaction is

$$Ag(s) + CN^-(aq) + O_2(g) \rightarrow Ag(CN)_2{}^-(aq) + H_2O$$

Balance this overall reaction.

Since the reaction occurs in basic conditions, we will follow the general rules stated earlier to balance an overall reaction and then add OH^- to eliminate the H^+.

1. The two half reactions are

$$Ag(s) \rightarrow Ag(CN)_2{}^-(aq)$$
$$O_2(g) \rightarrow H_2O$$

The oxidation state change for the silver half reaction is Ag(0) becoming Ag(I) and that for oxygen is from 0 to -2.

2. Now, balance each half reaction. The silver half reaction requires one electron on the right side. Since this half reaction involves cyanide ion, we must add two CN^- to the left side to balance the two cyanides in the complex.

$$Ag(s) + 2CN^-(aq) \rightarrow Ag(CN)_2^-(aq) + e^-$$

Since the O_2 molecule contains two oxygen atoms, we need two H_2O molecules on the right side of the half reaction.

$$O_2(g) \rightarrow 2H_2O$$

Since there are two oxygen atoms that go from O^0 to O^{2-}, we need a total of four electrons in this half reaction

$$O_2(g) + 4e^- \rightarrow 2H_2O$$

Now we balance the hydrogen atoms by adding four protons to the left side of the half reaction

$$O_2(g) + 4H^+(aq) + 4e^- \rightarrow 2H_2O$$

3. The individual half reactions are now balanced. We can combine them to get the overall reaction by eliminating the electrons. Multiply the silver half reaction by 4 and sum reactants and products to get the overall reaction:

$$4Ag(s) + 8CN^-(aq) \rightarrow 4Ag(CN)_2^-(aq) + 4e^-$$
$$\underline{O_2(g) + 4H^+(aq) + 4e^- \rightarrow 2H_2O}$$
$$4Ag(s) + 8CN^-(aq) + O_2(g) + 4H^+(aq) \rightarrow 4Ag(CN)_2^-(aq) + 2H_2O$$

This reaction occurs in basic conditions, so we add OH^- (four in this case) to each side of the reaction to eliminate H^+, producing H_2O on the left side of the reaction:

$$4Ag(s) + 8CN^-(aq) + O_2(g) + 4H_2O \rightarrow 4Ag(CN)_2^-(aq) + 2H_2O + 4OH^-(aq)$$

As a last step, we can eliminate two water molecules from each side of the overall reaction:

$$4Ag(s) + 8CN^-(aq) + O_2(g) + 2H_2O \rightarrow 4Ag(CN)_2^-(aq) + 4OH^-(aq)$$

Try balancing the following examples for practice. The main reactants and products are listed, but they are otherwise unbalanced. They are common enough for you to find the overall balanced reactions by an Internet search to check your result.

- Titration of Fe with permanganate: $MnO_4^-(aq) + Fe^{2+}(aq) \rightarrow Mn^{2+}(aq) + Fe^{3+}(aq)$
- Lead–acid battery: $Pb(s) + PbO_2(s) + SO_4^{2-}(aq) \rightarrow PbSO_4(s)$
- Alkaline battery (basic conditions): $MnO_2(s) + Zn(s) \rightarrow MnO(OH)(s) + Zn(OH)_2(s)$

You-Try-It 9.A
The balancing worksheet in you-try-it-09.xlsx contains additional un-balanced redox reactions. Identify the oxidation states of all atoms in these reactions and balance them.

9.4 BACKGROUND ON SPONTANEOUS REACTIONS AND EQUILIBRIUM

Before discussing reactions that occur in electrochemical cells, we need a little more background on the energetics of reactions. Reactant molecules will rearrange to form more stable, i.e., lower-energy, products. A reaction that produces lower-energy products is a spontaneous reaction. This does not mean that the reaction occurs immediately or rapidly; in fact, the rate at which a reaction occurs depends on other factors. The two following examples are spontaneous reactions:

$$CH_4(g) + 2O_2(g) \rightarrow CO_2(g) + 2H_2O(l) + \text{energy (mostly heat)}$$

$$2Fe(s) + \frac{3}{2}O_2(g) + 3H_2O(l) \rightarrow 2Fe(OH)_3(s) + \text{energy (mostly heat)}$$

Nothing happens when you first open the gas valve to a Bunsen burner. (Don't leave it open for too long!) The reason nothing happens is due to an *activation barrier* between the energy of the reactants and the products. This activation barrier makes the reaction of methane and oxygen very slow at room temperature. Applying a spark or lighted match to the burner head provides energy to overcome the activation barrier and we get a flame. The release of energy makes the reaction self-sustaining, that is, additional methane combusts immediately as it mixes with air in the hot flame.

Oxidation of iron to form $Fe(OH)_3(s)$ is the first step in the corrosion of iron and steel to rust or iron oxide, Fe_2O_3. The activation barrier for this reaction is much lower than for the oxidation of methane and metallic iron in contact with moist air will oxidize gradually at room temperature. We can't really compare the reaction rates of gases and solids. As an illustration, opening a hand warmer exposes iron powder to air and raises the temperature to $>50\ °C$ within a few minutes. The elevated temperature persists for several hours until all iron is oxidized. A balloon of methane and air at room temperature is unreactive indefinitely without a spark. Besides temperature, reaction rates are dependent on other reaction conditions. Rusting occurs much faster in the presence of salt, even with protective paint or zinc coatings.

The two reactions listed above are spontaneous and occur at a rate dependent on reaction conditions. They proceed because they produce products of lower energy than the reactants. This difference in energy is something that we can capture as heat or work. Electrochemical cells separate the two half reactions so we can capture that energy to do useful electrical work. We can also predict that mixing CO_2 and H_2O will not spontaneously form CH_4 and O_2. Likewise, oxidized iron will not spontaneously revert to metallic Fe. To make these reactions go in reverse, we must put energy into the system.

We discussed earlier that copper metal and silver ions undergo a spontaneous reaction to produce copper ions and silver metal.

$$Cu(s) + 2Ag^+(aq) \rightarrow Cu^{2+}(aq) + 2Ag(s)$$

Since we know that this reaction is spontaneous, we do not expect the reverse reaction to be spontaneous. No reaction occurs on adding a piece of Ag metal and to a solution that contains Cu^{2+}. Adding Cu wire to a room-temperature solution that contains Ag^+ initiates the reaction, but it is not obvious. We must wait some minutes or hours to be able to see formation of products. Having said that, the reaction is fastest at the beginning. As product forms, the rate gradually slows. The electroanalytical methods that we will discuss use cell designs that control the reaction rate so it is not a variable in making measurements. In the many measurement methods that we discussed in Part I of this text, reagents were chosen for rapid reaction. The redox titrations that are published in handbooks are pairs of reactants that react rapidly when they are mixed. An acid–base neutralization titration that is slow is performed as a back titration. Metal complexing agents that form UV/Vis-absorbing complexes must do so rapidly and completely.

Finally, will this reaction of Ag^+ and Cu continue indefinitely? No, at some point we reach equilibrium concentrations and the reaction stops. When we reach equilibrium, we write the reaction with arrows in both directions.

$$Cu(s) + 2Ag^+(aq) \rightleftharpoons Cu^{2+}(aq) + 2Ag(s)$$

We use the right arrow ($\rightarrow$) when we are talking about a reaction that is proceeding and concentrations are changing. We use the double harpoons ($\rightleftharpoons$) when a reaction has reached equilibrium. You will see the same reaction written both ways. The different notation indicates that we are considering either nonequilibrium or equilibrium concentrations of the reactants and products.

9.4.1 Really Big Concepts in Chemistry

To summarize the previous discussion, what do we mean when we say a reaction is complete? When a reaction is complete, the concentrations of the reactants and products reach equilibrium concentrations and do not change. On the microscopic level, chemical species continue to react. Reactant molecules, R, continue to form product molecules, P, and product molecules revert to reactant molecules. A dynamic model describes equilibrium as the condition when the rate of $R \rightarrow P$ equals the rate of $P \rightarrow R$. The net result is that there is no change in [R] and [P]. These really big concepts in chemistry are described by the following terms:

Thermodynamics describes the changes in the form of energy when a reaction occurs, for example, converting chemical energy to heat.
Kinetics describes how quickly or slowly a reaction occurs.
Equilibrium describes reactions in which the reactants and products coexist in unchanging concentrations.

Is a balloon filled with a mixture of H_2 and O_2 stable? Stable is a tricky term and I'll answer this question at the end of this discussion. You can fill a balloon with H_2 and O_2 and carry it around. So yes, the mixture is stable in practice. You might exercise some caution to avoid flames or sparks. Thermodynamically, we expect combustion of H_2 to be a spontaneous reaction (compare $E°$ values in Table 9.4). Why does it not react immediately on mixing? For the same reason we needed to apply a spark to a Bunsen burner.

We use the reaction coordinate model to graphically describe reaction energetics. Figure 9.2 shows this model, where the x-axis is the reaction coordinate, or the microscopic progress of the reaction, and the y-axis is energy.

E_a is the activation energy, the height of the activation barrier. At higher temperatures, a larger fraction of the reactant molecules will have enough kinetic energy to exceed E_a when they collide. ΔG is the difference in Gibbs energy between the reactants and the products.[4] Note that the relative

---

[4]General chemistry texts might use a diagram with ΔH, the heat of reaction, but the concept is the same.

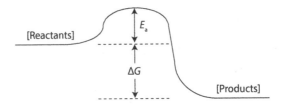

Figure 9.2 Reaction coordinate diagram for a spontaneous reaction.

energy between the starting reactants, [R], and the products, [P], are concentration dependent. This diagram is an energy snapshot of the reaction for a given set of concentrations of reactants and products. As concentrations change, the Gibbs energy difference between the reactants and products decreases. Figure 9.3 shows three snapshots as a reaction proceeds toward equilibrium. Figure 9.3(a) shows concentrations far from equilibrium, Figure 9.3(b) after the reaction has proceeded for some time, and Figure 9.3(c) is the model when reactant and product concentrations have reached equilibrium values.

To summarize, reactants that are mixed in nonequilibrium concentrations have chemical potential energy that drives the reaction toward equilibrium. This potential energy is the difference in Gibbs energy between the products and reactants, ΔG. The reactants will react to achieve equilibrium and reach the lower-energy condition. To answer the question about a mixture of H_2 and O_2, the reaction

$$H_2(g) + \frac{1}{2}O_2(g) \rightarrow H_2O(l) + energy\,(mostly\ heat)$$

is not stable thermodynamically, but we say that it is "metastable" kinetically. It is *not* an equilibrium situation. The reaction will not occur at room temperature until initiated with a spark. Once a spark provides enough energy for some of the H_2 and O_2 to overcome the activation barrier and react, the heat released sustains further reaction and the H_2/O_2 mixture can be explosive.

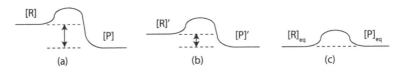

Figure 9.3 Reaction coordinate diagrams as a reaction goes to equilibrium.

9.5   REACTION ENERGIES, VOLTAGES, AND THE NERNST EQUATION

Reactants and products mixed in nonequilibrium concentrations have chemical potential energy that drives the reaction toward equilibrium. This potential energy is the difference in Gibbs energy between the products and reactants, ΔG. The reactants will react to achieve equilibrium and reach a lower-energy condition. For any arbitrary reactant and product concentrations, ΔG is given by

$$\Delta G = \Delta G^\circ + RT \ln(Q) \tag{9.1}$$

where

- ΔG° is the difference in Gibbs energy at standard concentrations
- R is the gas constant (8.3145 J/mol $\cdot$ K)
- T is absolute temperature (K)
- Q is the reaction quotient. (Q has the same form as any equilibrium constant expression, but the different symbol, Q vs K, indicates that the concentrations or partial pressures are not equilibrium values.)

This relationship shows that reactants at concentrations far from equilibrium will have greater Gibbs energy than reactants near equilibrium concentrations. At equilibrium, $\Delta G = 0$. We cannot extract energy from a system at equilibrium. In addition, note that

$$\Delta G^\circ = -RT \ln(K) \tag{9.2}$$

That these two quantities, ΔG° and K, are related should make sense. K tells us the relative concentrations of products to reactants at equilibrium, and ΔG° tells us how much energy we could get out of a reaction when they are at standard concentrations, that is, not at equilibrium concentrations.

9.5.1 E° and ΔG°

ΔG and E (or ΔG° and E° for standard concentrations) are both measures of the driving force of spontaneous reactions. The following relationships are given without proof:

$$\Delta G = -nFE \tag{9.3}$$

$$\Delta G^\circ = -nFE^\circ \tag{9.4}$$

where n is the number of moles of electrons transferred and F is the Faraday constant. The Faraday constant is the charge of one electron times the number of electrons per mole:

$$F = (1.60218 \times 10^{-19}\,\text{C})(6.02214 \times 10^{23}\,\text{mol}^{-1}) = 96485.3\ \text{C/mol}$$

You will also see it with equivalent units of 96485.3 J/(mol V). From Equation 9.3 we see that spontaneous reactions will have negative values of ΔG and positive values of E.

Recall that

$$\Delta G = \Delta G^\circ + RT\ \ln(Q)$$

Using Equation 9.3 and Equation 9.4 to substitute for ΔG and ΔG° we obtain

$$-nFE = -nFE^\circ + RT\ \ln(Q)$$

and rearranging gives

$$E = E^\circ - \frac{RT}{nF}\ln(Q) \tag{9.5}$$

This expression is the Nernst equation. We can use it to

- calculate the voltage of an electrochemical cell when the reactants and products are not at standard concentrations;
- find the concentration of one component in an equilibrium system from a voltage measurement if other components are not changing.

You will also see the Nernst equation written as

$$E = E^\circ - \frac{0.0592\,\text{V}}{n}\log(Q) \tag{9.6}$$

where RT/F at 298.15 K is 0.0257 V and $2.303 \times \log$ replaces ln.

The Nernst equation can be used to calculate E values for half reactions (see Example 9.6) or E values for overall reactions. Using the Nernst equation with overall reactions will tell us if a reaction is spontaneous or not. If it is positive, the reaction is spontaneous as written. If it is negative, the reaction occurs in the opposite direction as written. Note that for a reaction to occur the appropriate species must be present. One species must be in a form that can be reduced and another species must be in a form that can be oxidized.

Example 9.6 Nernst Equation.   Comparing E° values, we expect a spontaneous reaction when copper metal is placed in a solution of Ag^+. Will this reaction be spontaneous if the solution is at 100°C? You may use 0.1 M for ion concentrations in your calculation.

For this reaction condition, we can find E for each half reaction to predict if a reaction will occur spontaneously. The half reactions are

$$Ag^+(aq) + e^- \rightleftharpoons Ag(s) \qquad E° = 0.799\,V$$

$$Cu^{2+}(aq) + 2e^- \rightleftharpoons Cu(s) \qquad E° = 0.340\,V$$

Setting up the Nernst equation for each half reaction,

$$E_{Ag} = 0.799\,V - \frac{(8.3145\,J/mol\,K)(373.2\,K)}{(1)(96485\,C/mol)} \ln\left(\frac{1}{0.1M}\right)$$

$$E_{Ag} = 0.799\,V - 0.074\,V = 0.725\,V$$

Note that Q takes the form of an equilibrium constant expression for the half reaction as written, in this case as a reduction.

$$E_{Cu} = 0.340\,V - \frac{(8.3145\,J/(mol\,K)(373.2\,K)}{(2)(96485\,C/mol)} \ln\left(\frac{1}{0.1M}\right)$$

$$E_{Cu} = 0.340\,V - 0.037\,V = 0.303\,V$$

Looking at the E values for these half reactions at 100°C, we can predict that the reaction that occurs will be the same as at room temperature. The silver half reaction is significantly more positive than the copper half reaction. We can predict that placing copper metal in a boiling solution with Ag^+ will result in a spontaneous redox reaction.

Redox reactions eventually reach equilibrium and E becomes zero. So, analogous to the relationship between $\Delta G°$ and K, we have

$$E° = \frac{RT}{nF} \ln(K) \tag{9.7}$$

This gives us the relationship between $E°$ and K, so we see that either $E°$ or K values are useful in describing the extent to which a given set of reactants will go to products (see Example 9.7).

Example 9.7 K Calculation. Use the $E°$ values for the silver and copper half reactions to calculate K at 298 K.

We are jumping ahead a little for this example. To find K for the overall reaction, the $E°$ value for the overall reaction is needed. It is the difference between the $E°$ values of the half reactions:

$$E° = 0.799\,V - 0.340\,V = 0.459\,V$$

Now, we can use Equation 9.7 to find K, where $n = 2$ for the overall reaction

$$E^\circ = 0.459\,\text{V} = \frac{0.0592\,\text{V}}{2}\log(K)$$

$$\log(K) = \frac{(0.459\,\text{V})(2)}{0.0592\,\text{V}} = 15.5$$

$$K = 10^{15.5} = 3 \times 10^{15}$$

As you can see, a moderate difference in potential is equivalent to a very large equilibrium constant.

9.6   ELECTROCHEMICAL CELLS

Consider the spontaneous reaction of Zn metal in a solution of Cu^{2+}:

$$Zn(s) + Cu^{2+}(aq) \rightarrow Zn^{2+}(aq) + Cu(s) \quad \Delta G^\circ = -212.6\,\text{kJ/mol}$$

If we simply mix these reactants, that is, place a piece of Zn metal in a solution containing Cu^{2+}, they will react to produce Zn^{2+}, Cu metal, and heat. We can capture the energy that this reaction releases by setting up an electrochemical cell to separate the two half reactions. In such a setup, Zn is oxidized in one container, called a half-cell, and Cu^{2+} is reduced in a separate half-cell. A wire connects the two electrodes so that electrons can flow from the species being oxidized (Zn) to the species being reduced (Cu^{2+}).

As the reaction proceeds, the Zn electrode corrodes creating Zn^{2+} in the zinc half-cell solution. The Cu electrode is plated with fresh Cu(s), removing Cu^{2+} from the copper half-cell solution. This charge imbalance will stop the reaction and must be compensated. We do so with some type of salt bridge, which allows spectator ions to migrate to the two half-cells. Figure 9.4 shows a

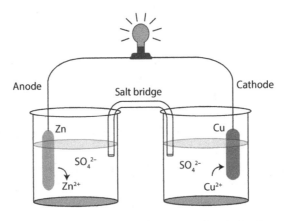

Figure 9.4   Schematic of an electrochemical cell.

schematic of this electrochemical cell. The salt bridge is a tube filled with a saturated KNO_3 solution. It has porous frits on the ends that prevent bulk mixing of the solutions. As the redox reaction proceeds, ions move through the salt bridge due to the electrostatic attraction to keep the two half-cells electrically neutral. NO_3^- will migrate into the left cell and K^+ will migrate into the right cell. A current of electrons moves through the wire and a current of ions moves via the salt bridge.

The driving force in the electrochemical cell is the same as for the case of just mixing the reactants. By separating the two half reactions, the electrons must travel through the wire so that we can capture the electrical energy to do work. In Figure 9.4, the electrical energy powers a light bulb. When we first set up this electrochemical cell and complete the circuit, we could measure a voltage through the wire. As the reaction proceeds, the system approaches equilibrium, the voltage drops, and given enough time goes to zero. Batteries are electrochemical cells that convert chemical energy to electrical energy. The reactants in fresh batteries are at nonequilibrium concentrations. As you use them, the reactants form products to approach equilibrium concentrations until the voltage drops to where the battery is no longer useful.

To refer to the electrodes in an electrochemical cell we follow these definitions:

Anode: Electrode where oxidation occurs (Zn in Figure 9.4).
Cathode: Electrode where reduction occurs (Cu in Figure 9.4).

The cell as diagrammed in Figure 9.4 operates as a *galvanic cell*, that is, it allows the spontaneous reaction to occur to capture the difference in Gibbs energy. If the light bulb is replaced by a power source, the reaction can be driven in reverse to "charge up" the cell. Zn^{2+} would be reduced to Zn metal, and Cu metal would be oxidized to Cu^{2+}. This process is called *electrolysis*. The cell is called an *electrolytic cell* and the electrode in the Zn half-cell is now called the cathode, and the electrode in the Cu half-cell is the anode. Using the definitions for anode and cathode as written above maintains consistency for both galvanic and electrolytic cells.

9.6.1 Shorthand Line Notation

To describe electrochemical cells, we use a shorthand notation with a vertical line for every phase boundary. Phase boundaries include electrode/solution boundaries and barriers between different solutions. We always start our notation with the anode on the left and the cathode on the right. For the example discussed earlier we abbreviate the cell as

$$Zn|ZnSO_4(aq, x\,M)\|CuSO_4(aq, y\,M)|Cu$$

The values in parentheses are the activities (in practice usually molar concentrations) for the species in solution. You will also see this shorthand notation used just for half-cells, for example,

$$Pt\big|H_2(g, 1.0\,atm)\big|H^+(aq, 1.0\,M)$$

which is the standard hydrogen electrode (SHE) written as the anode for an electrochemical cell. When partial pressures or concentrations are stated, the g and aq designations are somewhat redundant and not always written.

9.6.2 E_{cell} of Reactions

To find the voltage produced by an electrochemical cell, we simply sum all of the potentials in the circuit.

$$E_{cell} = \sum E$$

We will neglect small potentials that arise from the solution and salt bridge resistances and between electrical junctions. For a simple electrochemical system consisting of two half-cells, we simply sum the voltages produced by the two half reactions

$$E_{cell} = E_{cathode} + E_{anode}$$

When balancing reactions, we sometimes multiply one or both half reactions by a numerical factor so that the electrons cancel. The E or E° values do *not* get multiplied by these factors. For the example of the silver half reaction that we have used previously, try writing the Nernst equation for the half reaction with one electron and with two electrons. You will see that $n = 2$ has no effect because it cancels $1/[Ag^+]^2$ in the log term.

The E or E° of a half reaction that undergoes oxidation is the negative of the E or E° of the reduction half reaction. For the Cu and Zn example in Figure 9.4, we can calculate the cell voltage at standard concentrations. First, write each half reaction as a reduction. The more positive E° for the copper half reaction confirms our expectation that the reaction is spontaneous with copper being reduced and zinc being oxidized:

$$Cu^{2+}(aq) + 2e^- \rightarrow Cu(s) \qquad E^\circ = 0.340\,V$$
$$Zn^{2+}(aq) + 2e^- \rightarrow Zn(s) \qquad E^\circ = -0.763\,V$$

On writing the zinc half reaction in the direction of oxidation, we also change the sign of E°.

$$Cu^{2+}(aq) + 2e^- \rightarrow Cu(s) \qquad\qquad E^\circ = 0.340\,V$$
$$Zn(s) \rightarrow Zn^{2+}(aq) + 2e^- \qquad\qquad E^\circ = 0.763\,V$$

Now simply summing these $E°$ values gives

$$E°_{cell} = E°_{cathode} + E°_{anode} = 0.340\,V + 0.763\,V = 1.103\,V$$

Check that you get the same result (neglecting roundoff error) from the value of $\Delta G°$ and the relationship between ΔG and E_{cell}.

9.6.3   Example of a Concentration Cell

Consider two beakers containing silver nitrate solutions of different concentrations, say, 0.005 M and 0.2 M (Figure 9.5). If we open a valve between the two beakers, do we have a system at equilibrium? No, there is a concentration gradient and there is a driving force for diffusion to attain equal concentrations in both beakers.

We can set up an electrochemical cell to capture the chemical potential of this nonequilibrium system. Figure 9.6 shows the two containers connected by a salt bridge. Silver electrodes and a voltmeter complete the circuit. The two half reactions of this concentration cell are

$$Ag(s) \rightarrow Ag^+(aq) + e^-$$
$$Ag^+(aq) + e^- \rightarrow Ag(s)$$

which occur in the left and right half-cells, respectively, to ultimately achieve equal Ag^+ concentrations in each cell.

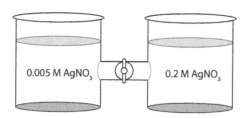

Figure 9.5   Two containers with a connecting bridge and valve.

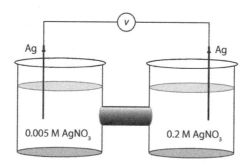

Figure 9.6   An electrochemical concentration cell.

The reactions in the two cells proceed to equalize the concentration in each cell. When the cell is connected, the silver electrode in the left half-cell will be oxidized to Ag^+. In the right half-cell Ag^+ will be reduced to silver metal onto the electrode. The reaction quotient for this reaction is

$$Q = \frac{[Ag^+]_{products}}{[Ag^+]_{reactants}}$$

and the Nernst equation for the overall reaction is

$$E = E^\circ - \frac{0.0592 \text{ V}}{1} \log(\frac{0.005 \text{ M}}{0.2 \text{ M}})$$

E° for the overall reaction is 0.0 V since it is the same half reaction on each side of the cell. Thus,

$$E = 0.0 \text{ V} - (-0.0950 \text{ V})$$
$$E = 0.0950 \text{ V}$$

The positive E value confirms that we expected a spontaneous reaction to occur since the system is not at equilibrium initially. If we set up this system and let the reaction occur, it would eventually reach an E of zero when the concentrations in the two cells became equal. In case you are wondering why batteries are not formed from simple concentration cells, this example is typical of the low voltage that such cells produce. Using different reactions in the two half-cells produces much larger voltages since E° will be larger than 0.0 V.

You-Try-It 9.B
The Nernst worksheet in you-try-it-09.xlsx contains examples of electrochemical cells. Predict cell voltages as a function of reactant concentrations.

9.6.4 Electroanalytical Methods

Before going on to the instrumental electroanalytical methods, some key points to remember about redox chemistry are as follows:

- When comparing half reactions, write them all as reductions, and the more positive will go as written.
- A spontaneous reaction will occur if the appropriate species are present.
- Use the Nernst equation when not working at standard concentrations.
- E_{cell} is the sum of all voltages in a circuit.
- Spontaneous reactions have positive E_{cell} and negative ΔG.
- When multiple redox reactions are possible, the reaction with the largest driving force, that is, the largest E_{cell}, will occur first.

This last point is the same as for other equilibria. When adding a strong base to a mixture of acids, the strongest acid is neutralized first. Similarly, when adding a strong base to a polyprotic acid, such as H_3PO_4, the base will neutralize the most acidic proton completely before removing other protons. A metal will form a complex with the ligand with the largest K_n. A precipitating agent will precipitate the ion of the least soluble salt before precipitating other ions. Adding an oxidizing agent to a sample will oxidize the most easily oxidized substance before oxidizing other substances.

Electrochemical or electroanalytical techniques are common in analytical chemistry owing to their simplicity and their ability to be miniaturized to create portable instruments. They also have a very significant advantage compared to other instrumental methods in that they can respond to one specific form of an analyte. Classical and spectroscopic methods often provide total elemental concentrations. Electrochemical methods can be specific for only one oxidation state, including different polyatomic ions, for example, NH_4^+ versus NO_3^-.

The following sections will describe the principles and instrumentation of potentiometry and voltammetry. After calibrating with standards, these electroanalytical methods determine an analyte concentration from an electrical voltage, V, or current, I, respectively. We already discussed the related method of coulometry in Section 3.7. Coulometry, including indirect coulometric titrations, uses the total charge, Q, required to react with all analyte to determine concentration. So, the common electrochemical methods of analysis (for an analyte A) can be classified as

coulometry: $Q \propto [A]$
potentiometry: $V \propto \log[A]$
voltammetry: $I \propto [A]$

These methods are electrochemical, meaning that a redox reaction or equilibrium is involved. Current is simply charge per unit time, that is, coulombs per second. Potential and current are related by Ohm's law:

$$V = IR \qquad (9.8)$$

where R is resistance. The electrochemical cells in the three methods are designed quite differently. Coulometry uses large electrodes to get complete reaction of the analyte. In potentiometry, the cell resistance is such that there is extremely low current. No redox reaction occurs, and the measured voltage is related to an equilibrium that is established at a sensor surface. In voltammetry, an applied voltage at a working electrode causes reduction or oxidation of the analyte. Unlike coulometry, experimental factors such as a small electrode area and diffusion to the electrode surface limit the measured current. The current is small so the bulk concentration of the analyte does not change, but $I \propto [A]$.

9.7 POTENTIOMETRY

Potentiometry is the measurement of a voltage to determine the concentration of an analyte ion. The electrochemical potential actually responds to analyte activity. Calibrating an instrument versus standard concentrations allows the measurement to be related to an unknown concentration. Such measurements can be made with an electrochemical cell as described in Section 9.6. The key difference for potentiometry is that the cell is constructed so that only one potential in the circuit changes, and this variable potential is sensitive to an analyte concentration. It is also important that the experimental apparatus does not change the analyte concentration as it is being measured. This aspect is accomplished by constructing the cell so that there is no appreciable electrical current during the measurement.

Potentiometric measurements are often not as sensitive as spectroscopic and other types of analytical methods, but they have the advantage of being sensitive to oxidation state. For example, potentiometric electrodes are available for NH_4^+, NO_3^-, and NO_2^-. Many other methods will provide only total nitrogen. The advantages of potentiometric measurements include

- species specificity (oxidation state and polyatomic ions)
- portability
- miniaturization.

Thus, they can be used for in situ applications such as the use of a microelectrode to measure ion concentrations in vivo and even in a single cell. They are also common in process applications for continuous monitoring of an industrial process or environmental system.

A key aspect of potentiometry is that a voltage develops at an electrode surface or across a sensing membrane. To measure that voltage, a potentiometric device requires the following:

- An indicator or test electrode or membrane that is sensitive to the activity of an analyte ion in solution (and thus relatable to analyte concentration).
- A reference electrode.
- One or more junctions between half-cells (equivalent to a salt bridge).
- A voltmeter for the readout.

These components are arranged in a complete circuit that includes the analyte solution (Figure 9.7). The simplest indicator electrode is a metal electrode in contact with a solution containing ions of the same metal. The metal ions in solution have some affinity to "stick" to the surface, leading to a surface charge. This surface charge will vary depending on the analyte concentration, providing the analytical signal. The reference electrode, analogous to the left half-cell in Figure 9.4, is designed to provide a constant voltage that does not

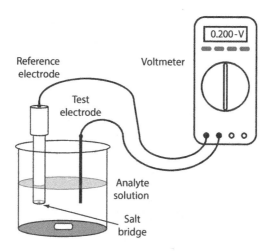

Figure 9.7 Voltage measurement using a metal indicator electrode.

change when the test solution changes. The circuit is completed in Figure 9.7 using some type of salt bridge between the analyte solution and the reference electrode. In practice this salt bridge or junction is accomplished with a pinhole or porous frit that prevents the solution in the reference electrode from mixing with the analyte solution.

Simple indicator electrodes are reproducible for only a few metals such as Ag. For example, you cannot place a Ca metal electrode in aqueous solution to measure Ca^{2+} because of its reactivity. More robust and selective sensing electrodes are usually necessary. As an example, a CuS crystal is a common sensing element to detect Cu^{2+}. For a generic working electrode, a Pt or Au metal electrode is often used to sense the overall redox potential of a solution. These electrodes are known as *oxidation-reduction potential (ORP)* electrodes. The redox potential will depend on both the concentration and type of ions in solution. A typical measurement range is $\pm$ 1100 mV, where solutions containing reducing agents produce negative potentials and solutions containing oxidizing agents have positive readings.[5] This approach is useful in detecting an endpoint in a titration or in monitoring the oxidizing or reducing power of industrial waste streams or natural waters. They are also commonly used to check the oxidizing power of swimming pools. ORP electrodes are not as useful for direct readings if the composition of a solution is complex, variable, or sample dependent. In these cases, electrodes with specificity for different analytes are needed.

Potentiometric electrodes are available for a wide variety of cations, anions, and gases in solution. Gases are measurable if they are in equilibrium with an ionic form such as NH_4^+ for NH_3 and HCO_3^- for CO_2. In general, the potential at an indicator electrode is given by the Nernst equation for the half

---

[5]When corrected to a standard hydrogen electrode reference electrode. Actual measurements depend on the specific reference electrode.

reaction that occurs at that electrode. For a generic half reaction between the oxidized, ox, and reduced, red, forms of an electroactive species

$$\text{ox} + \text{ne}^- \rightleftharpoons \text{red}$$

The Nernst expression for the half reaction is

$$E = E^\circ - \frac{RT}{nF} \ln \frac{a_{\text{red}}}{a_{\text{ox}}} \tag{9.9}$$

where the quantities in the ln term are the activities of the reduced and oxidized forms. For the typical case of an analyte ion interacting with the surface of an electrode or membrane, the equation for the half-cell potential simplifies to

$$E = E^\circ + \frac{RT}{z_i F} \ln(a_i) \tag{9.10}$$

where z_i is the charge, including sign, of the analyte ion and a_i is the activity of the analyte ion. This equation will give the potential of the half-cell in a circuit that is in contact with the test solution. It is often called the *Nernst–Donnan equation*, and membrane potentials are often called *Donnan potentials*. To avoid confusion, we will simply call them ISE (ion-selective electrode) equations for each example that we encounter.

Since potentiometric measurements are almost always calibrated with standards of known concentration, we use concentration notation in place of activity for the rest of the expressions in this chapter. The practical reason to do so is simply to have a notation that is easier to read. The experimental condition to make this approximation valid is that the measured standards and samples must have similar ionic strength. Experimentally, an ionic strength "buffer" is mixed with all standard and test solutions before measurements. A typical procedure is to dilute samples and standards 1:1 with an ionic strength adjustment buffer (ISAB). An ISAB typically consists of 1 M NaCl or other strong electrolyte to adjust the ionic strength. An ISAB solution might also contain a buffer to control pH and a metal complexing agent to remove interferences, depending on the target analyte and potential interferences. At 25°C our half-cell equation for analyte A becomes

$$E = E^\circ + \frac{RT}{z_i F} \log[A_i] \tag{9.11}$$

9.8 ION-SELECTIVE ELECTRODES (ISE)

9.8.1 Construction and Use

An "ion-selective" or "ion-sensitive" electrode is a sensing electrode that has a membrane surface that is involved in a chemical equilibrium with ions in solution. Different concentrations of the ion in solution produce different degrees of surface charge on the sensing membrane, and thus a change in the potential across the membrane. Any ionic analyte can be detected by potentiometry for which a robust ion-selective membrane is available. Membranes are made from glass, crystals, or polymers. The general concept is similar to an ion-exchange equilibrium, where ions in solution partition to some degree to the stationary phase.

As an example, fluoride can be sensed using a thin slice of a LaF_3 crystal to form a membrane. Both surfaces of the membrane are in equilibrium with the adjacent solutions

$$LaF_3(s) \rightleftharpoons LaF_2^+(s) + F^-(aq)$$

where $LaF_2^+(s)$ represents defects or "holes" of missing fluoride ions at the crystal surface. The LaF_3 membrane develops a potential between the inner and outer surfaces, depending on the extent of the defects on each surface. The defects on the inner surface depend on the $[F^-]$ of the internal solution, which is kept fixed. The defects on the outer surface depend on the $[F^-]$ of the test solution, which is variable for different solutions.

Figure 9.8 shows a schematic of an ISE measurement with the placement of the sensing membrane. Within the ISE body is the internal solution of fixed $[F^-]$ and an internal reference electrode. Measurements are made by reading the voltage between the ISE and reference electrode with a voltmeter. A junction or salt bridge on the end of the reference electrode completes the electrical circuit. Analyte solutions are stirred for uniformity so what contacts the electrode surface is representative of the bulk solution.

As a practical note, commercial ISEs often incorporate the working and reference electrodes into one unit (Figure 9.9). In this schematic, the outer surface of the LaF_3 membrane is visible on the end of the electrode. The other membrane surface is in contact with an inner solution and internal reference

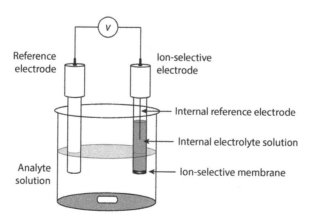

Figure 9.8   Set up to make a measurement with an ion-selective electrode.

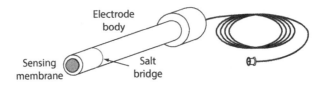

Figure 9.9 An integrated ISE assembly.

electrode in the center of the electrode assembly. A porous membrane on the side of the electrode serves as a salt bridge between the test solution and the external reference electrode, which is in an outer concentric compartment of the electrode body.

The voltmeter in Figure 9.8 measures E_{cell}, which is the sum of all individual potentials, E_i, in the electrical circuit:

$$E_{cell} = \sum E_i \tag{9.12}$$

The cell is constructed so that only the potential across the sensing membrane is variable as the analyte solution changes. For the LaF_3 equilibrium, the expression for the potential across the membrane, called the boundary potential, E_b, is

$$E_b = E° - (0.0592\,V)\log\,[F^-]$$

Where substituting the -1 charge of fluoride ion for z_i changes the sign before the log term. Adding all other potentials in the circuit gives us the voltage measured by the voltmeter:

$$E_{cell} = E_j + E_{ext.ref.} + E_{int.ref.} + (E° - (0.0592\,V)\log\,[F^-])$$

where

- E_j is one or more junction potentials, which are usually small, but develop across interfaces such as the salt bridge.
- $E_{ext.ref.}$ is the potential of the external reference electrode (the left electrode in Figure 9.8).
- $E_{int.ref.}$ is the potential of the internal reference electrode (the inner compartment of the ISE electrode in Figure 9.8).

Since $E°$, E_j, $E_{ext.ref.}$, and $E_{int.ref.}$ should be constant in a properly maintained cell, this expression simplifies to

$$E_{cell} = E_{const} - (0.0592\,V)\log\,[F^-]$$

where E_{const} is found by calibrating the cell with standard solutions. Similarly, a plot of E_{cell} versus log [F⁻] will produce a slope that should be near -0.0592 V, depending on the temperature of the cell. Example 9.8 illustrates the use of an ISE equation with a one-point calibration.

Example 9.8 Potential Calculation.   What is $[Ca^{2+}]$ in a water sample that produced a voltage of 0.0333 V using a Ca^{2+} ISE? Calibrating the cell with a standard solution of 0.163 mM Ca^{2+} produced a voltage of -0.0440 V. You may assume that the temperature was 25°C for each measurement. Prior validation experiments showed linear calibration line from 0.01–100 mM Ca^{2+}.

At constant temperature and no changes to other potentials in the circuit, E_{const} is the same when measuring the standard and test solutions. Since Ca^{2+} has a charge of +2, the ISE equation is

$$E_{cell} = E_{const} + (0.0296\,V)\log [Ca^{2+}]$$

Calibration with the standard solution allows us to determine E_{const} for this ISE:

$$-0.0440\,V = E_{const} + (0.0296\,V)\log(1.63\times10^{-4}\,M)$$
$$E_{const} = 0.0681\,V$$

Now we can use this E_{const} value to solve for the unknown $[Ca^{2+}]$:

$$0.0333\,V = 0.0681\,V + (0.0296\,V)\log[Ca^{2+}]$$
$$[Ca^{2+}] = 0.0667\,M$$

9.8.2 Reference Electrodes

A key factor in making repeatable potentiometric measurements is the stability of the reference electrode. To complete the electrical circuit of the cell, it must be in contact via a junction or salt bridge with the test solution. However, it must also provide an invariant potential even as the composition of the test solution changes. Potentiometric cells are designed to minimize the electrical current and the flow of ions. Doing so allows the concentrations in the reference cell to not change. Let us look at one reference electrode in detail to help us understand potentiometric instruments. A common reference cell is the silver/silver chloride half-cell:

$$\| \,AgCl(sat'd),\,KCl(x\,M)|\,Ag$$

The half reaction is

$$AgCl(s) + e^- \rightleftharpoons Ag(s) + Cl^-(aq)\quad E° = 0.222\,V$$

A convenient feature of this half reaction is that the silver is in a solid form when it is oxidized (left side) or reduced (right side). The voltage is then only dependent on [Cl$^-$], which is usually made relatively high so small losses or additions has a small effect on E.

A general concept of thermodynamic data is that we can determine values by combining reactions. You probably found quantities such as heat of fusion, ΔH_f, for molecular reactions using tables of elemental data in your general chemistry course. To illustrate this general concept, let us see how the $E°$ of the silver/silver chloride half reaction is obtained. The $E°$ of the Ag$^+$/Ag half reaction is well known at 0.799 V. Similarly, K'_{sp} for AgCl(s) is known to be 1.82×10^{-10}. We can use these known values to determine $E°$ for the Ag/AgCl half reaction:

$$Ag^+(aq) + e^- \rightleftharpoons Ag(s) \quad E° = 0.799\,V$$

Since $n = 1$ the Nernst equation for this half reaction is

$$E = E° - (0.0592\,V)\log \frac{1}{[Ag^+]}$$

Using $K'_{sp} = [Ag^+][Cl^-]$ to substitute for [Ag$^+$] gives us

$$E = E° - (0.0592\,V)\log \frac{[Cl^-]}{K'_{sp}}$$

Rearranging gives

$$E = 0.799\,V + (0.0592\,V)\log K'_{sp} - (0.0592\,V)\log [Cl^-]$$

and on inserting the value for K'_{sp}, we obtain

$$E = 0.222\,V - (0.0592\,V)\log[Cl^-]$$

At the standard concentration of 1 M Cl$^-$, $E = E° = 0.222$ V, as given above. Note that the potential at the Ag/AgCl reference electrode depends only on the chloride concentration at a constant temperature. The chloride solution in the reference electrode is isolated from contact with the analyte solution via a membrane or small junction so that there is no mixing of solutions. Thus, the reference potential will not change as the electrode is placed in different analyte solutions.

9.8.3   pH Electrode

The pH electrode is probably the most common example of an ISE, in this case for a proton or H$^+$. An ISE cell for pH is constructed similar to the schematic in Figure 9.8, but the LaF_3 membrane is replaced with a proton-sensitive membrane and a compatible internal reference solution.

The sensing element in a pH electrode is a thin glass membrane (<0.1 mm thick). Glass can be protonated, and the degree of protonation depends on the $[H_3O^+]$ in contact with the glass surface. An internal reference solution keeps the degree of protonation, and thus the electrical charge, constant at the interior surface. The potential across the membrane then depends only on the $[H_3O^+]$ in the analyte solution. A complete pH meter consists of an internal reference electrode, the H_3O^+-selective membrane, junction, external reference electrode, and a voltmeter. Similar to the fluoride electrode in Figure 9.9, commercial pH electrodes often combine all electrodes into an integrated probe that connects to a pH meter with control electronics and display.

The measured voltage is the sum of all potentials in the circuit:

$$\underset{E_{\text{ext.ref.}}}{Ag\,|\,AgCl(sat'd),\,KCl(1.0\,M)}\quad\underset{E_j}{\|}\quad\underset{}{[H_3O^+]}\quad\underset{E_b}{|glass|}$$

$$\underset{E_{\text{int.ref.}}}{H_3O^+(0.1\,M),\,KCl(0.1\,M),\,AgCl(sat'd)\,|\,Ag}$$

where

- $E_{\text{ext.ref.}}$ is the potential of the external reference,
- E_j is a junction potential across the salt bridge,
- E_b is the boundary potential across the glass membrane,
- $E_{\text{int.ref.}}$ is the potential of the internal reference electrode.

E_b is the only potential that varies, being dependent of $[H_3O^+]$ in the analyte solution. All other potentials in the circuit can be combined as E_{const}. The measured voltage, E_{cell}, is then

$$E_{\text{cell}} = E_{\text{const}} + (0.0592\,V)\log[H_3O^+]$$
$$E_{\text{cell}} = E_{\text{const}} - (0.0592\,V)pH$$

The temperature in the 0.0592 V term is assumed to be 25°C. The net effect is that the response of the pH electrode is linear with pH. If you have used a pH meter, you probably calibrated the meter with standard buffer solutions of known pH, often pH = 7 and pH = 4. As for any potentiometric measurement, the calibration corrects for day-to-day differences in room temperature and any changes in the junction potential.

9.8.4 Nikolsky Equation

The discussion of ISEs so far has neglected interferences, but no ion-selective membrane material is perfectly selective. Real membrane materials will have some small sensitivity to other ions besides the target analyte. The Nikolsky–Eisenman equation includes contributions to a measured potential from these other ions. A simplified form for the case of interfering ions of the same charge as the analyte ion and $T=25\,°C$ is:[6]

$$E_{ISE} = E_{const} + \frac{0.0592\ V}{z_i}\log\left(a_i + k_{i,j}a_j + k_{i,k}a_k + \ldots\right)$$

Where z and a have their usual meanings, the "i" subscript refers to the analyte ion, and "j", "k", etc. refer to interfering ions. The $k_{i,j}$ factors are potentio-metric selectivity coefficients. These coefficients are the relative sensitivity of a given ISE membrane to an interfering ion (j, k, etc.) versus the analyte ion, i. They are a number between 0 and 1, and a selective membrane will have a very small value. Manufacturer specifications usually provide the selectivity coefficients for major interferences for a given ISE. Example 9.9 shows a sample calculation.

The most common occurrence of an ISE interference is the alkaline or sodium error that occurs for pH electrodes. Commercial pH electrodes will generally give a measurement that is less than the true pH for very alkaline solutions. The exact pH where an electrode deviates depends on the specific glass membrane. Inexpensive pH meters may deviate starting at pH = 10. More specialized glass formulations can measure accurately up to pH = 12. The error occurs because the proton concentration is so low, $[H^+] < 10^{-12}$ M, that the glass senses the Na^+ and other cations that are in the test solution. In practice it is not a problem, solutions with a pH > 12 are easily titrated. Commercial pH electrodes also have an acid error at pH < ≈1. This error arises from the glass being saturated with H^+ and losing waters of hydration. As for the very high pH case, these solutions are so concentrated that they can be titrated with high accuracy.

Example 9.9 ISE Calculation with Interference. Calibrating an ammonium ISE with NH_4^+ standard solutions produced a calibration function of $E_{ISE} = 273.1$ mV + (53.5 mV)log[NH_4^+] with [NH_4^+] in ppm. The ISE has a selectivity factor, $k_{NH4,K}$, for K^+ of 0.069. Measurement of a river water sample gave a voltage reading of 280.5 mV. A separate potassium measurement of this water gave [K^+] = 12.4 ppm. Calculate [NH_4^+] for this water sample neglecting and including the presence of K^+.[7]

When the K^+ interference is present, the ISE expression will be:

$$E_{ISE} = 273.1\text{mV} + (53.5\ \text{mV})\log\left\{\left[NH_4^+\right] + (0.069)\left[K^+\right]\right\}$$

---

[6]Here I deviate from IUPAC conventions, Umezawa, Y. et al., *Pure Appl. Chem.* **2000**, *72*, 1851.
[7]Ion Selective Electrode measurements of NH4 and K in natural waters, *Nico2000 Ltd—Test Report CCR/20/11/07*, http://www.nico2000.net/analytical/ammonium/NH4inrealsamples.htm; accessed August 2022.

First neglect the potassium interference, substitute 280.5 mV for E_{ISE}, and solve for $[NH_4^+]$:

$$280.6 \text{ mV} = 273.1 \text{mV} + (53.5 \text{ mV}) \log [NH_4^+]$$

$$[NH_4^+] = 10^{0.14} = 1.38 \text{ ppm}$$

Now repeat the calculation with $k_{NH4,K}$ and $[K^+]$ included:

$$280.6 \text{ mV} = 273.1 \text{mV} + (53.5 \text{ mV}) \log \{[NH_4^+] + (0.069)(12.4 \text{ ppm})\}$$

$$10^{0.14} = [NH_4^+] + (0.069)(12.4 \text{ ppm})$$

$$[NH_4^+] = 1.38 \text{ ppm} - 0.856 \text{ ppm} = 0.52 \text{ ppm}$$

We can see in this example that an interference can have a significant effect on an ISE measurement.

You-Try-It 9.C

The **potentiometry** worksheet in **you-try-it-09.xlsx** contains ISE measurements of standards and test portions. Plot the standard data and use the calibration function to determine analyte concentration in the unknowns.

9.8.5   ISFETs and CHEMFETs

You might use pH or sensing electrodes that look different from the ion-selective membranes described already. Some of the more interesting progress in potentiometric measurements involves miniaturizing sensors and making sensor arrays. The principle is the same as for the ion-selective membranes we discussed, although many designs utilize solid-state transistor technology. Examples include ISFETs (ion-sensitive field effect transistors) and CHEMFETs (chemically modified field effect transistors). The basic principle is that an equilibrium occurs on a coating on an FET. Field-effect transistors operate through a gate voltage controlling the current that passes through the transistor. Surface charge on the coating of the gate electrode can provide this voltage, and thus the current through the transistor is dependent on an analyte concentration in solution. Figure 9.10 shows a simplified schematic of an ISFET.[8] In this example, the analyte ion will exist in an equilibrium with the electroactive membrane. The analyte concentration will affect the charge on the electroactive membrane, which will control the conductivity of the

---

[8]Adapted with permission from Wohltjen, H., Chemical Microsensors and Microinstrumentation, *Anal. Chem.* **1984**, *56*, 87A–103A. Copyright 1984 American Chemical Society.

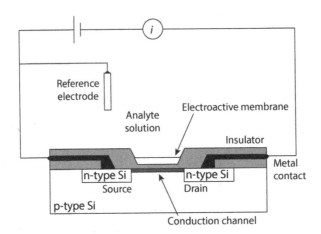

Figure 9.10 ISFET schematic. *Source:* Adapted with permission from Wohltjen, H., Chemical Microsensors and Microinstrumentation, Anal. Chem. 1984, 56, 87A-103A. Copyright 1984 American Chemical Society.

conductance channel. A higher analyte concentration will result in a larger current, *i*, measured between the source and drain. Note that this schematic is not drawn to scale; the ISFET sensing element can be made very small, on the order of 0.1 mm diameter. Such miniaturization also makes the ISFET technology amenable to creating solid-state devices containing arrays of sensors.

9.9   VOLTAMMETRY

In potentiometry, the Gibbs energy difference across a membrane generates the analytical signal. There is essentially no current flow. Electrolytic methods are an electroanalytical approach that uses an external source of energy to drive an electrochemical reaction, which would not normally occur. The externally applied driving force is either an applied current or an applied potential. When a constant current is applied, the resulting potential is the analytical signal. This *galvanostatic technique* is used to characterize materials and measure reaction rates. We will discuss a *voltammetric method* that uses applied potential to generate a measurable current. This current is the analytical signal and for many techniques is directly proportional to analyte concentration.

Voltammetry refers to a number of measurement techniques, including polarography, anodic stripping voltammetry, hydrodynamic voltammetry, and cyclic voltammetry in which an applied potential drives a redox reaction to produce a current, *I*. Unlike potentiometric measurements, which employ two electrodes on either side of a membrane, voltammetric measurements utilize a three-electrode electrochemical cell. The use of the three electrodes (working, auxiliary, and reference) along with a potentiostat allows accurate application of potential functions (waveforms) and the measurement of the resulting current. Figure 9.11 shows a simplified setup for a voltammetric measurement.

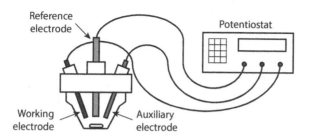

Figure 9.11 Simplified voltammetry experiment.

A stir bar is shown in the bottom of the cell, but stirring is only used for hydrodynamic and anodic stripping voltammetry (ASV). Not shown for clarity are ports to purge with inert gas to remove O_2 or to add reagents. Note that the cell is not drawn to scale; typical cells hold 10–100 ml of sample solution. Common configurations use an Ag/AgCl or saturated calomel (SCE) reference electrode, a platinum auxiliary electrode, and a glassy carbon working electrode. Gold, platinum, and silver are also used for the working electrode.

The three-electrode arrangement allows the current to flow between the working and auxiliary electrodes so that the composition of the reference electrode will not be affected as the measurement is made. Recall that in potentiometry the electrochemical cell was designed for essentially no current flow. In voltammetry, the current is the analytical signal and it must be generated and measured without affecting the applied potential, which is measured relative to the reference electrode.

A potentiostat is a voltage source that provides the applied voltage to the working electrode relative to the reference electrode. Commercial instruments are often a combination of a potentiostat and a galvanostat, a constant current source, and can apply the potential in a number of waveforms as described subsequently for specific techniques. Figure 9.12 shows a simplified schematic of a voltammetry experiment.

The different voltammetric techniques that are used for analytical measurements are distinguished from each other primarily by the potential function that is applied to the working electrode to drive the reaction and by the material used as the working electrode. Common techniques include the following:

- Hydrodynamic voltammetry
- Polarography: normal pulse polarography (NPP) and differential pulse polarography (DPP)
- Cyclic voltammetry (CV)
- Anodic-stripping voltammetry (ASV).

A common feature of all of these methods is that experimental conditions are designed so that the resulting electric current is proportional to analyte

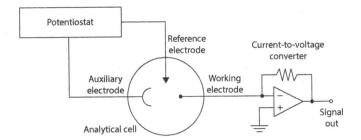

Figure 9.12 Schematic of a voltammetric cell.

concentration. This current is termed the faradaic current. When a potential is applied to an electrode, a capacitive current also occurs as ions in solution migrate to the electrode surface. Many of the specific experimental conditions and applied waveforms discussed are designed to maximize the measurement of the faradaic current versus the capacitive current background.

9.9.1 Polarography

Polarography is a specific type of measurement that falls into the general category of linear-sweep voltammetry. Linear-sweep voltammetry is a general term applied to any voltammetric method in which the potential applied to the working electrode is varied linearly in time. These methods would include polarography, CV, and rotating disk voltammetry. The slope of the applied potential ramp has units of volts per unit time, and is generally called the scan rate of the experiment. Figure 9.13 shows a simple linear sweep scan (b) and the resulting current (a). In this example, there are two analyte ions that cause an increase in the measured current. The two analytes have different redox potentials, and occur at different applied potentials.

The value of the scan rate may be varied from as low as millivolts per second (typical for polarography experiments) to as high as 1,000,000 V/s (attainable when ultramicroelectrodes are used as the working electrode). With a linear potential ramp, the faradaic current is found to increase at higher scan rates. This is due to the increased flux of electroactive material to the electrode at the higher scan rates. The amount of increase in the faradaic current is found to scale with the square root of the scan rate. This seems to suggest that increasing the scan rate of a linear-sweep voltammetric experiment could lead to increased analytical signal to noise. However, the capacitive contribution to the total measured current scales directly with the scan rate. As a result, the signal to noise ratio of a linear-sweep voltammetric experiment decreases with increasing scan rate.

The current versus potential response of a polarographic experiment, a linear sweep method controlled by combined convection/diffusion mass transport, has a typical sigmoidal shape. That is, it levels off against applied

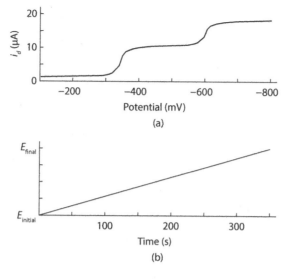

Figure 9.13   Linear sweep.

potential once the current is limited by mass transport of the analyte to the electrode surface. What makes polarography different from other linear-sweep voltammetry measurements is that it uses the dropping mercury electrode (DME). An electrochemical cell is usually a triangular glass vessel directly below a Teflon support that holds a tube of mercury. The tube extends into the sample solution and a mercury drop forms at the end of this tube to serve as the working electrode. The advantage of the DME is that it provides a clean electrode surface throughout the measurement and H^+ reduction requires a relatively high voltage on Hg.

A plot of the current versus potential in a polarography experiment will have current oscillations due to the drops of Hg falling from the capillary. If the maximum current of each drop is plotted, the result is the sigmoidal shape shown in Figure 9.13a. The limiting current (the plateau on the sigmoid), called the *diffusion current* because diffusion is the principal contribution to the flux of electroactive material at this point of the Hg drop life, is related to analyte concentration by the Ilkovic equation

$$i_d = 708nD^{1/2}m^{2/3}t^{1/6}c \qquad\qquad (9.13)$$

where D is the diffusion coefficient of the analyte in the medium (cm^2/s), n is the number of electrons transferred per mole of analyte, m is the mass flow rate of Hg through the capillary (mg/s), t is the drop lifetime in seconds, and c is the analyte concentration (mol/cm^3).

There are a number of limitations to the polarography experiment for quantitative analytical measurements. Because the current is continuously measured during the growth of the Hg drop, there is a substantial contribution from capacitive current. As the Hg flows from the capillary end, there is initially a large increase in the surface area. As a consequence, the initial current is dominated by capacitive effects as charging of the rapidly increasing interface occurs. Toward the end of the drop life, there is little change in the surface area, which diminishes the contribution of capacitance changes to the total current. At the same time, any redox process that occurs will result in a faradaic current that decays approximately as the square root of time (due to the increasing dimensions of the Nernst diffusion layer). The exponential decay of the capacitive current is much more rapid than the decay of the faradaic current; hence, the faradaic current is proportionally larger at the end of the drop life. Unfortunately, this process is complicated by the continuously changing potential that is applied to the working electrode (the Hg drop) throughout the experiment. Because the potential changes during the drop lifetime (assuming typical experimental parameters of a 2 mV/s scan rate and a 4 s drop time, the potential can change by 8 mV from the beginning to the end of the drop), the charging of the interface (capacitive current) has a continuous contribution to the total current, even at the end of the drop when the surface area is not rapidly changing. As such, the typical signal to noise ratio of a polarographic experiment allows detection limits of only approximately 10^{-5} or 10^{-6} M. Better discrimination against the capacitive current can be obtained using the pulse polarographic techniques described next.

9.9.2 Normal Pulse Polarography (NPP)

Pulse polarographic techniques are voltammetric measurements that are variants of the polarographic measurements, which try to minimize the background capacitive contribution to the current by eliminating the continuously varying potential ramp and replacing it with a series of potential steps of short duration. In NPP, each potential step begins at the same value (a potential at which no faradaic electrochemistry occurs), and the amplitude of each subsequent step increases in small increments (Figure 9.14). When the Hg drop is dislodged from the capillary (by a drop knocker at accurately timed intervals), the potential is returned to the initial value in preparation for a new step.

For this experiment, the polarogram is obtained by plotting the measured current versus the potential at which the step occurs. As a result, the current is not followed during Hg drop growth, and the normal pulse polarogram has

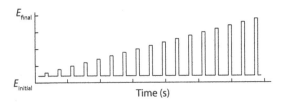

Figure 9.14   Normal pulse polarography waveform.

the typical shape of a sigmoid. By using discrete potential steps at the end of the drop lifetime (usually during the last 50–100 ms of the drop life, which is typically 2–4 s), the experiment has a constant potential applied to an electrode with nearly constant surface area. After the initial potential step, the capacitive current decays exponentially while the faradaic current decays as the square root of time. The diffusion current is measured just before the drop is dislodged, allowing excellent discrimination against the background capacitive current. In many respects, this experiment is akin to conducting a series of chronoamperometry experiments in sequence on the same analyte solution. The NPP method increases the analytical sensitivity by 1–3 orders of magnitude (limits of detection 10^{-7}–10^{-8} M) relative to normal dc polarography.

9.9.3 Differential Pulse Polarography (DPP)

DPP is a polarographic technique that uses a series of discrete potential steps rather than a linear potential ramp to obtain the experimental polarogram. Many of the experimental parameters for DPP are the same as for NPP (for example, accurately timed drop lifetimes with a potential step duration of 50–100 ms at the end of the drop lifetime). Unlike NPP, however, each potential step has the same amplitude, as shown in Figure 9.15. After each step, the potential returns to a slightly higher potential than the previous one.

In this manner, the total waveform applied to the DME is very much like a combination of a linear ramp with a superimposed square wave. The differential pulse polarogram is obtained by measuring the current immediately before the potential step, and then again just before the end of the drop lifetime. The analytical current in this case is the difference between the current at the end of the step and the current before the step (the differential current). This differential current is then plotted against the average potential (average of the potential

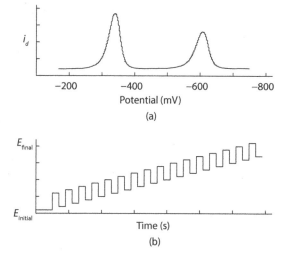

Figure 9.15 Differential pulse polarography.

before the step and the step potential) to obtain the differential pulse polarogram. Because the result is a differential current, the polarogram in many respects is similar to the differential of the sigmoidal normal pulse polarogram. As a result, the differential pulse polarogram is peak shaped.

DPP has even better ability to discriminate against capacitive current because it measures a difference current (helping to subtract any residual capacitive current that remains prior to each step). Limits of detection with DPP are 10^{-8}–10^{-9} M.

9.9.4 Cyclic Voltammetry (CV)

CV is an electrolytic method that uses microelectrodes and an unstirred solution so that the measured current is limited by analyte diffusion at the electrode surface. The electrode potential is ramped linearly to a more negative potential, and then ramped in reverse back to the starting voltage. The forward scan produces a current peak for any analytes that can be reduced through the range of the potential scan. The current increases as the potential reaches the reduction potential of the analyte, but then falls off as the concentration of the analyte is depleted close to the electrode surface. As the applied potential is reversed, it reaches a potential that reoxidizes the product formed in the first reduction reaction. Reaching this potential produces a current of reverse polarity from the forward scan, creating an oxidation peak that usually has a similar shape as the reduction peak.

Figure 9.16 shows a cyclic voltammogram of 5 mM potassium ferricyanide, $K_3Fe(CN)_6$, in 0.1 M KNO_3. First consider the forward scan, starting at the left of the upper part of the plot. The applied voltage starts at 0.7 V and is scanned to a more negative potential. The measured current changes little until the applied voltage approaches 0.3 V where the ferricyanide ion, $Fe(CN)_6^{3-}$, begins to be reduced to the ferrocyanide ion, $Fe(CN)_6^{4-}$. The current signal increases until it peaks, i_p, and begins to decrease. The dashed lines show how to determine i_p. Note that it is similar to the sigmoidal shape obtained in a polarography experiment (Figure 9.13), except that the current

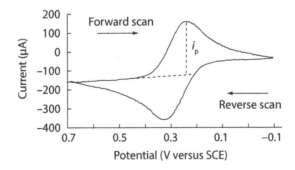

Figure 9.16 Cyclic voltammogram.

drops after reaching a peak. At the end of the forward scan, the applied voltage reverses direction and the current signal now appears below the forward scan. Since oxidation occurs, the current flows in the reverse direction and generates a negative peak. For a reversible reaction, the heights of the cathodic and anodic peaks are equal. Lines to find i_p are not shown for the anodic peak for clarity, but the process is the same as shown for the cathodic peak.

The peak current, i_p, at $T=25°C$ is described by the Randles–Sevcik equation

$$i_p = (2.69 \times 10^5) n^{3/2} A c D^{1/2} v^{1/2} \qquad (9.14)$$

The numerical value includes the Faraday constant and a kinetic factor. n is the number of moles of electrons transferred in the reaction. A is the area of the electrode (cm^2), c is the analyte concentration (mol/cm^3), D is the diffusion coefficient (cm^2/s), and v is the scan rate of the applied potential (V/s).

The potential difference between the redox peaks is theoretically 59 mV for a reversible reaction. In practice, the difference is typically 70–100 mV. Larger differences, or nonsymmetric redox peaks, are an indication of a nonreversible reaction. These parameters of cyclic voltammograms make CV most suitable for characterization and mechanistic studies of redox reactions at electrodes.

9.9.5 Anodic Stripping Voltammetry (ASV)

ASV is an electrolytic method in which a mercury electrode is held at a negative potential to reduce metal ions in solution and form an amalgam with the electrode. The solution is stirred to carry as much of the analyte metal(s) to the electrode as possible for concentration into the amalgam. After reducing and accumulating the analyte for some time, the potential on the electrode is increased to reoxidize the analyte and generate a current signal. The ramped potential usually uses a step function, such as in NPP or DPP.

The concentration of the analyte in the Hg electrode, c_{Hg}, is given by

$$c_{Hg} = \frac{i_l t_d}{n F V_{Hg}} \qquad (9.15)$$

where i_l is the limiting current during reduction of the metal, t_d is the duration of accumulation, n is the number of moles of electrons transferred in the half reaction, F is the Faraday constant (96485.3 C/mol), and V_{Hg} is the volume of the electrode. The expression for current produced by anodic stripping depends on the particular type of Hg electrode, but is directly proportional to the concentration of the analyte concentrated into the electrode. The main advantage of stripping analysis is the preconcentration of the analyte into the electrode before making the actual current measurement. Anodic stripping can achieve detection of concentrations as low as 10^{-10} M.

You-Try-It 9.D
The voltammetry worksheet in you-try-it-09.xlsx contains data from different types of measurements. Use the data to determine analyte concentration in the unknowns.

Chapter 9. What Was the Point? The first part of this chapter reviewed the terminology and conventions for working with redox reactions. These conventions include using standard reduction potentials and the Nernst equation and the energetics of electrochemical cells. The latter part of the chapter described a variety of electroanalytical methods for making quantitative measurements. The two main variations are

- potentiometry—measurement of a voltage under conditions of no current flow,
- voltammetry—measurement of a current induced by an applied potential.

PRACTICE EXERCISES

1. What reactions, if any, will occur for the following?
   (a) Cl_2 is bubbled through a solution of tin(II) chloride
   (b) Cl_2 is bubbled through an aqueous solution of HI
2. What metals listed in Table 9.4 could be used to produce $H_2(g)$ from 1.0 M acid solutions?
3. In which of the following mixtures will a spontaneous reaction occur?
   (a) $Ag(s)$ in 1.0 M $ZnCl_2$ solution
   (b) $Cu(s)$ in 1.0 M $ZnCl_2$ solution
   (c) $Ni(s)$ in 1.0 M $CuCl_2$ solution
   (d) $Zn(s)$ in 1.0 M $CaCl_2$ solution
4. Balance each of the following redox reactions:
   (a) $ClO_3^-(aq) + NO(g) \rightarrow Cl_2(g) + NO_3^-(aq)$
   (b) $CH_4(g) + O_2(g) \rightarrow CO_2(g) + H_2O(l)$
   (c) $Zn(s) + MnO_2(s) \rightarrow ZnO(s) + Mn_2O_3(s)$ (alkaline battery reaction)
   (d) $Zn(s) + NO_3^-(aq) \rightarrow Zn(OH)_4^{2-}(aq) + NH_3(aq)$ (basic solution)
5. Search the Internet to find examples of electron transfer reactions of importance in the following areas:
   (a) environmental
   (b) technological
   (c) biological

For each example specify the species that are oxidized and reduced, and the number of electrons transferred, and balance the overall reaction. Selecting a simple combustion reaction, the lead–acid battery, or anything from a textbook website is too easy. Ask a study partner check your selections.

6. What is the cell voltage of each of the following electrochemical cells:
   (a) $Ni|Ni^{2+}$(aq, 1.0 M) ‖ H^+(aq, 1.0 M)|H_2(g, 1 atm)|Pt
   (b) $Zn|Zn^{2+}$(aq, 1.0 M)‖H^+(aq, 1.0 M)|H_2(g, 1 atm)|Pt
   (c) $Ag|Ag^+$(aq, 0.5 M)‖Ag^+(aq, 0.5 M)|Ag
   (d) $Ag|Ag^+$(aq, 0.001 M)‖Ag^+(aq, 0.5 M)|Ag

7. A volume of 22.40 ml of 0.150 M permanganate titrant (MnO_4^-, which is reduced to Mn^{2+}) is required to reach the end point in a redox titration of Fe^{2+} ($Fe^{2+} \rightarrow Fe^{3+}$).
   (a) Write the balanced equation for this titration.
   (b) What is the stoichiometric factor between MnO_4^- and Fe^{2+} in this titration?
   (c) How many moles of Fe^{2+} were present in the analyte solution?

8. The amount of ClO^- in commercial bleach can be determined by adding excess I^- and then titrating the I_2 with $S_2O_3^{2-}$.
   (a) Write the balanced equation for the reaction of ClO^- and I^-.
   (b) Write the balanced equation for the titration reaction.
   (c) What is the stoichiometric factor between $S_2O_3^{2-}$ and ClO^- in this titration?

9. Find K from E for the electrochemical cell in Figure 9.4. You may assume that all concentrations are 1.0 M.

10. Describe the process that occurs, including changes in solution concentrations, when a lead–acid car battery is recharged.

11. Vitamin C is sold as an antioxidant dietary supplement. Explain the rationale for taking extra vitamin C in your diet.

12. What are the key differences between potentiometry and voltammetry? Include both the principles and practical considerations in your answer.

13. List and discuss the advantages and disadvantages of potentiometric measurements compared to wet chemical methods such as gravimetry and titration.

14. For each of the following samples, suggest sample preparation steps that might be necessary to measure the analyte using an ISE.
   (a) Mg^{2+} in seawater
   (b) NO_3^- in a freshwater stream
   (c) NH_4^+ in a sample of solid fertilizer
   (d) Cl^- in stainless steel

15. For each of the samples in the previous question, write the expression for an ISE that relates the measured potential, E_{cell}, to analyte concentration. You may use E_{const} for all potentials that do not change with analyte concentration and assume that temperature is 25°C.

16. Plot the following data, E versus $\log[F^-]$, and determine the calibration curve for this ISE.

[F−]	E (mV)
0.10 M	−96
0.010 M	−37
0.0010 M	18
0.00010 M	69
0.000010 M	87

(a) What is the linear range of this calibration data?
(b) On the basis of the slope of the calibration data, what was the temperature in the laboratory on the day that the calibration data was recorded?
(c) An unknown solution, which was treated in the same way as the solutions used to generate the calibration data, produced a measured voltage of 48 mV. What is the fluoride concentration in this unknown?

17. Why is it necessary to calibrate a pH meter with two different buffer solutions?

18. Describe how potentiometry with an ISE differs from CV when you have a mixture of ions in your sample solution.

19. Use Figure 9.13 to answer the following questions:
(a) Estimate how close the reduction potentials could be for two analytes and still be measurable separately.
(b) What was the scan rate for the voltammetry experiment?
(c) If the concentration of the second analyte, occurring at −600 mV, is known to be 1×10^{-4} M, estimate the concentration of the first analyte.
(d) What assumptions did you make to estimate the concentration of the first analyte in the previous question?

ATOMIC SPECTROMETRY

Learning Outcomes

- Describe the major instrument components in atomic absorption spectrometry (AAS), atomic emission spectrometry (AES), and inductively coupled plasma mass spectrometry (ICP-MS).
- Identify the characteristics that distinguish different atomic spectrometry methods.
- Identify potential biases and interferences for a given analyte and sample matrix.
- Use spectrometric data to calculate unknown concentrations.
- Describe the characteristics of different mass spectrometer designs and select an appropriate mass spectrometer design for a given application.

Elemental analysis is exactly what the name implies, determining the elemental composition of a sample. This term can refer to several different classes of methods depending on the concentration range being measured. Elemental analysis at the percentage level is accomplished using the classical methods discussed in Chapter 3. Analyses to determine the empirical formula of a molecule are performed in dedicated instruments called *elemental, combustion,* or *CHN analyzers*. CHN refers to carbon (C), hydrogen (H), and nitrogen (N) and variations include CHNO, CHNO/S, etc. In these instruments, a pure

Basics of Analytical Chemistry and Chemical Equilibria: A Quantitative Approach, Second Edition. Brian M. Tissue.
© 2023 John Wiley & Sons, Inc. Published 2023 by John Wiley & Sons, Inc.
Companion Website: www.wiley.com/go/tissue/analyticalchemistry2e

sample undergoes complete combustion and the elemental composition is determined by measuring the relative amounts of the combustion products (see Examples 3.1–3.4). Numerous instrumental methods exist for direct elemental analysis of solids, including X-ray fluorescence (XRF) and X-photoelectron spectroscopy (XPS). The details of these instruments are beyond the scope of this text, but the underlying principles are similar for any spectroscopic method.

This chapter describes the most common methods found in chemical laboratories for elemental analysis at trace-level concentrations, i.e., less than approximately 100 ppm. These instruments are based on the absorption or emission of light from gas-phase atoms or on the conversion of an atom to an ion for mass spectrometric detection. The mass spectrometers have the lowest detection limits, widest dynamic range, and they are capable of measuring most of the periodic table. The atomic absorption and emission spectrometers find use for more routine measurements at lower cost and complexity when a mass spectrometer is not needed. None of these instruments are amenable to measuring H, C, N, O due to their presence in the atmosphere. Detection limits vary for each element, but for many metals the LOD can be less than 1 ppb. Most measurements are made on samples that are, or can be dissolved, in liquid solution. For the most part, these are laboratory-based instruments, although some specialized portable instruments are available.[1]

Atomic absorption and atomic emission methods rely on the spectroscopic transitions of valence electrons between energy levels. The instruments can quantitate a given element by measuring the attenuation or intensity of light at a given wavelength. The principles of these measurements are the same as for molecular UV/Vis absorption and fluorescence spectroscopy, as discussed in Chapter 4. As in molecular spectroscopy, atomic emission is usually more sensitive than atomic absorption. Optical atomic spectrometry methods are most appropriate for metals and metalloids. Nonmetals have optical transitions in the vacuum UV spectral range (≤ 200 nm), and measurement using commercial atomic spectrometers is only possible through indirect methods.

A mass spectrometer (MS), operates by ionizing an analyte, separating different ions based on their mass-to-charge ratio (m/z), and counting the ions at each m/z. Since an ion is charged, a mass spectrometer detector amplifies the charge of the analyte ion to a measurable electrical current. This chapter will discuss the most common implementation of mass spectrometry for elemental analysis, which uses an inductively coupled plasma (ICP) as the energy source to atomize and ionize a sample.

Although the detection principles of optical and mass spectrometry are different, there is much overlap in the analytical applications and sample delivery. All atomic spectrometry methods convert the analyte to gas-phase atoms. This

---

[1]The most common portable instruments are based on XRF. Also available is a laser-induced breakdown spectrometer (LIBS), which is based on atomic emission.

process destroys the sample matrix and molecular species. Conversion of each analyte to a single elemental form provides the total concentration for a given element. However, any speciation information is lost. For example, measurement of chromium will give the total Cr concentration, but it will not determine if the analyte existed as Cr^{3+} or CrO_4^{2-} in a sample. Determining speciation requires separation of the different analyte forms before measurement or the use of other methods, such as an electroanalytical technique (Chapter 9).

The next section describes some general requirements that are common to all atomic spectrometry instruments. The subsequent sections describe atomic absorption, atomic emission, and inductively coupled plasma mass spectrometry (ICP-MS) in more detail. The last section describes other mass spectrometer designs, which have different mass ranges, precision, and resolving power that are needed for different applications in elemental and molecular analysis. To clarify some terminology, the term spectrometry refers to methods for making quantitative measurements.[2] Spectroscopy refers to methods that are used primarily to obtain a spectrum. In practice, the two terms are used interchangeably. On recording a UV/Vis absorption spectrum, we can use an absorbance value to determine an unknown concentration. Likewise, to quantitate the elements in a sample, we obtain a mass spectrum.

10.1 ATOMIZATION

Atomic spectrometry requires that the analyte exists in the gas phase. The most common method to convert analytes to gas-phase atoms is to introduce a test portion into a high-temperature environment. The most common continuous energy sources to create a suitable high-temperature environment are flames and plasmas. A plasma is a hot, partially ionized gas that excites and ionizes atoms effectively. An inductively coupled plasma (ICP) using argon support gas is the most common plasma source because of its very high operating temperature. AAS also makes use of heated cells for easily vaporized elements and electrically heated graphite furnaces for small sample sizes. All of these instruments require ventilation to exhaust the fumes from the high-temperature source.

Solid samples are usually dissolved in aqueous solution, but they can also be atomized directly by a spark or laser pulse. Flame atomic absorption spectrometry (FAAS) has a high gas flow and uses simple aspiration, which is the drawing of liquid through a tube via the suction created by flowing gases. Instruments that use a plasma source typically employ a peristaltic pump to deliver sample solution into the instrument. A peristaltic pump uses a loop of flexible tubing and one or more rotating rollers to squeeze the tubing and move the solution through the pump.

---

[2]Since atomic absorption instruments use a light source, the proper term is *spectrophotometry*.

10.1.1 Continuous Energy Source

The following processes are necessary to convert an analyte in an aqueous solution to gas-phase atoms in a flame or plasma:

Nebulization: suspension of solution droplets in the gas phase
Desolvation: evaporation of the solvent to leave salt particles
Atomization: decomposition of the salt particles to gas-phase atoms

The nebulization process converts a liquid sample to a form that can be introduced to the high-temperature source. Solvent removal begins during nebulization but mostly occurs in the high-temperature source. Desolvation converts the liquid droplets to salt particles. These salt particles are then atomized by the high energy of the flame or plasma.[3] Figure 10.1 shows a schematic of these processes to create gas-phase atoms. Each delivery step—nebulization, desolvation, and atomization—will have a certain efficiency that reduces the number density of gas-phase atoms. The efficiency of the atomization process will be the most susceptible to variations due to matrix effects in different samples. The subsequent processes of excitation, ionization, and molecule formation will depend on the temperature and can also be affected by interferences in the sample matrix.

The purpose of the nebulizer is to convert the liquid test solution into a fine mist or aerosol that can be transported into the high-temperature environment. There are two common types of nebulizers, pneumatic and ultrasonic. Figure 10.2 shows two designs of pneumatic nebulizers. In both the cross-flow (a) and concentric (b) designs, the flowing nebulizer gas aspirates the sample solution through the capillary and causes it to spray into a spray chamber as an aerosol. The spray chamber is not shown for the concentric design, but it encloses the end of the nozzle. The impact bead shown in the cross-flow nebulizer helps break large droplets into smaller droplets. The solution droplets must be less than approximately 5 μm for efficient delivery and atomization in

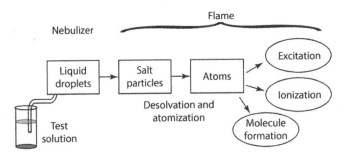

Figure 10.1 Atomization process for flame AAS.

[3]I avoid using the term vaporization, which can refer either desolvation or atomization.

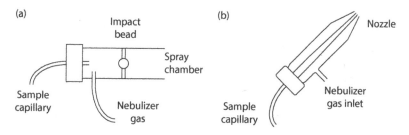

Figure 10.2 Schematics of (a) cross-flow and (b) pneumatic nebulizers.

the high-temperature source. A "flow spoiler" is often placed in the spray chamber to direct large droplets to the walls of the chamber and the waste drain. Fuel and oxidizer gases for a flame, or Ar for an ICP, flow through the spray chamber and entrain and carry the small droplets to the high-temperature source. The efficiency is only 1–5% due to loss of larger droplets going to the waste drain, but the method is simple, robust, and reproducible.

An ultrasonic nebulizer uses ultrasonic vibrations to create small uniform droplets as the test solution is delivered to the surface of a transducer.[4] These nebulizers often include a heated section and a cooling condenser to desolvate the aerosol droplets before they enter the high-temperature source. The ultrasonic nebulizer is more efficient than a pneumatic nebulizer and can achieve solution delivery into the flame of up to ≈80%. This greater efficiency can decrease the LOD for some elements by a factor of 10. The disadvantage of the ultrasonic nebulizer is higher cost and greater complexity. The choice of nebulizer design is often dependent on sample characteristics, such as the concentration of salt or suspended solids.

In measurements of nebulized liquids, samples are introduced sequentially with a rinse between each sample. The nebulizer must maintain a constant rate of sample introduction from sample to sample. The nebulizer, and other instrument components, must be inert to corrosion and retention of analyte. These characteristics are necessary to maintain a repeatable signal over time and to prevent the introduction of contaminants or erroneous measurements due to carryover. A typical procedure is to measure blank and calibration standards at the start of data collection. Due to the possibility of introducing bias from clogging or contamination, one of the calibration standards should be remeasured after every ten unknown test portions.[5] A change in the signal of this standard will alert the analyst to drift or other problem.

Table 10.1 lists approximate temperatures of different atomizer/excitation sources. Flames and plasmas have temperature gradients and zones, and Table 10.1 lists the typical temperature in the analytical zone. A higher temperature

[4]The ultrasonic transducer is a piezoelectric crystal driven by an RF electrical signal at >1 MHz.
[5]See, for example, EPA Method 200.8 Determination of Trace Elements in Waters and Wastes by Inductively Coupled Plasma Mass Spectrometry.

TABLE 10.1 Temperature of Different Excitation Sources

Atomizer/ Excitation Source	Approximate Temperature, °C
Acetylene/air flame	2300
Acetylene/N_2O flame	3000
Inductively coupled plasma	8000

source will, in general, have higher efficiencies for desolvating, atomizing, and exciting or ionizing the analyte. The plasma achieves a much higher temperature than the flames, and most atomic emission spectrometers and mass spectrometers for solution samples now utilize some type of plasma source. The high temperature of a plasma gives ICP-AES lower detection limits for most metals compared to flame AES or AAS. The exceptions are the easily ionized group 1 metals, Li, Na, K, etc., where there is little difference or even worse detection limits when using plasma excitation.

The schematic in Figure 10.1 also shows several processes—excitation, ionization, molecule formation—that occur in the high-temperature flame or plasma. Molecule formation reduces the concentration of gas-phase atoms and is a loss for any of the atomic methods. The excitation and ionization processes create the measured signal for atomic emission and mass spectrometry, respectively. These processes are losses when measuring ground-state atoms in AAS. Different flame conditions or solution additives are used for different analytes to minimize these losses and maximize the concentration of ground-state gas-phase analyte atoms. The difference in the efficiency of these different steps is the reason different elements have different detection limits. Easily atomized elements have low detection limits, and refractory elements, materials that are difficult to atomize, have higher detection limits.

10.1.2 Pulsed Energy Source

Pulsed energy sources include a graphite furnace, which is a graphite tube heated by a programmed electric current, an electrical spark between electrodes, and a high-energy laser pulse. The graphite furnace is used with AAS, which, along with other special sampling methods, is discussed in Section 10.2.

These pulsed energy sources do not require the sample to be in an aqueous solution, they can vaporize solids, slurries, or viscous liquids directly. The sample atomization and excitation or ionization occurs in one integrated step. The main advantages of these approaches are that there is little to no sample preparation, making the analytical process rapid. The disadvantage is that these approaches are more prone to errors due to interferences or matrix effects compared to dissolving and diluting a solid in an aqueous solution. Figure 10.3 illustrates two pulsed methods, a spark between electrodes (a) and laser ablation (b).

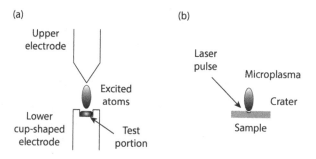

Figure 10.3 Schematic of pulsed vaporization methods for solid samples.

In the spark method, the solid sample is placed in or on the lower graphite electrode, sometimes mixed with additional powdered graphite. An electric pulse between the electrodes creates a spark that vaporizes some of the sample. The spark has sufficient energy to vaporize, atomize, excite, and ionize some of the solid. The spark can be interfaced to an atomic emission spectrometer or to a mass spectrometer. Owing to matrix effects, the spark source is used for repetitive analyses of similar samples, such as metals, minerals, concrete, or other solid materials. In these cases, there is little variation in the composition of the samples and standards, and measurements are repeatable and accurate.

In the laser ablation process, a short laser pulse of 3–5 ns is focused to a small spot on the sample surface. The short pulse is desirable because it concentrates the photon energy, resulting in efficient ablation and less sample heating. The laser pulse causes rapid heating of a very small volume of sample. The resulting "microexplosion" ejects solid particles, atoms, and ions into the gas phase. Each laser shot leaves a crater of typically 50 μm diameter. It essentially functions like a spark to provide the energy for vaporization and excitation. Measuring atomic emission from a laser microplasma is known as *laser-induced breakdown spectrometry* (*LIBS*). The advantage of LIBS is that it can be miniaturized for a field portable instrument.

The laser ablation process is also used for sampling material from a solid to deliver it to an ICP. Figure 10.4 shows a schematic of a laser ablation method for delivering sample to an ICP. These units are mostly used in geological or material science applications, but they can also be applied to sections from environmental and biological samples. A key advantage is that this method can sample material from different grains or microscopic portions of a solid sample to achieve spatial resolution in the elemental analysis. The laser can raster across features to make a two-dimensional map of elemental composition. The dashed lines in Figure 10.4 show the optical viewing path and the shaded dashed lines show the overlapping laser path. The dichroic mirror reflects the laser wavelength and transmits visible wavelengths for viewing and recording images. The flowing gas transfers ablated particulate matter to the

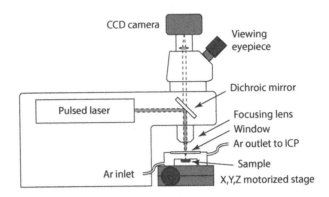

Figure 10.4 Schematic of laser ablation sampling of a solid sample.

ICP, where it is atomized and atoms are excited and ionized for either AES or mass spectrometric analysis.

10.2 ATOMIC ABSORPTION SPECTROMETRY (AAS)

The principles of optical transitions and spectroscopic instruments are the same for the measurement of both atoms and molecules. The major difference between atomic and molecular spectra is in the spectral width of the electronic transitions. The vibrational and rotational energy levels of molecules create broad absorption and emission bands in the electronic spectra (see the energy-level diagram in Figure 4.8). The absorption and emission lines in atomic spectra are very narrow because atoms lack these energy modes. This difference changes the requirements of the optical spectrometers, requiring a narrow bandpass to distinguish closely spaced absorption or emission lines. The advantage of the narrow atomic spectral lines is that one element can be measured in the presence of many others. In molecular spectra, overlap of the broad bands makes it difficult to measure more than two or three absorbing species in the same test solution.

AAS uses the absorption of light to measure the concentration of the gas-phase atoms. The gas-phase atoms absorb ultraviolet or visible photons to make transitions to higher electronic energy levels. The instrument measures the light power with the analyte atom present, P, and without the analyte present, P_o, to determine the absorbance, A:

$$A = -\log\frac{P}{P_o} \tag{10.1}$$

Commercial atomic absorption spectrophotometers are generally double-beam instruments, where the light source is split to measure P_o in a second beam path that does not transit the analyte region. The absorbance is directly proportional to analyte concentration (see Section 4.7 on the Beer–Lambert law).

The instrumentation for AAS is straightforward in principle, consisting of the following components:

- Light source
- Sample atomizer
- Wavelength separator
- Light detector and associated electronics

The performance of each component contributes to the limit of detection in a given instrument. Key aspects to obtain sensitive measurements in any absorbance measurement are a stable light source, a wavelength separator with efficient rejection of interfering light, and a low-noise detector. AAS also requires efficient conversion of the sample to gas-phase atoms. Because the intrinsic strength of electronic transitions and the efficiency of atomization processes vary, the LOD will be different for each element. Commercial instruments can achieve detection limits of 0.01–10 ppm for most metals. For perspective, Na and K occur in blood at 1000 ppm and Cu and Zn occur at 1–10 ppm. Toxic metals such as Cd or Pb can be present at concentrations on the order of 0.010 ppm or less. These metals are generally measured by a more sensitive ICP method.

The operational principles of each instrument component are described next, although many details are omitted for conciseness. Figure 10.5 shows a simplified schematic of the arrangement of components for an AAS spectrophotometer. Omitted from this schematic is the second beam path for measurement of P_0. Different manufacturers use various optical configurations, but a rotating chopper and beamsplitters is a common approach to alternately pass the light beam through the flame and reference path. Commercial instruments are computer controlled. On selecting the analyte metal, the computer sets an appropriate wavelength and bandpass and optimizes the optical alignment, detector gain, etc.

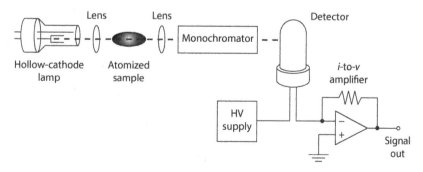

Figure 10.5   Schematic of an atomic absorption spectrophotometer.

10.2.1 Light Source

The most common light source for atomic-absorption measurements is a hollow-cathode lamp. These lamps contain a cup-shaped cathode made from, or coated on the interior with, the element to be measured. It is enclosed in a glass envelope containing an inert gas, usually Ne or Ar, at low pressure. An electrical discharge between the hollow cathode and a second electrode sputters some of the cathode material into the gas phase.[6] The energy of the electric discharge excites the sputtered atoms to higher electronic states, which decay by atomic emission. The lamp emission consists of narrow lines at the characteristic wavelengths of the element to be measured. These lamps are usually of a single element, but some are available with two or three similar metals, for example, Na and K.

The main advantage of this lamp design is that it generates a narrow-band light emission that will be resonant with the element being measured. Because the linewidth of the lamp emission matches the absorption line, the absorption of light is very sensitive, that is, there is not much extra light going to the detector. The disadvantage of these light sources is that only one element, or at most a few elements, may be measured without changing lamps. Modern instruments often contain a turret or carousel to house multiple lamps, typically two to eight. These instruments can sequentially measure multiple metal analytes in a given test portion by switching lamps and spectrometer settings via computer control. For elements with a relatively low melting point, the hollow-cathode design is not stable, and electrodeless discharge lamps have a longer life and more stable emission.

10.2.2 Atomizers

The general characteristics and requirements of atomizers were described in Section 10.1. This section describes the details of several types of atomizers for atomic absorption. The most common atomizers for AAS are a flame, graphite furnace, or a "cold-vapor" cell. In this context, "cold" is relative to a flame and the cell is usually heated above room temperature. Since atomic absorption occurs from ground-state atoms,[7] excited or ionized atoms are "lost" to the measurement. For some elements, the greater fraction of excited or ionized atoms at the high temperature of a plasma source offset the more efficient atomization efficiency of a plasma.

Flame AAS uses a slot burner, typically 10 cm long, to increase the path length and the measured absorbance per the Beer–Lambert law. Figure 10.6 shows a schematic of a slot burner in an AAS instrument. The cylindrical

---

[6]Sputtering is the vaporization of solid material due to collisions from atoms or ions that strike the solid surface.

[7]Ionized metals do have distinct absorption spectra. Ions generally are not measured in AAS because of variability in ionization efficiency due to sample matrix effects.

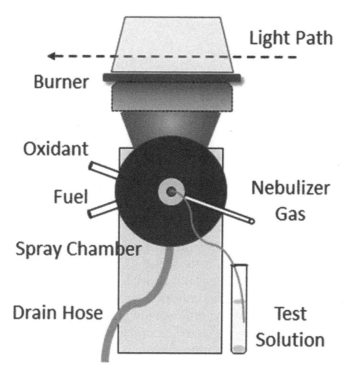

Figure 10.6   Schematic of a flame AAS slot-burner.

component below the burner is the spray chamber and the hoses are the nebu-lizer and gas supplies. The position of the light path is low in the flame. The flame is normally of a light blue color and not very visible. Adjustment knobs (not shown) move the burner assembly up and down, back and forth, and at an angle compared to the light path. These adjustments allow the flame to be positioned relative to the light path for the maximum absorption signal for a given element. Since a flame has different regions that vary in temperature and composition, the maximum concentration of gas-phase atoms can occur at a slightly different flame height for each element.

Analytical chemists generally like to work with linear calibration curves. Flame AAS is one method where nonlinear calibration curves are accepted. Figure 10.7 shows typical calibration curves for two different elements. Note that the x scales are different in these two curves. The Beer–Lambert law applies to AA spectrometry, but the nonlinearity arises from a combination of instrument factors and the efficiency of analyte atomization. Stray light intro-duces nonlinearity in a spectrophotometer at high absorbance (see Section 4.6). In atomic absorption it will depend on instrument settings such as the selected wavelength, slit width, and lamp current. The conversion of analyte to gas-phase atoms is also not constant versus concentration. Higher concentration solutions create larger salt particles after desolvation, that might not atomize

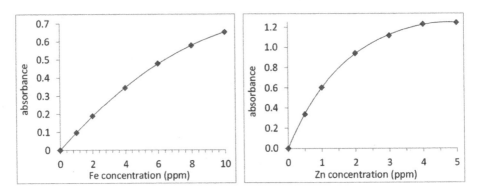

Figure 10.7   Flame AAS calibration curves for iron (left) and zinc (right).

completely. Calibration curves are measured daily for a given set of instrument settings. The experimental data is then used to find a calibration function that provides a good fit, generally a second-order or higher polynomial.

Besides being nonlinear versus concentration, the FAAS signal for an atom can be affected by physical, chemical, and ionization interferences. These sample matrix effects can be corrected, but they create a bias in a measurement if they are not recognized. A common physical interference is variability in the delivery of solution via the simple aspiration used in FAAS. Solutions that vary in density or viscosity will deliver analyte to the flame at different rates. Recognizing this problem is easy because the analyte can simply measure solution delivery with a graduated cylinder and a stopwatch. This problem is also easily corrected by matrix matching the standards to the typical test solution.

A well-known chemical interference is the suppression of calcium signal due to phosphate in a sample matrix. Calcium phosphate is a refractory material that is difficult to atomize in a flame. The presence of varying amounts of phosphate in samples will lead to varying biases in the calcium measurements. Phosphate could be removed from test solutions by precipitation, but an easier solution is to complex it in the gas phase in the flame. La^{3+} complexes phosphate more strongly than does Ca^{2+}. An excess of La^{3+}, known as a *releasing agent* in this role, is added to all standard and test solutions. Phosphate in the flame forms as lanthanum phosphate particles, releasing the calcium atoms for measurement without bias.

Barium has a relatively low ionization potential, and ionization of gas-phase Ba to Ba^+ is a loss when measuring a Ba absorption line. The degree of ionization can depend on the amount of other easily ionized atoms in the flame. In the following equilibria, the presence and ionization of potassium will produce free electrons, which will suppress the ionization of barium by Le Chatelier's principle:

$$Ba \rightleftharpoons Ba^+ + e^-$$

$$K \rightleftharpoons K^+ + e^-$$

Ionization of Ba becomes a problem when the amount of other ionized atoms is variable between standard solutions and different test solutions. As in the phosphate example, every unknown measurement has a variable bias depending on what is in the sample matrix. The solution is to add an excess of potassium to all test and standard solutions. Some fraction of Ba is still ionized, but that fraction is the same in all test and standard solutions. All of these matrix effects require some knowledge of sample composition and validation experiments to avoid biased measurements.

Graphite is an electrically conducting solid that is fairly inert with a high sublimation temperature (>3500°C). A graphite furnace consists of a graphite tube that has a small opening at the top of the tube to introduce a sample. The sample is often placed on a graphite sheet in the tube called a *L'vov platform*. After depositing a sample in the graphite furnace, an electric current is passed through the furnace via clamped electrical leads to rapidly raise the temperature. This process is often called *electrothermal vaporization* in product literature. The electrical current is applied stepwise over a time of 1–3 min as shown in Figure 10.8. The first step raises the furnace temperature above 100°C to evaporate water. The second step raises the temperature to 500–1000°C to ash or char organic matter. These drying and ashing steps remove steam and smoke that would add to the analyte absorbance if the measurement was done in only one step. The third step raises the tube temperature to 1500–2500°C to atomize the analyte. A final step raises the temperature even higher to do a burnout to avoid carryover to the next sample. The tube is purged with an inert gas to remove the steam, smoke, and atomized sample that is created during each heating step.

A graphite furnace has several advantages over a flame. It is a more efficient atomizer than a flame and it can accept very small quantities of sample directly. It also provides a reducing environment for easily oxidized elements. Unlike flame AAS (FAAS) which can only analyze solutions, graphite furnace AAS (GFAAS) can accept solutions, slurries, or solid samples. Because of the more efficient sample utilization, detection limits are 10–100 times lower compared

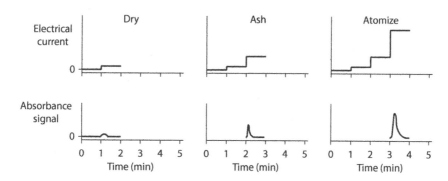

Figure 10.8 Sequence of steps in electrothermal vaporization.

to flame AAS. The price of the lower detection limits is more variability in the measurements, producing less precise results and requiring greater operator skill and instrument maintenance.

Flame atomic absorption measurements for most elements use a nebulizer and spray chamber to introduce analyte into the flame. Doing so introduces only a portion of the test solution into the flame. Analyte utilization can be much higher for some elements by replacing the nebulizer with a reaction vessel and flow cell. The analyte is converted to a volatile form and all of the analyte in a test solution is measured. Figure 10.9 shows a simple schematic of this set-up. The optical flow cell is usually glass with high quality windows to transmit UV and visible light. It is heated to create or maintain the analyte in the gas phase. Capturing all of a volatile species in a solution can decrease the LOD for a given element by a factor of 1000 or more compared to using FAAS.

Mercury is the one metal with an appreciable vapor pressure at room temperature. Mercury in different forms and types of samples can be dissolved in solution and oxidized to Hg^{2+}. It is then reducing with $SnCl_2$ to elemental mercury. Flowing an inert gas through the reaction solution sweeps the mercury into a flow cell in the light path. As in GFAAS, the absorbance signal is transient, but measures all of the mercury that was in solution.

The metalloids As, Se, Sn, Sb, Te, and Bi can form volatile molecular hydrides.[8] These hydrides are formed by reaction with sodium borohydride, $NaBH_4$, in acidic solution. After the reaction is complete, an inert gas sweeps the volatile hydride through the flow cell. The flow cell is heated to decompose the hydride to the atomic species, and the atomic absorption is again measured as a transient signal.

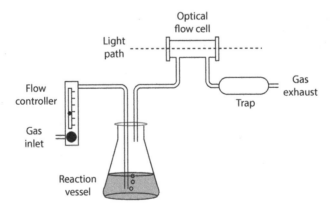

Figure 10.9   Schematic of a cold-vapor sample system.

[8] AsH_3, H_2Se, H_2Te, BiH_3, SbH_3, and SnH_4.

10.2.3   Wavelength Separator and Detector

The wavelength separator is usually a dispersive monochromator consisting of entrance and exit slits, mirrors, and a diffraction grating (refer to Section 4.6). The purpose of the monochromator is to isolate only one emission line from the hollow-cathode lamp and to reduce the amount of background light from the flame that reaches the detector. Manufacturer-provided manuals or operating software list typical instrument parameters for each element, including suggested wavelengths and potential interferences. Table 10.2 lists the relative sensitivity of three absorption lines of potassium. Measurements are normally done at the strongest absorbing line. If an interference is present at the strongest line, another line is chosen. Measuring a test solution at a weaker line is also an alternative to diluting a high concentration sample that is beyond the measurement range of a more strongly absorbing wavelength.

Instrument bandpass is controlled by the size of the slit width opening and is typically 0.2–2.0 nm. This bandpass is much larger than the linewidth of the atomic emission originating from the hollow-cathode lamp, which is approximately 0.002 nm. The narrow emission of the hollow-cathode lamp provides the resolution for the atomic absorption measurement. A wider monochromator slit allows more of the light beam to reach the detector for more sensitive detection. The disadvantage of the wider bandpass is an increase in any background absorption that is present.

Photomultiplier tubes are the most common detectors for AAS owing to their linear dynamic range and high sensitivity for UV and visible light. Researchers have built instruments with solid-state array detectors and continuum light sources for multielement AAS, but these designs have yet to be commercialized. The electronics consist of a current-to-voltage converter and amplifier to convert the PMT signal to a voltage signal. The electronics or software manipulates the P and P_0 signals from the detector to produce an absorbance measurement, which is displayed on screen and saved or printed by the instrument software.

10.2.4   Background Correction

Although interference from other elements can be avoided by selecting a line with no overlap, molecular species that form in the flame will have broad lines.

TABLE 10.2   Relative Strength of Potassium Absorption Lines at Different Wavelengths

Wavelength, nm	Relative Sensitivity
766.5	1.0
769.9	0.48
404.4	0.0055

These species can absorb some of the light, resulting in an erroneously high absorbance measurement. One method of background correction is to measure the absorption with a continuum source. Since the monochromator bandpass, ≥ 0.2 nm, is much larger than the hollow-cathode lamp linewidth, 0.002 nm, absorption of continuum-source light is primarily from the background. A typical continuum source is a deuterium arc lamp, which is useful through the UV wavelengths where molecular interferences are most common. The signal from the hollow-cathode lamp and the D_2 lamp are measured separately by pulsing, chopping, or modulating the light beam from one or both lamps.

There are other background correction methods that can be more accurate if the background absorption is not a smooth continuum or if working at wavelengths longer than the 400-nm limit of the D_2 lamp. The self-reversal method pulses the hollow-cathode lamp current to change the emission profile of the lamp.[9] At low current, the lamp emission is symmetric and measures both atomic and background absorption. At high current, the lamp profile consists of broadened wings with no intensity at the center wavelength. In this condition, the lamp is only resonant with background absorption. Another approach is to use the Zeeman effect by applying a magnetic field around the gas-phase atoms to shift the atomic absorption in and out of resonance with the hollow-cathode lamp emission.[10] Any measured absorbance when the atom is out of resonance is due to background absorption. These correction methods allow any background absorption due to molecular absorption or scattering to be subtracted from the measured signal to improve the accuracy of AAS measurements.

10.2.5   Quantitative Measurements

Unlike molecular UV/Vis absorption in a 1-cm cuvette, the proportionality constant between A and c can vary day-to-day due to changes in optical alignment, sample delivery, or other instrument factors. Quantitative AAS measurements are therefore always made using a recent calibration curve. Figure 10.7 showed typical nonlinear calibration curves, including very different linear ranges for different elements. Analysis by GFAAS can also have severe matrix effects for complex samples. In both GFAAS and FAAS, these issues can be minimized by using matrix-matched standards. Having pointed out numerous problems, AAS remains a workhorse analytical method for the more easily atomized metals in industrial and environmental applications. These applications tend to be repetitive analyses of similar sample types. Example 10.1 uses the calibration data from Figure 10.7 to illustrate a quantitative measurement with a manual background correction.

[9]Also called the Smith–Hieftje method.
[10]The process is a little more complicated than described. The magnetic field splits the atomic energy levels to create a triplet in the spectrum. The transition that remains resonant with the lamp wavelength is blocked by a polarizer.

Example 10.1 AAS Background Correction. You obtain the following measurements with an AAS spectrophotometer that does not have background correction capabilities:

$[Fe^{2+}]$, ppm	Absorbance
0.0 (0.1 M HCl blank)	0.000
2.0	0.188
4.0	0.345
6.0	0.479
8.0	0.578
10.0	0.652
Unknown	0.328

However, you notice that the flame appears "smoky" for the sample test solution compared to a clear flame for the blank and standard solutions. You know that your sample does not contain cobalt, so you replace the iron hollow-cathode lamp with a cobalt lamp (240.7 nm vs 248.3 nm for iron). You measure a signal of 0.10 1 with the cobalt lamp, which you attribute to scattering. Correct your measurement to determine $[Fe^{2+}]$ in the unknown.

The measurement with the cobalt lamp is due to particles that block some of the light, appearing as absorbance. Subtracting this background measurement gives

$$A_{corrected} = 0.328 - 0.101 = 0.227$$

Now you can use the calibration curve to determine the unknown concentration. Fitting the data to a polynomial:

$$A = -0.00360\left[Fe^{2+}\right]^2 + 0.101\left[Fe^{2+}\right]$$

Using the quadratic formula to solve for $A = 0.227$ gives $[Fe^{2+}] = 2.46$ ppm.

This example is a case where making another measurement with a known amount of Fe^{2+} spiked into the test solution could confirm if a bias is present or not. For example, spiking 1.0 ppm into the test solution gives a new concentration of 3.46 ppm. From the calibration curve, we expect a background-corrected absorbance of $A = 0.306$. If the actual measurement is different from this value, then more experiments are needed to find the source of error.

You-Try-It 10.A

The AAS worksheet in **you-try-it-10.xlsx** contains methods, calibration data, and measurement results. Plot the standard data and use the calibration function to determine analyte concentration. This data set includes a spike experiment to test for any bias in the measurement.

10.3 ATOMIC EMISSION SPECTROMETRY (AES)

AES, also called *optical emission spectrometry* (OES), uses quantitative measurement of the light emission from excited atoms to determine analyte concentration. In atomic emission, the analyte is converted to gas-phase atoms in a high-temperature energy source. The processes of nebulization, desolvation, and atomization are the same as in atomic absorption instruments. The fundamental difference between AES and AAS is that the source of the analytical signal comes from excited atoms rather than ground-state atoms (see Figure 10.1). The high-temperature atomization source provides sufficient energy to promote gas-phase atoms into higher energy levels. The excited atoms decay back to lower levels by emitting light. Some common examples of this process are flame tests for qualitative detection of metals in solution and the different colors in fireworks, which are produced by different elements. This process is exactly analogous to the molecular fluorescence described in Section 4.8. As for molecular fluorescence, atomic emission is more sensitive than atomic absorption for most elements.

The instrumental components for AES have some similarities to AAS. The sample must be converted to free atoms in some type of high-temperature excitation source. This energy source desolvates, atomizes, and excites the analyte atoms. Liquid samples are nebulized and carried into the excitation source by a flowing gas. Particulates from solid samples can be introduced by laser ablation of the solid sample in a gas stream. Solids can also be vaporized directly and excited by a spark between electrodes or by a laser pulse. Since the excited-state population depends on temperature, plasma sources provide higher atomic emission signals than does a flame. Figure 10.10 shows the assembly of a nebulizer and an ICP torch. A variety of excitation sources are used in experimental and commercial instruments:

- Flame: see Figure 10.6[11]
- Electrical spark or arc: see Figure 10.3
- ICP: see Figure 10.10
- Microwave-induced plasma: (MIP)
- Direct-current plasma (DCP)
- LIBS and laser-induced plasma.

The optical emission from the excited atoms is collected by optics and dispersed through a spectrometer. Figure 10.11 shows a simplified schematic. The excitation can occur via a flame, plasma, electrical spark, laser pulse, or other high-energy source. Since the transitions are between distinct atomic energy levels, the emission lines in the optical spectra are very narrow. The spectra of samples containing many of the heavier elements can be congested, and

[11]AAS instruments can usually be used in an atomic emission mode.

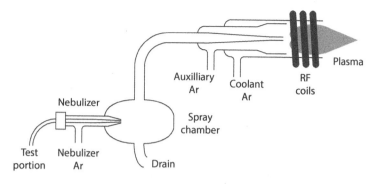

Figure 10.10   Schematic of an ICP torch.

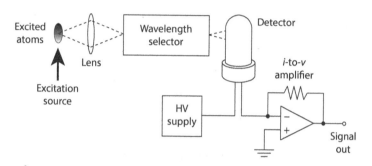

Figure 10.11   Schematic of an atomic emission experiment.

spectral separation of nearby atomic transitions requires a high-resolution spectrometer. The spectral bandpass of the ideal spectrometer will be narrow enough to discriminate closely spaced lines. In practice, a narrow bandpass requires a large instrument with small apertures and a sensitive detector.

The spectral bandpass of an AES spectrometer will usually be variable so that a larger aperture can be used for greater sensitivity when no spectral interferences are near an analyte emission line. Instrument *resolution* is the minimum spectral bandpass, and is typically less than 0.01 nm in commercial instruments. As an example, emission at 228.80 nm is the strongest line for cadmium. If arsenic is present in a sample, it can interfere with cadmium measurements owing to spectral overlap from the nearby As line at 228.812 nm. Figure 10.12 shows calculated emission lines for equal concentrations of Cd and As. The three spectra are offset vertically for clarity. This simulation shows that a spectral bandpass less than 0.005 nm is necessary to avoid the arsenic interference. Working at UV wavelengths is typical for most elements because the spectral widths will be narrower than at longer wavelength. The group I alkali metals are the exceptions; they have their strongest emission at visible wavelengths.

Since all atoms in the test portion are excited simultaneously, they can be detected simultaneously using a polychromator with multiple detectors.

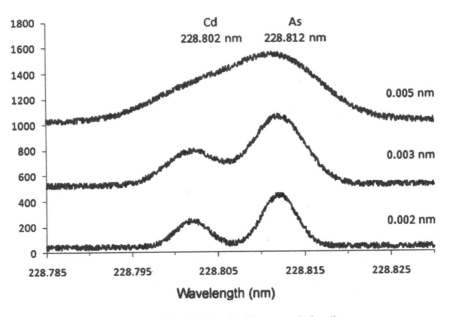

Figure 10.12 Resolution of adjacent emission lines.

This ability to simultaneously measure multiple elements is a major advantage of AES compared to AAS. It is not uncommon to measure 20 or more elements in an analysis. Figure 10.13 shows a schematic of a Rowland circle polychromator. The lens focuses the optical emission from the excitation region through the entrance slit and onto the concave grating. The concave grating has a radius of curvature and position so that it focuses the dispersed wavelengths of light around the image plane of the Rowland circle. Exit slits and detectors are positioned at the angles from the grating where strong emission from different elements occur. The example shows four detectors to monitor four different elements, two of which receive light (the dashed lines in the figure). Figure 10.13 shows an older design with discrete phototubes or photomultiplier tubes. Modern instruments now use solid-state array detectors, but the concept is the same in both cases.

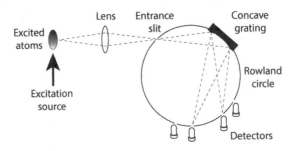

Figure 10.13 Schematic of a polychromator.

As in AAS, calibration of atomic emission measurements is done with standards and unknown concentrations are obtained from the calibration function. The linearity of ICP-AES is much better than in AAS, with calibration curves being linear over several orders of magnitude in analyte concentration. A disadvantage of AES compared to AAS is that the excited-state population of the atoms is more sensitive to the temperature of the excitation source. This sensitivity can reduce the precision of multiple measurements and reduce the accuracy between samples and standards.

The last topic of this section is to define atomic fluorescence spectrometry. *Atomic fluorescence spectrometry* is the measurement of atomic emission for gas-phase atoms that have been excited by a light source. There is no difference between the light of atomic fluorescence versus atomic emission, the only difference is the mode of excitation. Atomic fluorescence spectrometry works best for elements that are easily vaporized, with the primary example being Hg.[12] Lab methods use a cold-vapor delivery of the analyte after treatment in a reaction vessel. Dedicated commercial instruments can trap both free mercury and particulate mercury in air for environmental and workplace monitoring. These instruments sequentially desorb the two forms of mercury into a flow cell to measure them by atomic fluorescence.

You-Try-It 10.B
The AES worksheet in you-try-it-10.xlsx contains calibration data and repetitive measurements to validate the method. Plot the standard data and use the calibration function with the repetitive measurements to determine the limit of detection for each analyte.

10.4 INDUCTIVELY COUPLED PLASMA MASS SPECTROMETRY (ICP-MS)

Inductively coupled plasma mass spectrometry (ICP-MS) is the most powerful method for elemental analysis. ICP-MS can analyze almost all elements in the periodic table, with the exceptions of H, C, N, O, F, Cl, and the inert gases. H, C, N, and O are background gases from air and the other elements are not ionized efficiently in an Ar plasma. Typical detection limits are 1–0.001 ppb, with the best sensitivities occurring for the transition metals. The more electronegative nonmetals are not ionized as efficiently, but the detection limits are better than competing analytical methods. Since a mass spectrometer can scan the complete range of atomic masses, most of the periodic table can be determined at least semi-quantitatively in one measurement. In practice, fewer

---

[12]EPA Method 7474 Mercury in Sediment and Tissue Samples by Atomic Fluorescence Spectrometry.

elements are measured to achieve better levels of precision and accuracy in the results. ICP-MS can also measure the isotopic composition in a sample, which is not possible using optical spectroscopy. The bottom line is that the low detection limits and capability to analyze most of the periodic table makes ICP-MS the most powerful commercially available method for elemental analysis. The price of the better performance is a greater instrument complexity and greater operator skill needed to make the measurements.

Before discussing ICP-MS further, we must understand the nature of mass spectrometry. An excited state and an ion are not the same! For a crude analogy, consider that you have a positive charge and you are wearing a hat with a negative charge. With your hat on your head, you are in your ground state. Holding your hat in your hand and waving it over your head could be considered an excited state. Having a gust of wind blow your hat down the street leaves you as a positive ion. Being a charged ion now, you are acted upon by any electric field in your vicinity.

Mass spectrometers convert analyte atoms and molecules to charged species. Once charged, the ions can be manipulated by electric and magnetic fields. A variety of mass spectrometer designs use the difference in mass-to-charge ratio (m/z) of the ions to separate them from each other. The general operation of a mass spectrometer involves the following steps:

- Creating gas-phase ions
- Transfer the ions into a vacuum system that houses the mass analyzer
- Separating the ions in space or time based on their mass-to-charge ratio
- Measuring the quantity of ions at each mass-to-charge ratio

The instrument components to perform these steps are called the ion source, a vacuum interface, a mass-selective analyzer or mass filter, and an ion detector (Figure 10.14). Since mass spectrometers create and manipulate gas-phase ions, they operate in a high vacuum system at 10^{-5}–10^{-6} Torr. The mean free path at these pressures is on the order of meters, so the low pressure prevents collisions that would change ion trajectories. The low pressure also protects the mass spectrometer components from oxidation, which would degrade performance.

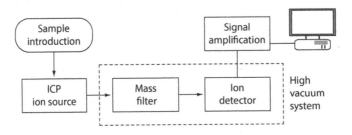

Figure 10.14   Schematic of mass spectrometer components.

Sample introduction methods for ICP-MS are the same as in ICP-AES. The difference is that analyte ions from the plasma source create the signal rather than photons from atomic emission. The plasma is interfaced to a high-vacuum mass spectrometer through an interface of small apertures called sampling and skimmer cones. There are usually two stages of vacuum pumps to make this transition from atmospheric to low pressure. The ions that pass through the last skimmer cone enter acceleration and focusing ion optics to send them to the mass analyzer. The details of these ion optics depend on the mass analyzer design, but for our purposes they prepare the beam of ions for the next stage in the mass spectrometer of separating ions based on m/z.

The most common mass analyzer for elemental analysis is the quadrupole mass filter. A quadrupole mass filter consists of four parallel metal rods arranged as in Figure 10.15. Ions travel down the length of the mass filter to the detector as shown by the dashed line. Constant (dc) and sinusoidal (ac) potentials are applied to the rods that affect the transverse motion of the ions. The ion velocity down the length of the mass filter is not affected. The rods function to eject ions out of the path so that they do not reach the detector. For a given set of voltages on the four rods, only ions of one m/z reach the detector. All ions of other m/z are ejected from their original path. A mass spectrum is obtained by measuring the ions that reach a detector as the voltages on the rods are varied. For elemental analysis, the ICP-MS must measure from the minor isotope of lithium, ^{6}Li, to the highest natural isotope of ^{238}U with unit resolution. The requirements on the mass spectrometer are therefore modest.

The motion of ions through the quadrupole mass filter follow erratic spiraling paths. The equations that describe the paths are beyond the scope of this text, but I will try to provide a qualitative description of the quadrupole operation. The two sets of rods have applied potentials of

x direction rods: $+(U + V_o\cos(\omega t))$
y direction rods: $-(U + V_o\cos(\omega t))$

where U is a dc voltage and $V_o\cos(\omega t)$ is an ac voltage of amplitude V_o and angular frequency ω. For a beam of positive ions traveling down the center of the quadrupole filter, the positive rods provide a repulsive force on average and the negative rods provide an attractive force on average. However, V_o is often a

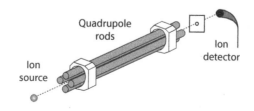

Figure 10.15 Schematic of a quadrupole mass analyzer.

factor of 5–10 greater than U, so both sets of rods are oscillating through plus and negative voltages.

An ion of mass m and charge q accelerated by a voltage, V, has a kinetic energy of

$$qV = \frac{1}{2}mv^2$$

At a given voltage, a heavier mass will have a lower velocity and travel a shorter distance. The transverse movement of heavier ions are therefore not affected by the ac potential as strongly as lighter ions. However, the heavier ions do respond to the average voltages on the rods. The average dc potential on the positive rods pushes the heavier ions to the center of the path and have little effect. The average dc potential on a negative rod attracts the heavier ions, causing them to drift out of the center path until they hit a rod and are removed. The net result is that the dc voltages function as a low-pass mass filter as shown by the shaded region in Figure 10.16a.

The effect of the ac potentials is less intuitive, but the lighter ions respond quickly to the changing ac potentials. The ac potential on the negative rods has a restoring force and the ac potential on the positive rods causes lighter ions to eventually be ejected from a stable path. The net result is that the ac voltages function as a high-pass mass filter as in Figure 10.16b. By setting U, V, and ω so the two mass cutoffs overlap, only one m/z has a stable path through the quadrupole filter to reach the detector. This one particular mass is shown in Figure 10.16c. This general configuration is known as a bandpass filter and is used in optical, electronic, and other applications.

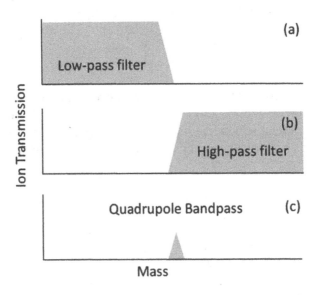

Figure 10.16   Mass filtering by the quadrupole rods for a given set of U, V, and ω.

The quadrupole rods are typically 10–15 cm long. The acceleration voltage from the ion source results in a transit time of 20–30 µs for high mass ions over this distance. This time allows ions to go through approximately 50 cycles of the ac voltages, which is enough to efficiently remove the ions with unstable paths. The U and V potentials are scanned more slowly than the transit time, but it is still possible to make multiple scans per second. For ICP-MS, the operator sets a time or total number of scans for signal averaging to improve precision in the measurement. The last stage of the mass spectrometer is an exit aperture and the ion detector. The ion detector converts the number of ions hitting it to a measurable voltage. It has a high internal gain and low intrinsic noise, similar to a photomultiplier tube. Details of ion detectors are discussed in the next section. For atomic spectrometry, the overall result is a mass spectrum of ion intensity versus m/z.

Although the inductively coupled plasma is very efficient at atomization, ICP-MS is not without interferences. Table 10.3 lists a few examples of isobaric interference.[13] The interferences can come from atmospheric gases, solution components, or from the sample matrix. Since Ar is used as the ICP support gas, argon species are particularly problematic. Several approaches to avoid, correct, or remove interferences in measurements are discussed below. Note that only a small number of interferences are listed in Table 10.3 for discussion. Any new ICP-MS analysis will require validation experiments and often significant method development to avoid interferences and obtain accurate results.

Since most elements have multiple isotopes, it is often possible to find one that is interference free. This approach is suitable for calcium, where measuring the minor $^{44}Ca^+$ avoids the interference from $^{40}Ar^+$. Table 10.3 shows that $^{44}Ca^+$ has a potential interference of the doubly charged $^{88}Sr^{2+}$. Since the mass analyzer separates ions based on m/z, $^{88}Sr^{2+}$ overlaps with $^{44}Ca^+$. The

TABLE 10.3 Examples of Isobaric Interferences in ICP-MS

Isotope	Relative Abundance	Interfering Ions
$^{40}Ca^+$	96.97	$^{40}Ar^+$
$^{44}Ca^+$	2.06	$^{88}Sr^{2+}$
$^{56}Fe^+$	91.66	$^{40}Ar^{16}O^+$, $^{40}Ca^{16}O^+$
$^{57}Fe^+$	2.19	
$^{58}Fe^+$	0.33	
$^{58}Ni^+$	67.77	$^{58}Fe^+$
$^{75}As^+$	100	$^{40}Ar^{35}Cl^+$
$^{112}Cd^+$	24.1	^{112}Sn, $^{40}Ca_2^{16}O_2^+$, $^{40}Ar_2^{16}O_2^+$, $^{96}Ru^{16}O^+$
$^{114}Cd^+$	28.7	$^{98}Mo^{16}O^+$, $^{98}Ru^{16}O^+$

[13]In thermodynamics isobaric refers to constant pressure. Here it refers to two ions of equal mass. Whether a given ion is an isobaric interference to an analyte ion depends on the resolving power of a mass spectrometer.

prevalence of doubly charged ions is element specific and only causes a problem in a few cases such as $^{44}Ca^+/^{88}Sr^{2+}$, $^{66}Zn^+/^{132}Ba^{2+}$, and $^{69}Ga^+/^{138}Ba^{2+}$. The fraction of $^{88}Sr^{2+}$ is generally on the order of a few percent of $^{88}Sr^+$. Measuring strontium simultaneously in a calcium determination can alert an analyst if this interference might be a problem. An accurate measurement of $^{44}Ca^+$ can then be made by correcting the peak at $m/z = 44$ using measurements of $^{88}Sr^+$ in the test solution and $^{88}Sr^{2+}/^{88}Sr^+$ from a validation standard.

Measuring all isotopes of a given element can also be useful to identify an interference. A mass peak that has an unusual intensity compared to the relative natural abundance for an element can indicate the presence of an interference at that isotopic mass. For example, tin has seven isotopes from ^{116}Sn to ^{124}Sn to choose for analysis even neglecting several minor isotopes. Where an interference can't be avoided, sample preparation steps can sometimes remove a matrix component. Ion-exchange extraction or chromatography can separate different species. In arsenic analysis, HCl should not be used in sample digestion or preparation to prevent the $^{40}Ar^{35}Cl^+$ interference. If chloride exists in the sample matrix, it can be removed before analysis. These steps add significant time and effort to an analysis, but they are often necessary for a problematic matrix.

For molecular interferences, most modern ICP-MS instruments contain a stage between the ion source and the mass analyzer that can remove polyatomic ions. This compartment is called a collision or collision/reaction cell. It consists of an enclosed cell within the mass spectrometer with a gas inlet and openings on the endplates that are collinear with the ion path. Inside the cell is a set of quadrupole rods that operate with a small ac voltage applied.[14] The ac voltage provides no ejection of ions, and the rods function only as an ion guide to keep the ions on their original path. The purpose of the cell is to allow the ion beam to pass through an elevated pressure of gas. The base pressure of a mass spectrometer is 10^{-5}–10^{-6} Torr. The mean free path at these pressures is on the order of meters, meaning that no collisions occur within an instrument. Raising the pressure to 10^{-2}–10^{-4} Torr allows an ion to undergo a small number of collisions over a 10–20 cm path, which is a typical length for one of these cells.

When operated in collision mode, the cell is maintained at an elevated pressure of a small inert gas such as He. As ions collide with the He gas, they lose kinetic energy in the direction of travel. For ions of a given mass, a polyatomic molecular ion will have a larger size or cross section than an atomic ion. The larger polyatomic ion therefore collides more often with He atoms, and loses a greater amount of kinetic energy, than does the atomic ion of the same mass. A small voltage barrier is set between the collision cell and the mass analyzer. This barrier prevents the polyatomic ions, which have lost more energy, from continuing to the mass analyzer. This method is called kinetic energy

---

[14]Hexapole and octupole ion guides are also common.

discrimination (KED) and the He gas pressure and voltage barrier are set to remove the polyatomic species without losing too many of the atomic ions.

When operated as a reaction cell, the cell contains an elevated pressure of a reactive gas such as ammonia, methane, or oxygen. There is a wide variety of chemical reactions to either remove interferences or to create cluster ions that shift the mass of an atomic ion to a higher m/z that is free of interferences. These approaches are not as general as the KED in a collision cell, but they can be very effective for specific analytes and interferences. Methane and ammonia are particularly effective at removing argon ions through reactions such as:

$$Ar^+ + CH_4 \rightarrow CH_3^+ + Ar + H$$

$$Ar^+ + CH_4 \rightarrow CH_4^+ + Ar$$

$$Ar_2^+ + CH_4 \rightarrow Ar + Ar + CH_4^+$$

$$Ar^+ + NH_3 \rightarrow Ar + NH_3^+$$

$$ArO^+ + NH_3 \rightarrow Ar + O + NH_3^+$$

Using a specific gas to remove interferences assumes that it does not react with the analytes of interest. The "mass shift" approach takes advantage of a reaction of a gas with one or more analyte ions. The following examples are just two cases of the many reactions that can be used to avoid interferences:

$$^{63}Cu^+ + 2\,NH_3 \rightarrow {}^{63}Cu(NH_3)_2^+ \text{ (mass-charge ratio of 97)}$$

$$^{64}Zn^+ + 3\,NH_3 \rightarrow {}^{64}Zn(NH_3)_3^+ \text{ (mass-charge ratio of 115)}$$

Applying which strategy to use depends on the specific analytes and the interferences that occur for a given sample matrix.

A very different, but less common approach due to the cost and complexity, is to use a high-resolution mass analyzer to measure an elemental ion separately from an interfering molecular ion. Various analyzer designs are discussed in the next section. Here we'll introduce the concept of peak resolution in a mass spectrum. Two adjacent peaks of equal intensity are defined as separated or resolved if the valley between them is 10% of the peak heights. The resolving power, R, of an instrument can be determined from these two peaks (real or hypothetical) using:

$$R = \frac{m}{\Delta m} \tag{10.2}$$

Here m is the mass of the lighter ion and Δm is the difference in mass between the two resolved peaks.[15] For example, a mass spectrometer with a resolving power of 500 can resolve an ion with $m/z = 500$ from an ion with $m/z = 501$.

---

[15]The definition of Δm should be specified when quoting a resolving power. Some workers prefer to define R using full-width at half maximum (FWHM) for Δm.

For unit resolution, $R = 500$ is more than sufficient for elemental analysis. Measuring an elemental ion independent from a molecular interference requires more resolution. Example 10.2 provides a specific case for the isobaric interference in arsenic analysis. Newer ICP-MS instruments are available that operate in low-, medium-, and high-resolution modes, where high resolution has an $R = 10,000$. The tradeoff is that increasing the resolving power leads to lower ion transmission and less sensitive measurements. The instruments are usually operated at the lowest resolution possible that can make accurate measurements of the elements of interest.

Example 10.2 Mass Spectrometer Resolving Power. Calculate the resolving power to measure arsenic, $^{75}As^+$ ($m/z = 74.9216$) independent of $^{40}Ar^{35}Cl^+$ ($m/z = 74.9312$).

$$R = \frac{m}{\Delta m}$$

$$R = \frac{74.9216}{74.9312 - 74.9216} = \frac{74.9216}{0.0096} = 7800$$

Based on this calculation, an ICP-MS will need to operate in high-resolution mode to make this measurement.

You-Try-It 10.C
Determining unknown concentrations from a calibration curve is essentially the same in both ICP-MS and ICP-OES measurements. For something different, the ICP-MS worksheet in you-try-it-10.xlsx contains ICP-MS data obtained via laser ablation to study an inclusion in a geologic sample. Make a two-dimensional plot of the data to visualize analyte distribution in the inclusion. The worksheet also provides calibration data as a function of mass to illustrate a semiquantitative analysis.

Before continuing to other mass analyzer designs, now is a good place to develop a sense of how we might choose between the instrumental methods discussed here and in the previous chapter for a given analytical problem. Electroanalytical methods and atomic spectrometry are both suitable for measuring ions in solution. Potentiometry is less sensitive than atomic spectrometry, but it is nondestructive and can detect analytes of different speciation. Likewise, electroanalytical methods can measure some anions for which AAS and ICP-OES are insensitive. The electroanalytical methods can also be portable or miniaturized more easily than spectroscopic methods. The bottom line is that the most appropriate analytical method will depend on the nature of the sample, the number of analytes, required detection limits, and the information that is desired. The continual development of these instrumental methods attests to their importance in a variety of applications.

When deciding between atomic spectrometry methods, the ICP methods have clear advantages in multielement measurements and lower detection limits. For measurement of one or a few easily atomized elements, the simplicity of flame AAS makes it a logical choice. AAS can provide better precision due to less sensitivity to temperature variations in the atomization source. For the plasma methods, ICP-MS can measure nonmetals, isotopes, and generally has lower detection limits than ICP-OES. For some elements, the difficulties caused by interferences can make ICP-OES a better choice. Even with the clear distinctions between the atomic spectrometry methods, there is still overlap in the capabilities of these instruments. Table 10.4 summarizes the characteristics of each atomic spectrometry method and Figure 10.17 provides an overview of selected instrument characteristics with approximate instrument costs. For a given type of instrument, more expensive instruments will have more capabilities such as multiple background correction methods,

TABLE 10.4 Characteristics of Different Atomic Spectrometry Instruments

Instrument	Main Advantage	Main Disadvantage
Flame AAS	Low cost, best precision	Least efficient atomization
Cold-vapor AAS	Low detection limits for certain elements	Low sample throughput
GFAAS	Very small sample size	Limited dynamic range
ICP-AES	Efficient atomization, multielement analysis, wide dynamic range	More expensive/complex than AAS
ICP-MS	Efficient atomization, multielement analysis, wide dynamic range, lowest detection limits, isotopic analysis	Most expensive/complex

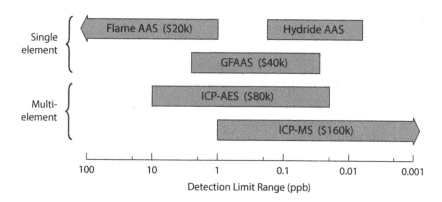

Figure 10.17 Summary of atomic spectrometry instruments.

variable resolution, a higher dynamic range, or a greater degree of automation and computer control.

10.5 OTHER MASS SPECTROMETER DESIGNS

Since we've been discussing instrumentation, now is a good place to introduce other mass spectrometer designs. The different designs are used for elemental analysis and molecular mass spectrometry applications, which are discussed in the next two chapters. In elemental analysis, the resolving power needed to separate an interference from an atomic analyte is easy to calculate and will be quoted as a numerical value. In molecular mass spectrometry, mass spectrometers are classified as having *nominal mass* resolution ($R \approx 1000$) or *accurate mass* resolution ($R > 10,000$). Resolving power, mass range, dynamic range, and precision are key parameters for selecting a certain instrument design for an analytical application.

We skipped discussing details of the detector in the ICP-MS. Commercial mass spectrometers use a variety of ion detector designs, and some instruments will have multiple detectors to measure both high and low signal levels. In all cases, ions striking the surface of a detector create an electrical current as the signal. The main difference in the different detectors is the degree of internal amplification that they provide and their time response. The current output from a Faraday cup is converted to a voltage with no internal amplification, which is known as *analog mode*. It is the least sensitive of the different detectors, but it is used for high ion signals that would saturate the more sensitive designs. The other detectors all produce internal amplification to convert a single ion to a measurable current pulse. These detectors are usually used in *counting mode*, where data acquisition electronics count and store the number of pulses or counts per second (cps). The upper limit for pulse counting varies, but generally pulse overlap causes nonlinearity at 10^8–10^9 counts per second. The following list describes some common ion detectors and their primary characteristics. The different mass spectrometer designs will use one or more of these detectors that best couple to the other characteristics of the mass analyzer.

Faraday cup: A Faraday cup is a metal cup that is placed in the path of the ion beam. It is attached to an electrometer, which measures the ion-beam current. Since a Faraday cup can only be used in analog mode, it is less sensitive than other detectors that are capable of operating in pulse-counting mode.

Electron multiplier tube (EMT): EMTs are similar in design to photomultiplier tubes, although without the photocathode. An ion hitting the first dynode ejects secondary electrons. These electrons eject more electrons as they cascade down a series of biased dynodes. The main advantage of an EMT is the high internal gain of 10^6 or more due to the amplification in the dynode chain. They can be used in analog or counting mode, but the higher gain results in a slower time response.

Channeltron: A channeltron is a horn-shaped continuous dynode structure that is coated on the inside with an electron-emissive material. An ion striking the channeltron creates secondary electrons that have an avalanche effect to create more secondary electrons and finally a current pulse. This design is compact and often used in quadrupole mass spectrometers, as represented by the horn-shaped detector shown in Figure 10.15.

Microchannel plate (MCP): An MCP consists of an array of glass capillaries (10–25 μm inner diameter) that are coated on the inside with an electron-emissive material. The capillaries are biased at a high voltage similarly to the channeltron. An ion that strikes the inside wall of one of the capillaries creates an avalanche of secondary electrons. This cascading effect creates a gain of 10^3–10^4 and produces a current pulse at the output. These detectors provide the fastest time response. Figure 10.18 illustrates an MCP detector and the general concept of amplification.

Daly detector: A Daly detector consists of a conversion dynode that emits secondary electrons when struck by an ion. The secondary electrons strike a scintillator that emits a light pulse that is measured by a photomultiplier tube, which can be outside of the vacuum system. The Daly detector is useful in a multidetector arrangement to achieve a high dynamic range. Removing a bias voltage allows ion to pass the dynode and reach a Faraday cup.

Table 10.5 lists some defining characteristics for the most common mass analyzer designs. The quadrupole mass filter is the most common mass spectrometer in ICP-MS and as a detector for gas and liquid chromatography. The magnetic-sector design is an older design, but it remains common in the earth sciences for isotope-ratio measurements. There are many variations on these analyzers, including the coupling of multiple analyzers in tandem for various applications and ionization sources. The time-of-flight (TOF) design provides

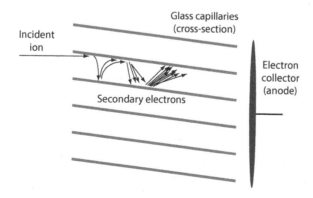

Figure 10.18   Cut-through schematic of a microchannel plate detector.

TABLE 10.5 Common Mass Analyzer Designs

Design	Typical R	Characteristic	Applications
Quadrupole	1,000	Compact	Elemental analysis and small molecules
Magnetic-sector	Variable*	High dynamic range High precision	Isotope ratios
TOF	10,000–50,000	Unlimited mass range	Accurate mass of small molecules, mass of large biomolecules
Ion trap	1,000	Compact Can accumulate ions	Used to interface ion source to TOF, MS^n
Orbitrap	>100,000	Highest resolution	Very complex samples

*400–10,000 depending on instrument design and settings.

accurate mass analysis of small molecules and has the highest mass range for analyzing biomolecules such as proteins and DNA/RNA fragments. The ion trap is similar in performance to a linear quadrupole, but it is capable of accumulating and storing ions for tandem mass spectrometry experiments. The orbitrap has the highest resolving power and is used for the most challenging mass spectrometric problems. It uses a different type of detection than the other designs, and we won't discuss it further.

The quadrupole is the most common small molecule mass analyzer because of its compact design. The operating principles of the quadrupole filter were described in Section 10.4. In elemental analysis using ICP-MS, the quadrupole filter is scanned to measure each element that is present in the test portion. This same operation, referred to as *full scan mode*, is possible as a chromatographic peak elutes from a column. Full scan operation is typical when trying to identify unknown components in the analytical sample. When the target analytes are known and the primary goal is quantification, the mass analyzer does not scan all m/z but can measure only one or a few selected fragment ions. This mode is known as *selected-ion monitoring* (SIM). Since more time is available to count only the signal at the selected m/z, SIM provides better sensitivity and precision in a quantitative measurement. For the greatest sensitivity and universal detection, the mass spectrometer sums the total current due to all ions. This mode is called *total-ion current* (TIC). The limitations of most commercial quadrupole analyzers include an upper limit of $m/z = 2000$ Da and moderate resolution. The upper mass limit can often be extended by a factor of two with loss of the lower mass region.

Because of the bulky magnet, the magnetic-sector analyzer is less common now than in the past. It has the advantage of being able to accommodate multiple detectors at the exit of the analyzer. This capability gives it the highest dynamic range of the common mass spectrometer designs. The ability to acquire simultaneous data on different channels also helps maximize sample utilization and improve precision in quantitative measurements. These advantages make it common in the earth sciences to measure isotope ratios for dating (geochronology) and other applications.

Figure 10.19 shows a schematic with one detector and the dashed line shows the path for ions of one m/z. The ion optics in the ion source chamber extract and accelerate ions to a kinetic energy, E_k, given by

$$E_k = \frac{1}{2}mv^2 = zeV \tag{10.3}$$

where m is the mass of the ion, v is its velocity, z is the charge of the ion, e is the elementary charge, and V is the applied voltage of the ion optics. The ions enter the flight tube between the poles of a magnet and are deflected by the magnetic field, B. In the schematic in Figure 10.19, the magnetic field is perpendicular to the page. The dashed line indicates the path of ions of one specific mass-to-charge ratio, m/z, that reach the detector. These ions have a mass-to-charge ratio so the Lorentz force due to the magnetic field matches the inertial force due to the ion mass and velocity for the radius of curvature, r, of the ion path:

$$Bzev = \frac{mv^2}{r} \tag{10.4}$$

Combining and rearranging Equations (10.3) and (10.4) to eliminate v gives:

$$\frac{m}{z} = \frac{B^2r^2e}{2V} \tag{10.5}$$

This equation shows that the m/z of the ions that reach the detector can be varied by changing B or V. For an isotope-ratio measurement, it is common to hop between two voltage settings to alternately measure each isotope. For an instrument with multiple detectors, V and B are set so that the appropriate ions are striking the different detectors.

The time-of-flight mass spectrometer (TOF-MS) does not have an intrinsic mass limit. In practice, special detection methods are needed when ions approach 1 MDa. However, the TOF is the most useful analyzer for large

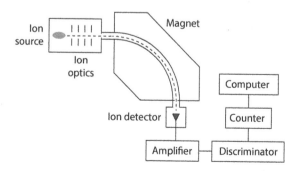

Figure 10.19   Top-down schematic of a magnetic-sector mass spectrometer.

molecules such as proteins and DNA or RNA fragments. The TOF analyzer design uses the difference in transit times through a flight tube to separate ions of different masses. It operates in a pulsed mode, so ions must be produced or extracted in "bunches." A fixed electric field, V, accelerates all ions into a drift region that is free of electric fields. An ion of charge z will have a kinetic energy, E_k, as given in Equation (10.3). For ions of equal charge, the lighter-mass ions have a higher velocity than heavier ions and they reach the detector at the end of the drift region sooner. The time, t, for an ion to transit the mass spectrometer is d/v, where d is the distance through the drift tube to the detector. Since d and V are fixed, the transit time depends on m and z:

$$t = \frac{d}{v} = \sqrt{\frac{d^2}{2eV}} \sqrt{\frac{m}{z}} \qquad (10.6)$$

Figure 10.20 shows the schematic of a TOF-MS instrument that includes a reflectron in the flight tube. The reflectron is a series of rings or grids with applied voltages that act as an ion mirror. This ion mirror compensates for the spread in kinetic energies of the ions as they enter the drift region and improves the resolution of the instrument. The output of an ion detector is recorded with an oscilloscope or digitizer as a function of time to produce the mass spectrum. Typical flight tubes are 1–2 m in length, and accelerating voltages in the kilovoltage range result in flight times in the microsecond range. To achieve the best spectral resolution, the detector is usually a fast microchannel plate detector. The detector output goes to a digitizer that operates at approximately 1 GHz to provide sufficient speed to obtain resolutions of 10,000 or greater.

The arrow in Figure 10.20 represents a laser pulse that creates a plume of ions. The pulsed operation of the TOF design couples seamlessly to pulsed ionization methods. Large biomolecules can be vaporized and ionized intact using matrix-assisted laser desorption ionization (MALDI). In this method, the analyte is dispersed in a matrix that absorbs the energy of the laser pulse. Analyte molecules are carried into the gas phase during ablation of the matrix and remain intact. They become ionized by proton transfer or other acid–base types of interaction. Figure 10.21 shows as an example the mass spectrum of a fullerene sample. The mass of the C60 fullerene ion is 720 Da.

As a final comment on mass spectrometry, the different component designs will be used in different applications depending on the analytical goal and the

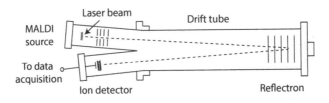

Figure 10.20 Schematic of a reflectron TOF-MS.

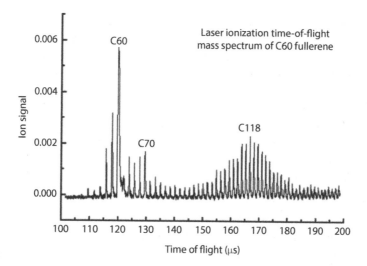

Figure 10.21 Example of a TOF mass spectrum (courtesy of R. Craig Watson, Virginia Tech).

required resolving power, mass range, precision, and analysis speed. In addition to quantitative measurements, mass spectrometers are used to determine isotopic composition, aid in structural analysis of organic molecules, and to sequence large biomolecules. As a chromatography detector, a mass spectrometer provides sensitive detection and the ability to confirm the identify analytes if they have a unique fragmentation pattern.

Chapter 10. What Was the Point?   This chapter described a variety of spectrometric methods for making quantitative measurements in elemental analysis. There are various and overlapping aspects for selecting a method of analysis and interpreting data.

- AAS: simplest and best precision for easily atomized metals.
- AES: low detection limits and multielement capability.
- ICP-MS: lowest detection limits, multielement capability, nonmetals, and isotopic analysis.
- Other mass spectrometer designs: stable isotope ratios, accurate mass analysis, high-mass biomolecules.

PRACTICE EXERCISES

1. List and discuss the advantages and disadvantages of atomic absorption versus AES.
2. List and discuss the advantages and disadvantages of potentiometric measurements compared to AAS.

3. For each of the following analytes, choose an appropriate electrochemical or spectroscopic instrumental method. If multiple methods might be useful, choose the one that will be most precise and rapid. Give the rationale for your choice in your answer and suggest compatible sample preparation methods.
   (a) Ca^{2+} and Mg^{2+} in leafy vegetables,
   (b) NO_3^- and NH_4^+ in surface water,
   (c) Pb, Cd, Zn, and Cu in soil from an old landfill,
   (d) ClO_4^+ in seawater,
   (e) Co in stainless steel.

4. What mass spectral resolution is necessary to quantitate ^{112}Cd (exact mass 111.90276 u) in the presence of $^{40}Ar_2^{16}O_2^+$ (exact mass 111.91460 u)?

5. You have the following data for measurement of Pb in water using AAS. The instrument is known to have a linear response for Pb from 0 to 10 ppm. What is the lead concentration in the unknown test solution?

$[Pb^{2+}]$, ppm	AAS Signal, AU
0.0	0.033
10.0	1.284
Unknown	0.471

6. Compare the advantages and disadvantages of using an ICP excitation source rather than a flame in AES?

7. Plot the following data, AES signal versus $[Ca^{2+}]$, and determine the calibration curve for this measurement. Discard any points that do not fall on the expected line.
   (a) What is the linear range of this calibration data?
   (b) An unknown solution produced a signal of 0.455. What is $[Ca^{2+}]$ in this test portion?
   (c) Another unknown solution produced a signal of 0.952. What is your course of action before calculating the concentration in this sample?

$[Ca^{2+}]$, ppm	AES Signal, AU
0.0	0.0
1.0	0.011
2.5	0.047
5.0	0.082
10.0	0.149
25.0	0.364
50.0	0.692
100.0	0.833

8. Electronic counting instruments can set a discriminator level (DISC) to preferentially count signal pulses and not count noise pulses, which have lower peak intensity. Plot the following data and determine the discriminator level that gives a maximum in the signal-to-noise ratio.

DISC, mV	Blank, cps	Test Solution, cps
5	8,120,000	8,970,000
10	618,000	671,000
20	33,800	288,000
30	6,290	210,000
35	6,040	190,000
40	5,920	179,000

9. Discuss why background correction is necessary in atomic spectroscopy and why the background might vary for different sample matrices.

10. What is the main advantage of using GFAAS compared to other atomic spectrometry methods? What are the disadvantages of GFAAS?

11. Refractory samples are difficult to dissolve, even in strong acid solutions. List the different sample introduction and vaporization methods to analyze such samples by atomic spectrometry. Discuss the relative advantages and disadvantages of these approaches.

12. Compare molecular and atomic spectra as shown in Chapters 4 and 10. Discuss the differences in the specifications for the spectrometers that are used in these two types of spectroscopy.

13. Explain why cold-vapor AAS is the preferred AAS method for the measurement of mercury.

14. Explain why nonmetal elements are not usually measurable using atomic absorption or emission spectrometry.

15. What is a stable isotope and what instrumental method is required for measuring the ratio of stable isotopes? Explain why.

16. What characteristics of a magnetic-sector mass spectrometer make it suitable for isotope ratio measurements?

17. What characteristics of a time-of-flight mass spectrometer make it suitable to measure the mass of large biomolecules?

CHAPTER 11

MOLECULAR STRUCTURE DETERMINATION

Learning Outcomes

- Describe the information that you can extract from a mass spectrum.
- Describe the principles and instrumentation of Fourier-transform infrared (FTIR) spectroscopy.
- Determine the functional groups in a molecule from an infrared spectrum.
- Describe the principles and instrumentation of nuclear magnetic resonance (NMR) spectroscopy.
- Predict the structure of a molecule based on an NMR spectrum.

11.1 INTRODUCTION

Most of this textbook focuses on methods and issues in quantitative analysis, i.e., determining the amount or concentration of an analyte in a sample. This chapter is slightly different in that it describes instrumental methods for qualitative analysis. Qualitative analysis includes determining the structure or other physical property of a molecule or material. These measurements often produce numerical results, but we classify them as qualitative methods because their main aim is to determine a property rather than an amount. For example,

Basics of Analytical Chemistry and Chemical Equilibria: A Quantitative Approach, Second Edition. Brian M. Tissue.
© 2023 John Wiley & Sons, Inc. Published 2023 by John Wiley & Sons, Inc.
Companion Website: www.wiley.com/go/tissue/analyticalchemistry2e

the melting point range of an organic compound is a numerical result that provides information about the identity and purity of a solid sample. A melting point determination does not provide a quantitative measure of the impurities.[1] Likewise, describing the tensile strength or compressibility of a material requires numerical values, but these measures describe a physical property and not the amounts of chemical species in the sample.

Mass spectrometry, IR, and NMR spectroscopy can be used for quantitative analysis. In fact, mass spectrometers are becoming a common detector for the gas and liquid chromatographic separations that are discussed in Chapter 12. Unlike atomic and molecular spectroscopy in the UV/Vis, IR and NMR methods are not capable of ultra-trace analysis. Most IR and NMR applications are limited to quantitating components at a level of approximately 1% or more. Common quantitative applications for IR and NMR spectroscopy are monitoring the progress of a synthetic reaction and determining impurities in pharmaceuticals.

Rather than discussing quantitative applications, I will describe the instrumental methods in this chapter from the perspective of determining or confirming a molecular structure. This activity to identify or "ID" a molecule is very common in chemistry. In these measurements, the sample must be one pure compound to obtain a spectrum that is representative of the molecule. It is possible to record spectra of a mixture, but the spectral interpretation is much more difficult. Applications include determining the composition of a mixture, identifying the active agent that has been purified from a natural product, and determining the structures of intermediate and final products when synthesizing a new molecule. For an analyte that you expect to isolate from a sample, confirming its presence can often be based on only one type of spectrum. In these cases, the spectrum of the pure compound serves as a reference.

To determine the structure of a newly synthesized molecule, a chemist will often use all of the tools listed in Table 11.1. Chapter 3 discussed combustion analysis using CHNO analyzers. These instruments provide the ratio of C, H, O, N (and S) in a sample. Having the empirical formula of a molecule is often the first step to then use the spectral methods to determine the structure. We will assume that we have the results of CHNO analysis in any of the examples that we discuss. We won't discuss crystallography using X-ray diffraction in any

TABLE 11.1 Summary of Structure Determination Tools

Method	Information
Combustion analysis	Empirical formula
Mass spectrometry	Molecular mass and fragmentation pattern
Infrared spectroscopy	Functional groups
NMR spectroscopy	Atomic and functional group proximity
X-ray diffraction	Crystal structure

[1]Freezing point depression can be used in some specialized measurements where the sample matrix does not vary.

depth. For molecules and macromolecules that form good-quality crystals, this tool provides the most definitive information to determine a crystal structure.

Part of the reason that multiple methods are needed for structure determination is that spectra can be similar for similar molecules. Many molecules have the same nominal mass, and the mass spectra of geometric isomers can be nearly identical. Infrared (IR) spectroscopy identifies functional groups in a molecule, but not their three-dimensional arrangement. Nuclear magnetic resonance (NMR) spectroscopy provides the relative proximity of atoms in a molecule, but these spectra can be sufficiently complicated that other spectra are needed to help interpret the NMR spectrum.

IR and NMR spectroscopy share basic concepts with UV/Vis absorption spectroscopy discussed in Chapter 4. The instrument components and designs are quite different in the very different spectral regions. As a reminder, Table 11.2 lists energy and wavelength ranges for these different spectroscopies. The different units that are used in the different techniques are simply to have a convenient scale. X-ray wavelength is often expressed in angstroms, Å, which is 10^{-10} m. A single X-ray wavelength is directed onto a solid sample and the instrument records the angles at which the X-rays are diffracted. The UV/Vis region uses wavelength in nm. Wavenumbers, cm^{-1}, is the common x-scale for IR spectra. Wavenumbers is simply the inverse of wavelength and is thus an expression of energy. The ranges listed for UV/Vis and IR spectroscopy are typical upper and lower bounds for the spectra. The lower energy side of each of these ranges will vary depending on instrument components. The range listed for NMR instruments is the RF radiation that excites proton nuclei for different instruments with different size magnets. The actual scan range is a small fractional range (ppm) at the NMR base frequency.

Each section in this chapter will describe the principles of the method and provide a sample spectrum. Mass spectrometer designs were described in the previous chapter. Two sections here will describe the basic instrument components of FTIR and NMR spectrometers. Being familiar with the instrument components makes it much easier to select data acquisition parameters when you record a spectrum. I have left out much of the theory behind these techniques. Likewise, the examples in this chapter are rather simple molecules.

TABLE 11.2   Spectral Range of Spectrometers for Structure Determination

EM Region	Energy or Frequency Range	Wavelength Range	Type of Transition
X-ray	≈10 keV	0.7–1.5 Å	Bragg diffraction
UV/visible	50,000–12,500 cm^{-1}	200–800 nm	Valence electrons
Infrared	4000–400 cm^{-1}	2.5–25 μm	Molecular vibrations
NMR (radio waves)	60–800 MHz	5–0.3 m	Nuclear spin flips*

*Transitions occur between energy levels that are split by a magnetic field. State-of-the-art NMR spectrometers operate at > 1.0 GHz.

When you encounter more complicated spectra, there are numerous library and online resources available on spectral interpretation for each of these methods.

11.2 MOLECULAR MASS SPECTROMETRY

Mass spectrometry was introduced in Chapter 10 for elemental analysis using the inductively coupled plasma mass spectrometry (ICP-MS). The general principles of a mass spectrometer are the same for either elemental or molecular analysis. The sample introduction, ionization mechanisms, instrument characteristics, and mass spectral data can be quite different. Since a high-temperature plasma destroys polyatomic species, it is obviously not suitable for structure determination of organic molecules. ICP-MS determines each element that was present in a test mixture. For structure determination, molecular mass spectrometry requires a pure substance. If multiple analyte molecules are present, the overlap of the many fragments in the mass spectrum makes it difficult to interpret. A molecule of interest can be purified using the methods discussed in Chapter 2 and introduced into the mass spectrometer vacuum system on a probe that is heated to vaporize it. It has become more common to separate an analyte from other sample components using chromatography and to record the mass spectrum recorded as it elutes from a separation column.

As described in the previous chapter, obtaining a mass spectrum involves:

- Creating gas-phase ions (and transfer to vacuum system if needed)
- Separating the ions in space or time based on their mass-to-charge ratio
- Measuring the quantity of ions of each mass-to-charge ratio

In elemental analysis, there was one m/z for each elemental isotope. In molecular mass spectrometry, a given molecule can produce multiple ions of different m/z. The term *molecular ion*, M^+, is reserved for the ion that has not lost any structural moieties. The molecular ion gives us the molecular mass, which can help confirm the identity of the molecule. Smaller ionized pieces of a molecule are called *fragment ions*. We call the relative distribution of different molecular fragments in a mass spectrum the *fragmentation pattern*. The y-scale in a mass spectrum is generally scaled to the most abundant peak, known as the *base peak*. The y-scale, or *relative abundance*, will have units of 0–1000, 0–100%, or something similar. The fragmentation pattern of a given molecule will be repeatable for some ionization methods. It is thus extremely useful to help identify a molecule because the experimental mass spectrum can be compared to reference spectra in search libraries.

We start here with *electron ionization* (EI). Other interfaces and ion sources for gas and liquid chromatography are discussed in the next chapter. An EI source consists of a heated tungsten or rhenium filament that is at a negative voltage compared to a ground plate called the electron trap. Electrons that transit from the filament to the ground plate create an electron beam. Operating the electron beam at an energy of 70 eV produces a de Broglie wavelength that matches the bond length of organic molecules. The 70-eV energy therefore provides the optimum ionization efficiency, which is approximately 0.1%. Analyte molecules, M, in the ion source that interact with the electron beam gain internal energy and can lose an electron to form a radical cation:

$$M + e^- (70\,eV) \longrightarrow M^{\cdot +} + 2e^-$$

This ion will undergo fragmentation, often to the point that the peak for the molecular ion is very small or not observed in the mass spectrum if $M^{\cdot +}$ is very unstable. Operating the electron beam at a lower energy can reduce fragmentation for some molecules, but it provides a lower sensitivity due to less efficient ionization. It is important to note that fragmentation and rearrangements occur very rapidly in the ion source before entering the mass analyzer. Fragmentation in the mass analyzer is not common, but it will change the energy of the ions and contribute to the baseline noise.

The main advantage of EI is that the fragmentation pattern at a given electron beam energy is reproducible in different instruments. Comparing an unknown mass spectrum to a library of EI mass spectra provides confirmation of the identity of the molecule. Instrument software will often include search libraries. There are also several freely available online databases with mass spectral and other data:[2]

- MassBank: https://massbank.eu/MassBank
- NIST Chemistry WebBook: https://webbook.nist.gov/chemistry
- Spectral Database for Organic Compounds, SDBS: https://sdbs.db.aist. go.jp/sdbs/cgi-bin/cre_index.cgi

For a new molecule that is not in the mass spectral libraries, the fragments in the mass spectrum can provide structural information. Table 11.3 lists some of the common fragment ions that appear in mass spectra of organic molecules. The low mass region, $m/z < 44$, is not as useful due to interferences of atmospheric gases. However, seeing a series of peaks that differ by 14 is indicative of an alkyl chain in the molecule. Many fragments are the result of rearrangement reactions in the gas phase, but most of these reactions are known and tabulated in reference sources. I list the tropylium ion in Table 11.3 as an example. The presence of a peak at $m/z = 91$ is indicative of a phenyl group with at least one alkyl carbon attached. Toluene is the simplest example, and the toluene molecular ion rearranges to the seven-member ring of tropylium.

The mass of the molecular ion and the fragmentation pattern provides a lot of information to confirm the identity of a molecule, especially if it is coupled

---

[2]These databases are searchable by name, formula, etc. The MassBank and SDBS databases also allow you to enter a list of mass spectral peaks for searching. URLs accessed August 2022.

TABLE 11.3 Common Molecular Fragments

Fragment	Empirical Formula	m/z
Methylene	Loss of $-CH_2-$	14
Methyl	Loss of $-CH_3$	15
Propyl	$C_3H_7^+$	43
Butyl	$C_4H_9^+$	57
Pentyl	$C_5H_{11}^+$	71
Hexyl	$C_6H_{13}^+$	85
Chlorine	Fragment containing Cl	35, 37 (3:1 ratio)
Phenyl	$C_6H_5^+$	77
Bromine	Fragment containing Br	79, 81 (1:1 ratio)
Tropylium	$C_7H_7^+$ (rearrangement)	91

with a chromatographic retention time. For a new molecule, i.e., one for which a reference compound is not available, the mass spectrum alone is usually insufficient to determine the geometric structure. Mass spectrometry does not provide information about the proximity of functional groups in a molecule. A mass spectrum that has a peak at $m/z = 57$ suggests that the molecule contains a $C_4H_9^+$ fragment ion, but it doesn't tell us if that group is an n-butyl group or some other fragment on the original molecule. Likewise, a molecule might contain a certain functional group, but that group does not form a fragment ion in the mass spectrum. The mass spectra of n-butyl and tert-butyl benzene have a strong tropylium peak at $m/z = 91$. Due to the stability of this fragment, there is almost no significant peak for either molecule at $m/z = 57$.

As described in Section 10.5, different mass analyzer designs have different mass ranges and resolving power. The most common analyzers in chemical analysis are the quadrupole, which provides nominal mass resolution, and the time-of-flight, which provides accurate mass resolution. The need for accurate mass capability arises because the masses of isotopes are not integer values. Table 11.4 lists the isotopes of the elements that are common in organic molecules with their exact mass and natural abundance. Carbon-12 is assigned a value of exactly 12.0000 and all other isotopes are measured relative to ^{12}C. A given peak in a mass spectrum will be the summation of the isotopes in the ion.

For small molecules and small fragment ions, the minor isotopes have little significance in the mass spectrum. For larger molecules, the probability that the molecular ion or a large fragment ion contains one of the minor isotopes increases significantly. As an example, Figure 11.1 shows the mass spectrum of caffeine at nominal mass resolution. Interpreting this particular mass spectrum is not easy due to the rearrangements that occur. The peak at $m/z = 137$ occurs from loss of CH_3NCO and the peak at $m/z = 109$ occurs from the additional loss of CO. Based on the structure of caffeine, see if you can draw structures of these two fragment ions at $m/z = 137$ and $m/z = 109$.

TABLE 11.4   **Exact Mass and Isotopic Abundance of Select Elements**

Element	Nominal Mass	Exact Mass	Abundance
Hydrogen	1	1.007825	99.99
	2	2.014102	0.01
Carbon	12	12.00000	98.93
	13	13.00335	1.08
Nitrogen	14	14.00307	99.64
	15	15.00011	0.36
Oxygen	16	15.99491	99.76
	17	16.99913	0.04
	18	17.99916	0.20
Fluorine	19	18.99840	100.00
Chlorine	35	34.96885	75.76
	37	36.96590	24.24

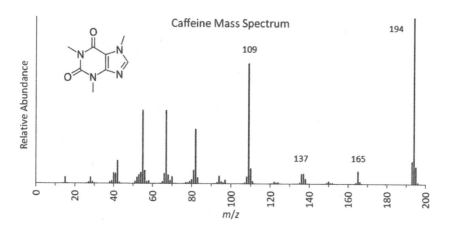

Figure 11.1   Electron-ionization mass spectrum (EI-MS) of caffeine.

The formula weight of caffeine, $C_8H_{10}N_4O_2$, is 194.194 g/mol. Formula weight is a weighted average that includes all isotopes. You won't see a peak in the mass spectrum at this mass. The nominal mass of caffeine is

$$8 \times 12 + 10 \times 1 + 4 \times 14 + 2 \times 16 = 194\,Da$$

The exact mass for the molecular ion that contains the primary isotopes is

$$8 \times 12.00 + 10 \times 1.007825 + 4 \times 14.00307 + 2 \times 15.99491 = 194.0804\,Da$$

The molecular ion is the base peak for caffeine and occurs at $m/z = 194$, but there is also a significant peak at $m/z = 195$. For simplicity we will consider only the carbon-13. For eight carbon atoms in the molecule, there is an ≈9%

probability that the molecular ion will contain one ^{13}C. Doing a simple calculation for an $[M + 1]^+$ ion that has seven ^{12}C atoms and one ^{13}C gives

$$7 \times 12.00 + 1 \times 13.00335 + 10 \times 1.007825 + 4 \times 14.00307$$
$$+ 2 \times 15.99491 = 195.0837 \, Da$$

This calculation overestimates the exact mass for caffeine, but it illustrates the point. Figure 11.2 shows the partial mass spectrum of the caffeine base peak at accurate mass resolution.[3] Besides the $[M + 1]^+$ ion at 195.0829 Da there is a smaller peak for the $[M + 2]^+$ ion at 196.0850 Da with an intensity that is ≈1% of the base peak.

The exact mass and isotope-pattern calculations become more useful as molecules get larger. You can find numerous online calculators that predict the isotope pattern when you enter the empirical formula. For a newly synthesized molecule, the experimental accurate mass determination of the molecular ion and a match to the predicted isotope pattern are strong evidence that the molecule has the expected empirical formula. Analysis of the fragmentation pattern can then help to identify structural motifs in the molecule. Further structural confirmation is then facilitated using FTIR and NMR spectrometry.

You-Try-It 11.A

The **mass-spec** worksheet in **you-try-it-11.xlsx** contains examples of mass spectra. Tabulate the major fragments in each mass spectrum. From the table of fragments, identify stable fragment ions and fragments that were lost. Propose the structure of the analyte molecule. Use one of the databases listed above to confirm your answer.

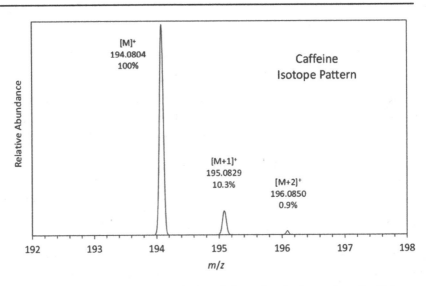

Figure 11.2   Accurate mass resolution of the molecular ion region of caffeine.

[3]Note that we refer to a calculated value as an *exact mass* and the measured value as an *accurate mass* determination.

11.3 FOURIER-TRANSFORM INFRARED SPECTROSCOPY

Chemists divide the infrared region of the electromagnetic spectrum into the near-infrared (NIR), mid-infrared (IR), and far-infrared (FIR) ranges. This energy classification is based on the usefulness in chemical analysis. Other disciplines, e.g., astronomy, will use slightly different cutoffs between regions. Likewise, classifications in thermal imaging are short-wavelength, mid-wavelength, and long-wavelength infrared: SWIR, MWIR, and LWIR, respectively. Chemists use NIR-absorption spectroscopy for quantitative analysis, especially in industrial quality control applications. I've not discussed the NIR region in this text due to the advanced nature of the data analysis for deconvoluting the overlapping absorption bands. Likewise, FIR spectroscopy has some niche uses for chemists with materials and proteins, but it is not a routine method in most labs. This section discusses the mid-infrared region, generally called simply the IR, of 4000–400 cm^{-1}.[4] An IR spectrum is useful to determine the functional groups in a molecule and it can help identify a substance by comparing the spectrum of an unknown to spectra in a reference library.

The basic principle of infrared absorption is the same as discussed for UV/Vis absorption spectroscopy in Chapter 4. Energy of a photon is converted to internal energy of a molecule. The major difference is that UV and visible photons put electrons into higher electronic excited states. Photons in the mid-IR region have enough energy to increase the vibrational energy in a molecule. Vibrations include stretching and bending motions between the atoms in the molecule. The much lower energy of the infrared photons means that instrument components must be made from IR reflecting and IR transmitting materials. We'll see that the overall spectrometer design is also quite different due to the noise characteristics of IR detectors.

Figure 11.3 shows the lower portion of the potential well for the ground electronic state of a diatomic molecule. This example is the simplest case for a vibration as the only possible motion is stretching and compression of the bond, i.e., a change in interatomic distance (x-axis). A bond has a zero-point vibration, meaning that it is in periodic motion even in its lowest energy state. This zero-point motion is indicated by the width of the horizontal line that is labeled as $v'' = 0$ in Figure 11.3. The other horizontal lines represent the allowed excited vibrational states. You can see that adding energy to a bond causes the average interatomic separation to increase. The population of these vibrational states depends on the ambient temperature as governed by the Boltzmann distribution. In recording routine IR spectra at room temperature, we will make the simplifying assumption that only the lowest vibrational state is populated. That is, we will not observe a $v'' = 1$ to $v'' = 2$ transition. Not shown in the figure are rotational levels, which occur on top of each vibrational state. High-resolution spectra of gas-phase molecules will show

---

[4]The low-energy cutoff of this range depends on instrument components.

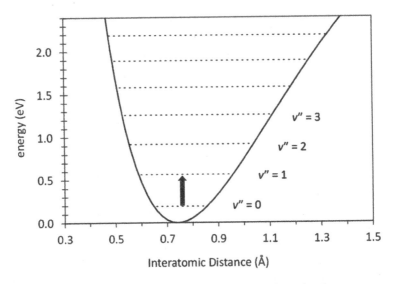

Figure 11.3 Vibrational levels in a diatomic molecule.

extensive fine structure for each IR peak, but rotational transitions are not resolved in the IR spectra of liquids and solids.

Given the previous discussion, the vertical arrow in Figure 11.3 shows the transition that is observed in the mid-IR region. Transitions to the higher vibrational levels, $v'' = 0$ to $v'' = 2$, 3, and so on, are called overtones. Overtone transitions are much weaker than the $v'' = 0$ to $v'' = 1$ absorption, which is referred to as the fundamental transition.[5] Another type of absorption is called a combination band, which occurs at an energy that simultaneously excites two fundamental transitions. Overtone and combination bands are generally weak and are not very useful in interpreting IR spectra. These bands will be labeled in reference spectra to prevent confusion with the bands due to the fundamental transitions.

Figure 11.3 showed the simplest case for a molecular stretching vibration between two atoms. Larger molecules will have more vibrational modes, including asymmetric and symmetric stretches, rocking, scissoring, twisting, and wagging. Each of these fundamental motions will have a progression of vibrational levels like Figure 11.3, but with a different energy spacing. A fundamental transition at 1000 cm^{-1} will have a first overtone transition near 2000 cm^{-1}. Recognizing an overtone or combination band is important so that one of these absorptions is not assigned as a fundamental band that does not actually exist in the molecule. For the rest of this discussion, we will only consider the fundamental absorptions.

In the IR region, the electric field of a photon couples to the dipole moment between two atoms in a molecule. For absorption to occur, there must be a

[5]Fundamental and overtone vibrations are a general concept that apply to sounds, vibrating strings, etc.

change in dipole moment during the vibration. This condition leads to a selection rule of transitions being allowed or not allowed, which are said to be IR active or inactive, respectively. These selections rules can be useful for highly symmetric molecules or complexes, but they are less relevant for most organic compounds. We are neglecting the more advanced topics such as selection rules and symmetry. Our goal is simply to use IR spectra to determine the functional groups in a molecule. More advanced analysis can provide more useful information.

In the following discussion, we refer to the strength of a bond. Stronger bonds require higher-energy photons to promote them to an excited state. These energies are what we plot on the x-axis of a spectrum. We also refer to the strength of a transition. In this regard, we are referring to the intensity or absorption value of a peak relative to the y-axis. Be sure to keep these two concepts separate to avoid confusion.

The energy of a fundamental absorption will depend on the mass of the two atoms attached by a bond and the type of the bond. We can use a simple spring analogy to understand some trends in IR absorption energies. Consider a spring attached to a support that has a weight of mass, m, hanging on its free end. Displacing the weight by a small amount sets it in motion with a periodic oscillation. The fundamental frequency, f, of the motion is given by Hooke's law

$$f = \frac{1}{2\pi}\sqrt{\frac{k}{m}} \tag{11.1}$$

where k is the force constant of the spring. This mechanical analogy does not have a zero-point motion and the amplitude of the displacement is not quantized, but we see a similar square root dependence in trends in IR spectra. Multiplying the frequency by Planck's constant and replacing m with reduced mass for the two moveable atoms gives us:

$$E = \frac{h}{2\pi}\sqrt{\frac{k}{\mu}} \tag{11.2}$$

The reduced mass, μ, is found from the masses, m_1 and m_2, of the two atoms that form the bond:

$$\mu = \frac{m_1 m_2}{m_1 + m_2} \tag{11.3}$$

Table 11.5 shows two examples that are consistent with the simplified spring model for a bond. The increasing bond strength in the carbon–carbon single, double, and triple bonds shows an increase in IR energy for these absorption bands. This trend is analogous to an increasing force constant in Equation 11.2. The reduced-mass examples in Table 11.5 show decreasing band energy.

TABLE 11.5 **Trends in Energy of Select Stretching Vibrations**

Trend	Bond	Approximate Energy (cm^{-1})
Bond-strength	C–C	1100
	C=C	1650
	C≡C	2200
Reduced-mass	C–H	3000
	C–C	1100
	C–Cl	700
	C–Br	600

The energy of the C–C versus C–H varies as $1/\sqrt{\mu}$ for these two cases. The carbon–halogen band energies are lower than predicted based only on reduced mass, but the bond strength also decreases for the heavier halogens.

Since the energy of vibrational excitation depends on the bond strength and the atoms in the bond, the IR spectrum allows us to identify these functional groups. The simplified model above does not give us information to predict the intensity of an absorption band, the linewidth, or the variability in the absorption wavelength. Numerous absorption bands have characteristics that are similar in different molecules and representative of the corresponding functional groups. For example, the C=O stretch of a carbonyl group tends to have a large intensity. Similarly, the O–H stretching absorption band will be very broad if hydrogen bonding occurs.

These general characteristics can vary depending on the details of a molecule. The C–H absorption bands might be the strongest lines in the spectrum of an alkane because no other functional groups are present. In the spectrum of a molecule that has an aldehyde or ketone group, the C–H will appear smaller due to the C=O band being the most intense. Some spectra can also appear different if measured as a dilute solution or pure substance. The O–H stretch will have the characteristic broad peak if measured neat (pure), but it can show a sharp line if measured as a dilute solution. The relative peak intensities can also appear different if a solution is too concentrated. Bands that are clipped or "flat-topped" indicate a suppressed peak signal due to detector saturation. Some of the instrument sampling issues that can affect spectra are described in the next section.

The IR spectral region is divided into two ranges: the functional group region (4000–1500 cm^{-1}) and the fingerprint region (1500–400 cm^{-1}). Interpreting a spectrum is usually done by starting with the functional group region. The position of bands in the functional group region is less variable than the fingerprint region, and the presence of a given peak can confirm that a functional group is present in a molecule. Table 11.6 lists the typical energies for common functional groups. Although there is a range for each

TABLE 11.6   IR Absorption Bands of Common Functional Group

Group	Bond Vibration	Energy (cm^{-1})	
Hydroxyl	O–H stretch	3600–3650	Moderate, sharp
Hydroxyl (H-bonded)	O–H stretch	3500–3300	Strong to moderate, broad
Amine	N–H stretch	3300–3500	
Alkyne	C–H stretch	3300	Moderate, sharp
Alkene	C–H stretch	3000–3100	Moderate
Alkane	C–H stretch	2850–2960	Moderate
Aldehyde	C–H stretch	2750 and 2850	Weak
Nitrile	C≡N stretch	2210–2260	Strong, sharp
Carbonyl	C=O stretch	1650–1750	Very strong
Amine	N–H bend	1580–1650	Moderate
Alkanes (methylene)	C–H bend	1470	
Alkanes (methyl)	C–H bend	1450	
Amine	C–N stretch	1180–1360	

group, it is small and separated from overlap with most other groups. A general strategy is to identify as much as possible about the hydrocarbon backbone and then to assign other functional groups. More extensive tables and charts can be found in online and reference sources for this type of spectral interpretation.

The position of the absorption bands in the fingerprint region are variable depending on the exact structure of a molecule. The variability makes this region less useful for identifying functional groups, but it is useful in comparing the spectrum of an unknown to a reference spectrum. This region is called the fingerprint region because if the spectrum of an unknown matches a reference spectrum, there is a good chance that your unknown is the same as the reference. Instrument software often comes with libraries of reference spectra, and a mathematical algorithm can find the best match to an experimental spectrum.

The Figure 11.4 shows the example of the IR absorption spectrum of isopropanol. Note the strong broad band of the OH group. The C–H region can identify methylene and methyl groups between 2970 and 2880 cm^{-1}. The lack of peaks between the C–H stretch and approximately 1500 cm^{-1} reveals the absence of other functional groups.

You-Try-It 11.B

The FTIR worksheet in you-try-it-11.xlsx contains examples of FTIR spectra. Tabulate the major peaks in each spectrum. Use the Table 11.5 or other sources to identify the functional groups that are present in each molecule.

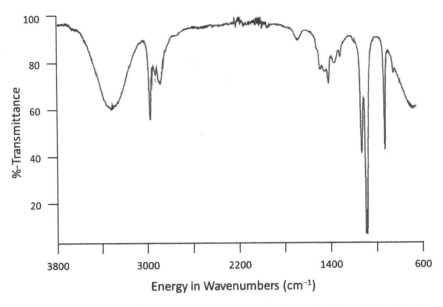

Figure 11.4 IR absorption spectrum of isopropanol (neat liquid on ATR prism).

11.4 FTIR INSTRUMENTATION

First, review the schematics of the dispersive UV/Vis absorption spectrophotometers that are shown in Section 4.6. The dispersive design uses a diffraction grating to separate the different wavelengths of light by diffracting them at different angles. The spectral bandpass is determined by the diffraction grating, the spectrometer focal length, and the size of the entrance and exit slits. Reducing the spectral bandpass for a given spectrometer and diffraction grating is accomplished by decreasing the slit widths. Doing so allows us to resolve more narrow spectral features, but at the cost of reducing the light that reaches the detector. In the ultraviolet and visible spectral regions, the photomultiplier tube detectors operate via the photoelectric effect and they have very low intrinsic noise. We can reduce the light passing through a UV/Vis spectrometer and still make the measurement. Detectors for the infrared spectral region are solid materials that operate by different mechanisms. We needn't go into the details, but they all create an electrical signal due to absorption of infrared photons. The significant factor, compared to the UV/Vis region, is that the energy of IR photons is only a factor of ≈ 10 higher than the thermal energy at room temperature. The result is that there is a significant background signal from the detectors due to thermal energy. If we reduce the intensity of the IR light too much, we are not able to measure it above this background.

Due to the detector characteristics, an interferometer performs better than a dispersive spectrometer. The interferometer measures all wavelengths simultaneously, which is known as the multiplex advantage. This advantage has a time advantage compared to a scanning dispersive instrument. Multiple scans are recorded to improve the signal-to-noise of the resulting spectrum via signal averaging. We won't go into the details, but resolution in an interferometer does not depend on narrowing the size of an aperture. That gives FTIR an advantage in a higher amount of light that goes through the sample and reaches the detector. This throughput advantage further improves the signal-to-noise in a measurement.

Fourier transform infrared (FTIR) spectrophotometers consist of a continuum light source, an interferometer, a sample holder or sampling device, and an IR detector. Figure 11.5 shows a top-down schematic of an FTIR spectrometer, where the dashed lines represent the light path. The Michelson interferometer in the shaded region of the figure consists of a beamsplitter and two flat mirrors.[6] The beamsplitter transmits 50% of the light and reflects 50% of the incoming light 90° along a different path. The light that travels via the two arms of the interferometer are reflected by flat mirrors back to the beamsplitter. Again 50% is transmitted and 50% is reflected. It is true that 50% of the light entering the interferometer is returned to the light source, but the absolute magnitude of light reaching the detector is not important.

One mirror is fixed in position and the other mirror is on a track or rails to be moveable. The moveable mirror scans back and forth in the direction of the light path. When the two mirrors are at the same distance from the beamsplitter, they have zero difference in the distance that photons travel. The result is complete constructive interference and the detector signal is a maximum. When the movable mirror is at any other position, the degree of destructive interference depends on the difference in path length for each different infrared wavelength. The net result, as the moveable mirror scans in position, is a detector signal that forms an interference pattern, called an interferogram. Since all wavelengths are passing through the interferometer, the interferogram is a complex pattern. However, it contains the intensity information at all

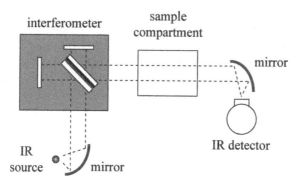

Figure 11.5   Schematic of an FTIR instrument (top-down view).

[6]A beamsplitter is similar to mirrored sunglasses that attenuate the transmitted light by simply reflecting some fraction of it.

wavelengths, which can be extracted mathematically. The most common manipulation of the interferogram is the Fast Fourier Transform (FFT). We won't discuss the details of this manipulation, but it converts the interferogram with units of mirror distance (cm) to a spectrum with units of energy (cm^{-1}).

Recall from Chapter 4 that absorbance is the $-\log$ of transmittance, which is the ratio of light power transiting the sample divided by the light power transiting a reference:

$$A = -\log(P / P_o) \tag{11.4}$$

The light power at each wavelength (or energy) for the reference depends on the emission spectrum of the continuum light source and losses due to reflection, solvent absorbance, etc. The light power at each wavelength (or energy) for the sample depends on the same things as for the reference, plus attenuation due to absorbance by the analyte of interest.

Since the interferometer has only one light path, the measurement of P_o is made by recording a reference interferogram in the absence of sample. This interferogram is stored in the software memory for subsequent measurements.[7] The instrument software can usually display this reference spectrum, which is useful to check that the sampling device is clean and to detect changes during subsequent measurements. Sample spectra are then recorded by scanning the interferometer with the absorbing analyte present in the light path.

The absorption spectrum due to the analyte is extracted by taking the Fourier transform of the reference and sample interferograms and applying Equation 11.4 at each wavelength. The resulting data can be displayed as absorbance or transmittance versus wavelength or energy. Using a transmittance y-scale is common to make it easier to compare experimental spectra to older spectra in the literature and reference sources.

Modern FTIR instruments are easy to operate, with the user generally only setting spectral resolution, the number of scans to average, and preferred scale for the y-axis. The operator does need to consider and choose an appropriate sample holder. The sample compartment in an instrument is simply empty space to accommodate sample holders. Due to IR absorption by water vapor and CO_2, the sample compartment might be purged with dry nitrogen or contain desiccant to reduce background absorbance. Obviously, water is not used as a solvent in FTIR spectroscopy. Plastic films are simply placed in the beam path. They must be thin enough to not completely block the light at which they absorb.

---

[7]See graphic in *Advantages of a Fourier Transform Infrared Spectrometer*, Thermoscientific technical note 50674 (2018); https://www.thermofisher.com/ftir; accessed August 2022.

Liquid or solid compounds can be dissolved in solvents and supported between two IR-transmitting salt plates in the light path. The solvents are limited to simple molecules with few IR absorption bands. Carbon tetrachloride, chloroform, and tetrachloroethylene are common, but each has some blacked-out spectra regions. The analyte concentration in solution must be adjusted so the absorption bands are in a reasonable absorbance range. The sample solution is introduced between two salt plates using a syringe via a port on the sample mount. Solid samples can also be ground in a mortar and pestle with KBr salt. This mixture is then pressed into a pellet and the pellet is placed in the light path. Moisture causes problems for both the salt plates and the KBr pellet method.

Due to difficulty with salt plates, a popular sampling method on modern instruments uses attenuated total reflection (ATR). There are numerous designs, and Figure 11.6 shows one variation. The liquid or solid sample is placed on the surface of a prism. For solids, pressure is applied by screwing down an anvil on the sample to get good contact. The prism is an IR-transmitting material such as diamond, ZnSe, or Ge, and the infrared light transits the prism via total internal reflection. Some fraction of the light power extends past the prism face, known as the evanescent wave. The evanescent wave interacts with the sample material, which absorbs at its characteristic energies. There is a difference in the penetration of the evanescent wave as a function of wavelength, and the spectrum using an ATR sample device can be different from a spectrum of a sample in KBr or a solvent. Some instrument software will correct for this wavelength dependence, but it is a factor of which the instrument operator should be aware.

You-Try-It 11.C
There is no calculational exercise for FTIR instrumentation. If your institution has one or more FTIR spectrophotometers, determine their x-scale range, resolution, and types of sampling accessories. Often once you know the instrument model number, you can find this information from manufacturer product brochures.

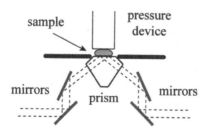

Figure 11.6 Schematic of an attenuated total reflectance sampling accessory.

11.5 NUCLEAR MAGNETIC RESONANCE SPECTROSCOPY

Along with mass spectrometry and infrared spectroscopy, nuclear magnetic resonance (NMR) spectroscopy is the third routine instrumental tool that chemists use to determine the structure of molecules. A key advantage of an NMR spectrum is that it can provide insight about the proximity of different structural motifs in a molecule. Where an FTIR spectrum might show that a molecule contains –CH$_3$ and –CH$_2$– groups, the NMR spectrum can show if the methyl and methylene groups are adjacent. For example, is the methyl group attached to an alkyl chain,–CH$_2$CH$_3$, or is it isolated from other protons in an acetate group –COOCH$_3$.

We will only discuss proton NMR spectroscopy, but it is possible to use nuclear magnetic resonance for any element that has an isotope with a nonzero nuclear spin. Beside protons, chemists often use NMR spectroscopy for boron-11, carbon-13, fluorine-19, and phosphorous-31. We are also omitting discussion of advanced techniques such as two-dimensional NMR, solid-state NMR spectroscopy, and magnetic resonance imaging (MRI).

Electrons and nuclei have spin angular momentum. We will neglect the details and simply say that nuclei have a property called spin. The important thing for spectroscopy is that nuclear spin gives a nucleus a magnetic moment so that it acts like an atomic-level bar magnet. For protons, the nuclear spin can be +1/2 or −1/2, which we also refer to as spin up or spin down. In the absence of an external field, these magnetic moments are oriented in random directions and the energies of the two spin states are equal, i.e., degenerate. When nuclei are placed in a magnetic field, they can align with their spin in either the same direction or the opposite direction as the external magnetic field. Nuclei that have their spin aligned with the magnetic field have a lower energy. Nuclei that have their spin aligned against the magnetic field have a higher energy. A physical analogy is two bar magnets with north (N) and south (S) poles. When brought together, the two magnets will orient as [N-S][N–S] rather than [N–S][S–N]. The net result is that in a magnetic field, the nuclei have two different orientations of different energy.

Given two energy states, the population of each state will depend on the Boltzmann distribution and the ambient temperature. The energy difference is small and the higher energy state is just slightly less populated than the lower energy state. The difference in population does allow an absorption transition to "flip" a spin from the lower energy to the higher energy state. The energy difference depends on the magnitude of the applied magnetic field and Figure 11.7 shows this energy splitting as a function of field strength. For reference, 1.41 T equals 14,100 Gauss and the magnetic field at the surface of the Earth is less than 1 Gauss. Transitions between the two levels occur in the radiofrequency portion (RF) of the electromagnetic spectrum. The 60 and 300 MHz labels are the RF frequencies for proton transitions for the two listed magnetic fields.

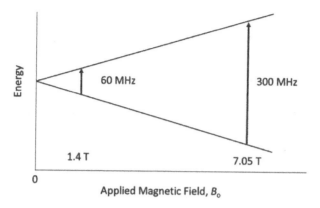

Figure 11.7 Splitting of nuclear spin states as a function of applied magnetic field B_0.

Based on Figure 11.7, absorption of RF radiation by nuclei will simply tell us that protons are present. One factor that makes NMR spectroscopy useful for structural analysis is that the magnetic-field strength will vary slightly for each unique local environment in a molecule. All atoms in a molecule are surrounded by bonding electrons to varying extents. The applied magnetic field induces the electrons to produce a small opposing magnetic field. This opposing magnetic field *shields* a given nucleus from the applied field. The degree of shielding depends on the type of bonding near a proton and the proximity of electron-donating or electron-withdrawing functional groups. The net result is that each unique proton in a molecule has a slightly different energy splitting and the spin–flip transition occurs at a slightly different frequency.

Before looking at the extent of shielding, it is useful to know how spectra are displayed. The instrument software takes the absorption spectrum as a function of frequency and normalizes the x-scale relative to a reference compound. Doing so makes it much easier to compare spectra obtained on instruments that operate at different frequencies. The x-scale in an NMR spectrum is expressed as a chemical shift, δ, and calculated using

$$\delta = \frac{\nu_{\text{peak}} - \nu_{\text{ref}}}{\nu_0} \times 10^6 \text{ ppm} \tag{11.5}$$

where ν_{peak} is the resonance frequency for a specific atom position, ν_{ref} is the resonance frequency of a reference standard, and ν_0 is the operating frequency of the spectrometer. The frequency difference due to differences in shielding is quite small. When all frequencies in Equation 11.5 are in units of MHz, this relative measure is multiplied by 10^6 to express δ in units of ppm. We are only discussing proton NMR spectroscopy, but if the type of data is ambiguous, the nuclei is stated explicitly as δ_H, δ_C, $\delta(^{19}F)$, and so on.

The most common reference compound for proton and carbon-13 NMR spectroscopy is tetramethyl silane, $Si(CH_3)_4$, or TMS. It has 12 equivalent protons to produce a single strong peak.[8] Silicon is less electronegative than carbon. Considering only one methyl group:

$$- \overset{|}{\underset{|}{Si}} \leftarrow CH_3 \qquad - \overset{|}{\underset{|}{C}} \Leftarrow CH_3$$

An Si atom pulls less electron density in its direction compared to a C atom. The methyl proton in $Si-CH_3$ experiences less shielding than protons in most hydrocarbons and the resonance for TMS occurs at a higher frequency. The proton peak due to the TMS reference is set as $\delta = 0.0$. Almost all hydrocarbon proton resonances will occur at a lower frequency and have a positive value of δ. Approximately 1% TMS will be added to the solvent, which is usually a deuterated compound, such as chloroform-d, acetone-d6, or benzene-d6, to avoid overlap with ^1H peaks. In practice, an NMR instrument might use the signal from deuterium, ^2D, or residual $CHCl_3$ in chloroform-d as the reference signal.[9] For any reference method, the software will still plot the spectrum assuming a TMS reference at $\delta = 0.0$.

Table 11.7 lists typical chemical shifts for different types of protons. The chemical shifts will vary slightly for a given type of proton depending on the

TABLE 11.7 Chemical Shifts of Common Proton Groups

Proton Group		Chemical Shift
Tetramethyl silane (reference)	$Si(CH_3)_4$	0.0
Methyl (alkane)	$-CH_3$	0.9–1.1
Methylene (alkane)	$-CH_2-$	1.2–1.5
Acetylene	$-C\equiv C-H$	2.5–3.0
Methyl or methylene (next to carbonyl)	$CH_3-C=O$	2.0–2.7
Methylene (next to alcohol)	$-CH_2OH$	3.3–4.0
Bromo-	$-CH_2Br$	3.4–3.6
Chloro-	$-CH_2Cl$	3.6–3.8
Methyl or methylene (next to ether)	$-O-CH_3$	3.3–3.9
Methyl or methylene (next to ester)	$COO-CH_3$	3.7–4.1
Vinyl	$>C=CH_2$	4.6–5.0
Alcohols and amines	$-OH, -NH_2$	1–5 (variable)
Aromatic	C_6H_6	6.5–8.0
Aldehyde	$-CO-H$	9.1–10
Carboxylic acid	$-COOH$	10–12

[8]The protons are equivalent due to rapid rotation of the bonds. This condition is generally true for any molecule in solution at room temperature.

[9]Modern instruments measure the ^2D signal continuously as a feedback mechanism to maintain a constant magnetic field. This process is known as locking.

exact structure of the molecule and nearby substituents. More electronegative groups will create greater shielding and have a higher δ value. Pi electrons provide significant shielding as seen in the large shift for protons on an aromatic ring. Peaks at higher δ are said to be shifted *downfield*. The chemical shift for protons on –OH and –NH$_2$ groups is variable depending on solvent and concentration due to hydrogen bonding.

As described so far, the chemical shifts do not tell us a lot more than the functional group region in an infrared absorption spectrum. There is one additional factor that changes the shape of a given NMR peak. Nonequivalent proton nuclei on adjacent carbon atoms can interact with each other. The interaction is called spin–spin or J-coupling and is mediated by the bonding electrons between the two nuclei. The coupling decreases as the number of bonds between two nuclei increases, and we will only consider protons interacting on two adjacent carbons. J-coupling results in a peak splitting into multiple smaller peaks, which we call a multiplet. The number of smaller peaks in a multiplet is determined by the number of adjacent nonequivalent protons. One adjacent proton splits a peak into two, called a doublet. Two adjacent protons split a peak into three peaks, called a triplet. A methyl group will split a peak into four peaks, called a quartet. The J-coupling for two adjacent protons is a fixed value, it does not depend on the applied magnetic field strength. One advantage of using a higher-frequency instrument is to spread out multiplets that might overlap otherwise.

These splitting patterns are illustrated in Figure 11.8 for ethanol, where the numbers above the peaks are the integrated area for each multiplet. There are three types of protons. The alcohol proton peak at 2.61 ppm has no adjacent protons and is a singlet. The two methylene protons are coupled to a methyl group and split into a quartet centered at 3.687 ppm. The three methyl protons are adjacent to the two methylene protons and split into a triplet at 1.226 ppm.

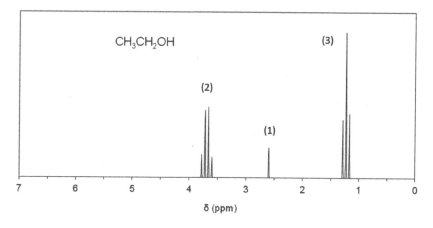

Figure 11.8   NMR spectrum (simulated) of ethanol in CDCl$_3$.

If there are nonequivalent protons on each side of a methylene proton, the result is a multiplet of a multiplet, e.g., a triplet of doublets. This splitting phenomenon is what allows the analyst to determine the proximity of different types of protons. Numerous examples are in the NMR worksheet in you-try-it-11.xlsx. Example 11.1 illustrates the process of spectral interpretation.

Example 11.1 NMR spectrum. Figure 11.9 shows the NMR spectrum of a molecule with an empirical formula of $C_4H_8O_2$. The numbers in parentheses give the integrated area of each multiplet. What is the structure of this molecule?

Given the empirical formula, $C_4H_8O_2$, it is worthwhile to first think about possible isomers. There are no peaks at high δ where aldehyde and carboxylic acid resonances occur, so those functional groups can be eliminated. The two oxygen atoms in the formula suggest a diol or an ester. Since the singlet in the spectrum has an integration of 3 we can also rule out the diol possibility. We can infer that the molecule has ether and/or carbonyl groups in some combination.

First, we see that there are three types of protons. The singlet with an integration of 3 is not split. It is due to a methyl group that has no adjacent protons. The triplet near 1.2 ppm also has an integration of 3 and can be assigned as another methyl group. Since it is a triplet, we can infer that it is adjacent to a methylene group. The quartet at $\delta = 4.12$ with an integration of 2 is consistent as being adjacent with this methyl group. Our molecule contains an isolated methyl group and an ethyl chain that we can deduce are separated by an ester group (R_1COOR_2). That narrows it down to either ethyl acetate ($R_1 = CH_3$ and $R_2 = CH_3CH_2$) or methyl propanoate ($R_1 = CH_3CH_2$ and $R_2 = CH_3$). The telling feature is the quartet at $\delta = 4.12$. This methylene group is shifted downfield due to the adjacent ester linkage, i.e., $R_2 = CH_3CH_2$. We can conclude that this NMR spectrum is that of ethyl acetate.

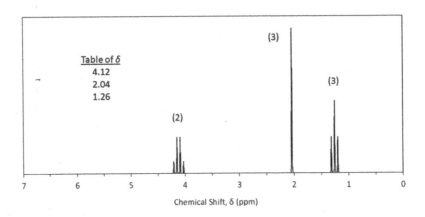

Figure 11.9 NMR spectrum (simulated) of $C_4H_8O_2$.

You-Try-It 11.D

The NMR worksheet in you-try-it-11.xlsx contains examples of NMR spectra. Identify the multiplets in each spectrum. Use the peak positions and splittings to determine the structure of each molecule.

11.6 NMR INSTRUMENTATION

A given instrument will have a magnet of a fixed strength, which splits the levels of different nuclei to different degrees. Instrument are therefore labeled by the frequency of the RF radiation for proton NMR spectrometry. Small benchtop instruments use permanent magnets and are available at 60–100 MHz. Typical instruments with superconducting magnets are available at 300–800 MHz. State-of-the-art instruments will operate at or above 1 GHz. The high-frequency instruments require a larger magnet, but they separate the energy levels to a greater extent. Doing so provides better spectral resolution, which is especially important for larger molecules that have many lines in the spectrum. The higher-field instruments also have greater sensitivity since the population difference between the lower and upper levels will increase.

The magnetic field must be very homogeneous so that all nuclei experience the same external field, B_0. The position of a given peak in a spectrum is then due only to differences in the local environment and the peak will be narrow. Before recording a spectrum, an instrument is "shimmed" to improve the field homogeneity. The shims are coils of wire that produce small magnetic fields when an electric current is applied. The computer software generally optimizes the shim currents based on a reference peak signal. The sample, dissolved in a solvent, is held in a tube in the bore of the magnet. The sample tube is often physically spun in the bore of the magnet to further average inhomogeneities in the applied field. Spinning is not required and is turned off for some advanced experiments.

The following description is oversimplified, but it introduces the general concept of an NMR experiment. As noted in the previous section, the population difference between the upper and lower energy states is quite small due to the thermal energy that is available at room temperature. Due to this small population difference, performing a direct absorption experiment is not very sensitive. Once populations are equal, the probability of stimulating emission in a downward transition is the same as stimulating absorption in an upward transition.

Modern NMR spectrometers use a time-pulse method to improve sensitivity. The slightly higher population in the ground state results in a net magnetization of the sample in the direction of the magnetic field, which we define as the z-axis. A pulse of RF radiation excites some nuclei to reorient this net magnetization into the x-y plane. Figure 11.10 shows a simplified schematic of a superconducting magnet and the sample tube that drops into the bore of the

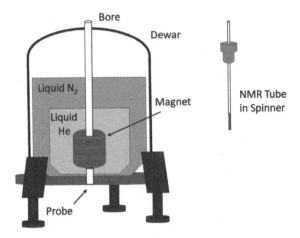

Figure 11.10 Side-view schematic of an NMR magnet and sample tube (enlarged).

magnet. A probe is inserted into the bore from the bottom. It has electrical connections from an instrument console and contains the two wire coils that function as antenna. One coil applies the pulse of RF radiation and the other coil, called the pick-up or receiver coil, records the signal. At the end of the excitation pulse, the nuclei begin to relax to return the net magnetization to equilibrium. As they do so, the net magnetization precesses or rotates around the z-axis. The movement of the net magnetization induces a current in the pick-up coil, which is the signal. The absorption measurement is now more like an emission experiment, where the signal is recorded on a nominally zero background.

The short-duration RF pulse has a broad frequency distribution so that it excites protons in all local environments simultaneously. Each of the non-equivalent proton nuclei precess at a slightly different frequency after excitation. These signals are also decaying as the magnetization returns to equilibrium along the z-axis. The signal picked up by the receiver coil is a complicated interference pattern.

Figure 11.11 shows a typical signal, which is known as a free induction decay (FID). The free induction decay has an x-scale with units of time, s. Performing a Fourier transform on the FID converts the detector signal to the inverse, s^{-1} or Hz, which is frequency. As discussed in the previous section, the instrument software converts the frequency scale to the relative δ scale in ppm.

With UV/Vis and FTIR absorption instruments, there are minimal instrument parameters for the operator to set, generally only spectral resolution and signal averaging. NMR spectrometers have numerous parameters for the operator to choose and a few common ones are noted here. The FID is recorded multiple times and averaged to improve the signal-to-noise ratio. The number of scans to average will depend on analyte concentration, with more averaging needed if concentration is low.

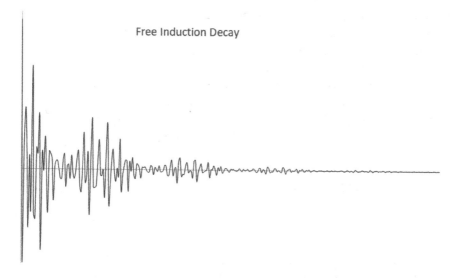

Figure 11.11   Free-induction decay of an NMR signal (simulated).

An acquisition time is chosen to acquire nearly the full FID. A delay or repetition time allows the magnetization to recover before applying the next excitation pulse. Any one of these settings might not be optimized, but they are chosen to obtain spectra of good signal-to-noise in the shortest time. For example, some instruments let you vary the RF pulse that shifts the magnetization vector from the z-axis. Nominally the best signal will occur when the magnetization is shifted 90°. However, this experiment requires a longer repetition time before the next pulse. The operator will often choose a smaller angle to be able to do more signal averaging and shorten the time required to obtain a spectrum.

There are also numerous post-processing routines for NMR data. Mathematically correcting a sloping baseline can improve the accuracy in peak integration. Apodization applies various functions to the FID to minimize the impact that the data is truncated at the end of the acquisition time. This data manipulation can often improve the resolution in the spectra. Finally, the instrument software will integrate the peaks and display the relative signal from each multiplet. The software might do this integration automatically or it might allow you to choose the x-range for integration.

You-Try-It 11.E
There is no calculational exercise for NMR instrumentation. If your learning institution has one or more NMR spectrometers, determine their operating frequency, types of nuclei that can be measured, and capabilities for advanced experiments. Once you know the instrument model number, you can find some of this information from manufacturer product brochures.

Chapter 11. What Was the Point?   This chapter presented a very brief overview of the most common instrumental methods to determine the structure of a molecule. The goal was to introduce the terminology and basic principles of these methods and to put them in perspective with the other spectroscopies that we've discussed. One chapter only scratches the surface of these topics. You can find more detail in introductory and advanced texts on each of these instrumental methods. Unambiguous determination of a molecular structure usually requires a combination of several of these methods.

PRACTICE EXERCISES

1. Why are different types of electromagnetic radiation used in different types of spectroscopy?
2. How will the presence of two or more species affect using mass spectrometry for structure determination?
3. How does a mass spectrum of a molecule help to determine the structure of the analyte?
4. Describe the principles and major components of a mass spectrometer.
5. Define mass spectrum terms such as m/z, molecular ion, fragmentation pattern, and isotope pattern.
6. Which mass analyzer designs (Section 10.5) are suitable for accurate mass determinations?
7. How will the presence of an impurity affect the IR spectrum of a substance? Does the concentration of the impurity matter?
8. What is an overtone absorption in IR spectroscopy?
9. Look up the structure of the following molecules and predict the energy in wavenumbers for each major peak that will occur in the functional group region of the IR spectra.
   (a) isooctane
   (b) benzene
   (c) phenol
   (d) polyvinyl alcohol
   (e) isopentyl acetate (smells like bananas)
   (f) benzaldehyde
10. What is the purpose of the two coils in an NMR probe?
11. How does the size of the magnetic field, B_0, affect an NMR spectrum?
12. Try predicting the NMR spectrum for each compound listed in question 9.

CHAPTER 12

ANALYTICAL SEPARATIONS

Learning Outcomes

- Choose appropriate chromatographic methods (TLC, HPLC, GC, electrophoresis) to match analytical goals.
- Predict elution order of solutes for different types of separation columns.
- Describe chromatographic instrumentation and chromatograms using correct terminology.
- Use manufacturer literature to choose appropriate chromatographic columns and detectors for a given class of analytes.
- Use data from chromatograms and electropherograms to identify, characterize, or quantify the constituents in a test sample.

This chapter describes the principles, experimental procedures, and equipment for thin-layer chromatography (TLC), gas chromatography (GC), high performance liquid chromatography (HPLC), and electrophoresis. These separation methods are coupled to some type of visualization or detection to create an integrated system for qualitative and quantitative analysis. I refer to these methods as *analytical separations* to differentiate them from column chromatography for sample cleanup or purification, as described in Section 2.8. Analytical separations encompass a wide range of sample types and applications and I do not attempt to provide a comprehensive review of all methods.

Basics of Analytical Chemistry and Chemical Equilibria: A Quantitative Approach, Second Edition. Brian M. Tissue.
© 2023 John Wiley & Sons, Inc. Published 2023 by John Wiley & Sons, Inc.
Companion Website: www.wiley.com/go/tissue/analyticalchemistry2e

This chapter introduces the principles and the most common implementations of analytical separations, proceeding from the simpler to the more advanced methods.

Placement of this chapter at the end of the text does not reflect a lack of importance of separations. The power to simultaneously separate and quantitate analytes in complex mixtures makes chromatographic instruments ubiquitous in analytical laboratories. We discuss these methods last because they depend on the concepts of partitioning and the various detection methods introduced in the earlier chapters.

As a refresher, Chapter 2 described liquid–liquid extractions, solid-phase extractions (SPE), and column chromatography to isolate analytes from interferences. The extractions operate in a stepwise or batch mode. In SPE, analytes that interact strongly with the stationary phase are retained and other solutes are washed out. The analytes on the stationary phase are then eluted for collection by changing the mobile phase. SPE separates very different classes of solutes, for example, neutral organics from salts or nonpolar compounds from polar compounds. The individual analytes are often separated and quantitated in a subsequent GC or HPLC analysis.

Many of the same solvents and stationary phases are used in both SPE and liquid column chromatography. The difference is that chromatography is operated as a continuous process with solutes eluting from the column at different times. The principles for TLC and HPLC are completely analogous to column chromatography (illustrated in Figure 2.11). Solutes in the liquid mobile phase interact with a stationary phase to differing degrees, travel through the column at different rates, and elute at different times. The difference is that column chromatography can handle mg to g quantities of analytes. Analytical separation columns have much greater efficiency, but they are separating nanogram to picogram amounts of analyte.

The challenge in an analytical separation is to choose an appropriate stationary phase and develop conditions that separate similar solutes, relying on small differences in elution time. Very weak or very strong interactions lead to rapid or slow elution, respectively, and are not desirable in chromatography. What is common in all chromatographic separations is that the separation of the components in a mixture will be reproducible under constant conditions. A mixture component that elutes at the exact same time as a pure reference compound is probably that same compound.

12.1 THIN-LAYER CHROMATOGRAPHY

Planar or *thin-layer chromatography* (TLC) uses a 0.1–0.2-mm thick coating of solid adsorbent on a flat substrate as the stationary phase and a liquid solvent as the mobile phase.[1] The principles of a TLC separation are analogous to

[1]Planar chromatography methods that achieve higher resolution and sensitivity are called *HPTLC* or *ultrathin-layer chromatography* (*UTLC*).

those in column chromatography. Different sample components will partition between the mobile and stationary phases to different degrees and travel up a plate at different rates. The different components in the test sample are now separated in different spots or zones. Separated spots are detected visually, and the size and darkness can indicate the relative amount of each component. The obvious difference in TLC is that the analytes remain on the stationary phase and do not elute as in column chromatography. TLC is useful to separate the major components in a mixture, which is generally less than ≈ 10 compounds.[2]

The most common stationary phases are alumina, Al_2O_3, for nonpolar solutes and silica gel, porous SiO_2, for more polar solutes. These adsorption stationary phases have polar functional groups on the solid surface. Solutes in the mobile phase will interact with the stationary phase to varying degrees depending on the functional groups of each solute. The more polar solutes in a mixture will have a stronger interaction with the stationary phase and are retained to a greater extent. They progress up the plate at the slowest rate. Less polar solutes interact less with the stationary phase. They remain in the mobile phase to a greater extent and progress up the plate faster. Other stationary phases are available, such as cellulose for polar biological analytes and reversed-phase coated adsorbents. The reversed-phase adsorbents are more expensive, but they are useful to test different solvents when developing a reversed-phase HPLC method.

Figures 12.1 and 12.2 illustrate the preparation and separation process in TLC. A small amount of test solution is "spotted" on the lower part of a TLC plate using a capillary tube. The test solution should contain $\approx 1\%$ of the analyte mixture in a volatile solvent. Too much analyte overloads the stationary phase and results in smeared spots. Too little analyte makes the spots difficult to visualize. The spotted test solutions are allowed to dry before continuing.

The spotted TLC plate is placed in a lidded container with the bottom edge, but not the spots, in a running solvent.[3] The illustration in Figure 12.2 shows

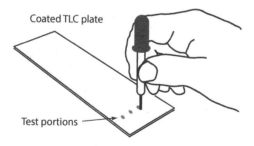

Figure 12.1   Spotting a TLC plate with sample.

[2]Two-dimensional TLC can increase this number approximately fourfold by running a TLC plate with two different solvents after rotating 90° between solvent runs.
[3]A piece of filter paper is often added to the container to help maintain a saturated atmosphere of the running solvent.

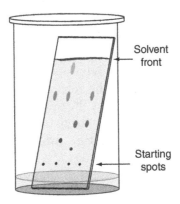

Figure 12.2   Separation of sample components on a TLC plate.

four different test solutions spotted on the TLC plate to run simultaneously. The solvent moves up the plate by capillary action and "carries along" the sample components. In Figure 12.2, the slightly darkened area on the plate is where the solvent is moving upward. Separated components are visible above each of the original spots. This process is called *running* or *developing* the TLC plate.

The running solvent is selected to match the polarity of the analytes. Table 12.1 lists common solvents that are chosen for their low toxicity and low expense. Using a mixture of miscible solvents makes it easy to adjust the mobile phase polarity by varying the ratio of the solvents. Suitable solvent mixtures for specific classes of solutes can be found in laboratory handbooks and manufacturer literature.

When the solvent front nears the top of the TLC plate, the plate is removed from the container and the solvent front is marked with a pencil. The solvent on the TLC plate is allowed to evaporate and separated spots are visualized depending on their properties. Colored substances will appear as distinct spots as in Figure 12.2. For colorless components on a TLC plate, the analyte spots are visualized by irradiating with ultraviolet light or staining. Many commercial

TABLE 12.1   Common TLC Mobile Phases

Polarity	Solvent
Less polar	Hexane
	Cyclohexane
	Toluene
	Diethyl ether
More polar	Ethyl acetate
	Acetone
	Methanol
Polar	Water
	Acetic acid

TLC plates contain a fluorescent dye that glows when irradiated with short-wavelength UV light (254 nm). Organic compounds will absorb the UV light and appear dark compared to the glowing TLC plate. Visualization systems will also include a long wavelength UV light (366 nm), which can excite fluorescent compounds that will appear as bright spots on a dark background.

Common stains are I_2 vapor, applied by placing the TLC in a closed jar with solid I_2, and spray solutions of phosphomolybdic acid, potassium permanganate, cerium(IV) sulfate, or *p*-anisaldehyde. Commercial vendors supply numerous visualization reagents for specific classes of analytes. Ninhydrin is used to visualize amino acids and other primary amines. A plate is sprayed with ninhydrin solution, and by warming, the ninhydrin reacts to form a purple complex to visualize the spots.

Figure 12.3 shows a schematic of a developed TLC plate that was used to check three test samples for the presence of a reference substance. The labels show where the test samples were spotted and the extent of the solvent front. The TLC plate was developed until the solvent front neared the end of the TLC plate, removed from the running solvent, and stained to visualize the spots. In this example, all samples contained more than one component and samples 1 and 3 contained the reference substance.

The distance that a given component travels along a TLC plate is described by a ratio, R_F, called the *retardation factor*. It is calculated as the distance that one solute has traveled divided by the distance the solvent has traveled:

$$R_F = \frac{\text{spot distance}}{\text{solvent distance}} \tag{12.1}$$

The spot distance is measured from the starting point to the center of a spread spot after the TLC plate is removed from the solvent. Using the ratio provides a simple measure for comparing how far different substances migrate for a given solvent and stationary phase combination (see Example 12.1). A good solvent or a solvent mixture will produce an R_F of approximately 0.5 for the target compound.

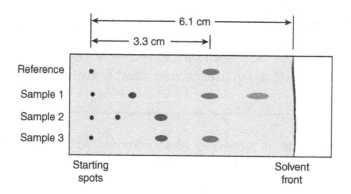

Figure 12.3   Visualized spots on a TLC plate.

The R_F value for a given substance will be approximately the same for a given solvent, stationary phase, and temperature. Owing to day-to-day and plate-to-plate variations, a reference substance should be run on the same TLC plate with the test samples.

Example 12.1 R_F Calculation. What is the R_F value for the reference substance in Figure 12.3?

The distance traveled by the reference substance is 3.3 cm and the distance traveled by the solvent is 6.1 cm. The ratio is

$$R_F = \frac{3.3\,\text{cm}}{6.1\,\text{cm}} = 0.54$$

The simplicity and speed of TLC (5–30 min depending on plate size) makes it very useful for a number of applications. Some examples are

- monitoring the progress of a reaction
- evaluating the effectiveness of a purification procedure
- rapidly determining the number, nature, and relative amount of impurities
- testing solvent mixtures for other chromatographic methods
- semiquantitatively analyzing the components in a mixture.

In monitoring a reaction, TLC can show the disappearance of the starting material, appearance of the desired product, and the generation of side products. TLC can determine the chemical nature of impurities relative to the polarity to the target analyte. In the case of using a nonpolar mobile phase with an alumina or silica stationary phase, solutes that migrate farther than the desired product are more nonpolar and solutes that migrate less are more polar than the compound of interest.

TLC is less useful than other chromatographic methods in quantitating analytes because of the difficulty in getting a reproducible and quantitative signal with the spotting procedure. However, it is possible to make semiquantitative measurements, and TLC is used in many applications that do not require analysis at trace levels. Besides measuring the area or darkness of a TLC spot, the spot can be scraped from the plate and the analyte extracted for quantitation or structural confirmation by another analytical method.

You-Try-It 12.A

The TLC worksheet in you-try-it-12.xlsx contains data from TLC separations. Tabulate R_F values for each reference and test portion and determine the composition of the samples.

12.2 CHROMATOGRAM TERMINOLOGY

Section 2.8 discussed the basic principles of chromatography, which apply to TLC, GC, and HPLC separations. Although the principles are the same, the experimental details of analytical separations differ from the methods for sample purification. The most significant difference is the addition of a detection system to quantitate the solutes as they exit the column. Since test samples are loaded onto a GC or HPLC column using a syringe, the start of a chromatographic separation is called the *injection*. In both cases, a small plug of the test sample is injected into the flowing mobile phase, and the start time of this injection is set as $t = 0$. The detector registers an increase in signal whenever a solute elutes from the column. The result is a *chromatogram* consisting of a series of peaks versus time after injection.

The terminology to describe chromatograms is the same for both GC and HPLC. The time from injection to elution for a single component in the test sample is called the *retention time* or *total retention time*, t_R, for that component. A component that interacts strongly with the stationary phase will be "retained" on the column for a longer period of time and have a longer retention time compared to components with weaker interactions. A related measure is the *adjusted retention time*, t_R':

$$t_R' = t_R - t_M \tag{12.2}$$

where t_M is the hold-up time. The *hold-up time* is the time for unretained substances to transit the column. It depends on the mobile phase flow rate and the column void volume. Observing a peak at t_M is common in chromatograms because the detector registers a signal from unretained solvent in HPLC or from air in GC, both of which are injected with the test sample.

The peak resolution, R_s, between two chromatographic peaks is defined by

$$R_s = \frac{t_{R(2)} - t_{R(1)}}{0.5\left(w_{b(2)} + w_{b(1)}\right)} \tag{12.3}$$

where w_b is the width of a peak at its base. Subscripts 1 and 2 refer to peaks from two different components, with component 1 eluting before component 2. Figure 12.4 shows an example of a chromatogram with two peaks that are nearly baseline separated. Dotted lines overlap the peak edges in the figure to help determine the baseline widths. Baseline separation requires a peak resolution of approximately 1.5. Baseline separation is desirable to reduce uncertainty in determining integrated peak areas for quantitative measurements. Peak overlap can introduce errors in quantitation, especially for a small peak that overlaps with a larger peak.

Calculating R_s is only useful for two adjacent peaks (see Example 12.2). The two peaks in a chromatogram with the greatest overlap are called the *critical*

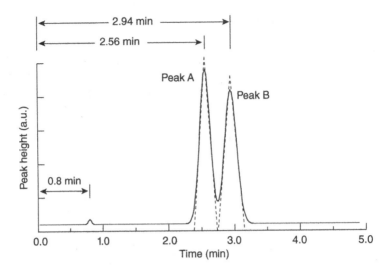

Figure 12.4   Chromatogram retention times.

pair. These two peaks serve as the benchmark to optimize a separation. $R_s \propto \sqrt{N}$, where N is plate number to be discussed below. Calculating R_s for a critical pair can let you predict the degree to which you must increase plate number to improve the separation efficiency.

Example 12.2 Peak Resolution.   Calculate the resolution of the two peaks in Figure 12.4.

The baseline widths of the two peaks are approximately 0.33 and 0.38 min, with an average of 0.36 min. Inserting values into the equation for R_s gives

$$R_s = \frac{2.94\,\text{min} - 2.56\,\text{min}}{0.36\,\text{min}}$$

$$R_s = 1.06$$

Although they are not baseline separated, these two peaks can be quantitated by peak fitting to get integrated areas or by using the peak heights. If peak B in a different chromatogram was ¼ of that shown in Figure 12.4, do you think that we could make the same statement for quantitating peak B?

Two additional descriptors for chromatographic peaks are in common use. Since the mobile phase is flowing at a constant rate, for example, in milliliters per minute, retention time and the volume of mobile phase entering the column are related by

$$V_R = F_c t_R \tag{12.4}$$

where F_c is the volumetric flow rate of the mobile phase through the column at a given column temperature.[4] Chromatographic data is often phrased as total retention volume, V_R, and hold-up volume, V_M. Since t_R and V_R are directly related, working with either time or volume is acceptable. For simplicity, we will only work with time data in this text.

Finally, to compare chromatographic data for different columns and conditions, the retention of a given substance on a given stationary phase is described by a retention factor, k:

$$k = \frac{t_R - t_M}{t_M} = \frac{t_R'}{t_M} \qquad (12.5)$$

The retention factor provides a relative measure versus the hold-up time; for example, $k = 10$ indicates that a given solute takes ten times longer to elute than a solute that is unretained. A solute that is not retained has $k = 0.0$. A solute that elutes at a time that is twice t_M has $k = 1.0$. This solute spends half of its time in the mobile phase and half of its time retained on the stationary phase. Note that retention factors do not depend on the column length, flow rate, or other experimental parameters (see Example 12.3). Manufacturer literature will often provide retention factors, which help determine if a given column is suitable for the solutes in a given application.[5]

Example 12.3 Chromatographic Retention Factors. What are the retention factors of the two peaks in Figure 12.4?

The retention factors for peaks 1 and 2, respectively, are

$$k_{(1)} = \frac{t_{R(1)} - t_M}{t_M} = \frac{2.56 \text{ min} - 0.80 \text{ min}}{0.80 \text{ min}} = 2.20$$

$$k_{(2)} = \frac{t_{R(2)} - t_M}{t_M} = \frac{2.94 \text{ min} - 0.80 \text{ min}}{0.80 \text{ min}} = 2.68$$

There are two common means of quantitating analytes from chromatograms—comparing the chromatogram of a test portion to a standard or reference chromatogram or using an internal standard. A reference chromatogram is usually needed in either case to know the retention times of the analytes of interest. If experimental conditions change, the reference must be rerun because retention times generally do not vary linearly with most separation parameters.

---

[4]The temperature dependence is a greater factor in GC than in HPLC since column temperature is often varied during a gas chromatogram. Volumetric flow rate at ambient temperature has the symbol F_a.

[5]The term capacity factor with the symbol k' is also common.

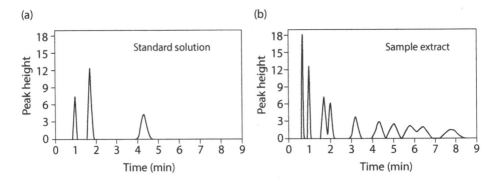

Figure 12.5 (a) Reference and (b) test portion chromatograms.

When using an external reference, the standard mixture is run the same day under the same conditions as the test portions. As an example, Figure 12.5(a) shows the chromatogram of a reference solution containing 1.0 μg/ml each of caffeine, theobromine, and theophylline. Figure 12.5(b) shows the chromatogram of an extract from cocoa powder run under the same conditions as the reference solution. Comparing the two chromatograms allows the analyte peaks to be identified on the basis of retention time. Each analyte can be quantitated by comparing peak heights or peak areas between the test portion and the standard solution. As Figure 12.5(a) shows, the detector sensitivity can vary for equal amounts of different analytes. Likewise, the peak width and sometimes peak shape is not constant for all peaks.

Using an internal standard is the more common means of calibrating a chromatogram. Since chromatographic runs can take up to two hours for a complicated sample, eliminating a separate reference run saves a great deal of time. Of greater significance, the internal standard reduces calibration errors due to drift in detector sensitivity or if injection amounts vary between different chromatograms. The requirement for a chromatographic internal standard is that it be compatible with the separation column and detector and that the internal standard peak not overlap with any of the analyte peaks. The detector sensitivity for the internal standard relative to each analyte is determined by running a reference mixture of the analytes. The relative sensitivity will not change for variations in injection amount or detector sensitivity.

You-Try-It 12.B
The quantitation worksheet in you-try-it-12.xlsx contains data from chromatographic separations. Use the calibration data to quantitate each target analyte in these data sets.

12.3 SEPARATION EFFICIENCY

The challenge in an analytical separation is to separate or "resolve" every component of interest in the test portion. Peaks that are not resolved can be difficult or impossible to quantitate. The main factor in achieving an acceptable separation is to choose an appropriate column. The following terminology provides descriptors to compare the performance of different columns for given classes of analytes.

The ability of a chromatographic column to resolve closely spaced peaks is called the *column efficiency*. It is described by two related terms, the *plate number*, N, and the *plate height*, H. The plate number is found from a peak in the chromatogram using

$$N = \left(\frac{t_R}{\sigma}\right)^2 = 16\left(\frac{t_R}{\omega_b}\right)^2 \tag{12.6}$$

where σ is the standard deviation assuming a Gaussian-shaped peak. σ can be replaced with the easily measured baseline width, w_b, using the numerical factor of 16.[6] The plate height is inversely related to plate number by

$$H = \frac{L}{N} \tag{12.7}$$

where L is the length of the column. In older literature, N is called the number of theoretical plates and H is called the height equivalent to one theoretical plate (HETP). When using the adjusted retention time, t'_R, in Equation (12.6), these terms are called the effective plate number, N_{eff}, and the effective plate height, H_{eff}. The plate number and plate height terminology originate from distillation theory, where a greater number of physical disks or plates along the length of a distillation column produced higher resolution in a fractional distillation process. We encountered a similar effect with sequential extractions in Section 2.5. Increasing the number of sequential extractions, n, even for a small volume of extracting solvent greatly increases the fraction of solute that is transferred from one phase to another. A very efficient chromatographic column will have large N and small H to produce narrow peaks. Closely spaced peaks will be clearly separated or resolved. N and H provide useful measures to compare the performance of different columns for a given class of analytes.

Plate height is related to the mobile phase flow rate and several column parameters, which are described below. The following discussion of column efficiency uses linear flow rate, u, which is related to the volumetric flow rate, F_c, by

$$u = \frac{F_c}{A_c\varepsilon} = \frac{L}{t_M} \tag{12.8}$$

[6]The baseline width is approximately 4σ for a symmetric peak.

The subscript c in F_c denotes flow rate corrected to column temperature. F_c can be found from the flow rate measured at the column exit at ambient temperature, F_a, using $F_c = F_a(T_c/T_a)$, where T_c and T_a are the column and ambient temperatures in kelvin, respectively. A_c is the cross-sectional area of the column and ε is the fraction of open space between the packing particles. In practice, u is found by dividing the column length, L, by the hold-up time, t_M. In gas chromatography, the mobile phase is compressible, and u is replaced with the average linear flow rate, $\bar{u}$. In liquid chromatography, the compressibility is very small and $\bar{u} = u$.

There are several sources of peak broadening that limit the resolution in a chromatogram. These broadening mechanisms are due to physical phenomena and cannot be changed for a given column. The following van Deemter equation applies to "packed" columns where the stationary phase is immobilized on the surface of small particles that fill or pack the column. Packed columns are used in all HPLC and some GC methods. A simplified version of the van Deemter equation can be written as:

$$H = A + \frac{B}{u} + Cu \tag{12.9}$$

where u is the linear flow rate of the mobile phase and the A, B, and C terms are described below. This expression allows us to understand why choosing different columns can improve separations. Column selection is often a tradeoff between acceptable performance and the cost of more expensive instrumentation.

Figure 12.6 shows plots of H versus u for several different columns. For any given column, there is an optimal flow rate that provides the smallest H and the best separation efficiency. This optimal flow rate is often available in column specifications from the manufacturer or can be determined experimentally for a given mobile phase. In Figure 12.6, the column represented by the solid line

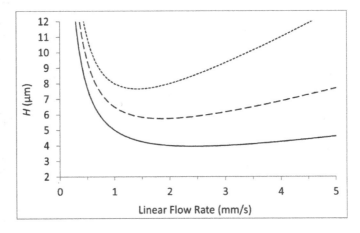

Figure 12.6 Plate height versus mobile phase flow rate.

has the best efficiency for $u = 2.3$ mm/s. In practice, the separation can be run at a faster rate to shorten analysis time with only a slight increase in H.

The multipath A term broadens peaks due to the random paths that identical solutes can take through a packed column. As an analogy, predict the outcome of rolling ten identical balls down a hill, starting them all at the same time. In theory, the ten balls should reach the bottom of the hill at the same time. In practice, they will each hit different bumps and depressions and travel slightly different paths, arriving at the bottom of the hill at different times. The smoothness of the ground will affect the spread in arrival times at the bottom of the hill. At the microscopic level, A is directly proportional to the diameter of the packing material, with smaller particles producing a smaller A term. Smaller particles are analogous to a smoother hill in our analogy and have a smaller spread in the multiple paths that are available to solutes. The open-tubular design of capillary GC columns eliminates this broadening mechanism and provides much higher separation efficiency than packed columns.

The B term arises from molecular diffusion. For a given solute, molecules that diffuse in the direction of the mobile phase elute faster than average and molecules that diffuse in the opposite direction elute later. Solute molecules also diffuse perpendicular to the mobile phase direction, but this does not contribute to peak broadening. Because of its dependence on flow direction, this term is often called *longitudinal diffusion*. In the extreme case of very low flow, solutes would diffuse lengthwise to significantly broaden the chromatographic bands. The large increase in H at low flow rate, $u < 1$ mm/s in Figure 12.6, is the result of the B/u term. The effect of diffusion decreases with increasing flow rate. The less time that solutes are in the column, the less time they have to diffuse and broaden the peaks.

The C term includes the equilibration times of a solute in both the stationary and mobile phases. This broadening results from the macroscopic nature of the mobile and stationary phases and is also known as the resistance to mass transfer. The time for a given solute to transit the column depends on the degree to which the stationary phase retains the molecule as it is carried by the mobile phase. In the ideal case, all molecules will be in constant contact with the mobile and stationary phases. In practice, solute molecules will spend some amount of time in the mobile phase too far from the stationary phase surface to interact. These molecules will be swept ahead of molecules that are interacting with the stationary phase. Likewise, solute molecules that diffuse into the stationary phase are not at the interface and will lag behind the average solute. These effects can be reduced by reducing the volume of the stationary and mobile phases, for example, by using a very thin stationary phase coating. Unlike the B term broadening, the C term increases with u and is the cause of the rising slope at higher u in the plots in Figure 12.6. When solutes transit the column faster, the "out-of-action" time delays due to resistance to mass transfer have a greater effect on peak broadening.

The van Deemter equation only considers broadening mechanisms in the column and assumes that the peaks are symmetrical. Peak broadening can occur due to solute dispersion outside of the column if the volume of connecting tubing, detector cells, and so on, is larger than necessary. Commercial systems are usually well tested to minimize this broadening. Asymmetric peaks can occur for certain analytes that interact with the packing material or if the column is overloaded with too much test sample. Peaks can be "stretched" on the leading edge of a peak, called *fronting*, or on the trailing edge, called *tailing*. Either asymmetry can degrade resolution due to greater overlap of adjacent peaks.

The peak resolution can be broken into several components and is often expressed by the following resolution equation:[7]

$$R_s = \frac{\sqrt{N}}{4}\left(\frac{k}{k+1}\right)\frac{\alpha-1}{\alpha} \tag{12.10}$$

N and k are the parameters defined earlier and α is the separation factor:[8]

$$\alpha = \frac{t_{R(2)}}{t_{R(1)}} = \frac{k_{(2)}}{k_{(1)}} \tag{12.11}$$

where $k_{(2)}$ and $k_{(1)}$ are retention factors for two adjacent peaks. Since $k_{(2)}$ has the longer retention time, α is always 1.0 or greater. An α of 1.1 is usually sufficient to produce baseline separation in HPLC (for a typical column with $N = 10,000$). The two adjacent peaks with the smallest value of α are another measure of the critical pair. These two peaks serve as the benchmark to optimize a separation. In the simplest case, separating the critical pair can be accomplished by using a longer column. Since the peak resolution increases as the square root of N and the time to complete a chromatogram increases linearly, this approach is not the most efficient. Choosing a more advanced, and usually more expensive, column with greater N will improve R_s.

The resolution equation provides several guidelines to help choose a suitable analytical column. For solutes with large k, the resolution does not vary significantly because the term $k/(k + 1)$ approaches 1. A rule of thumb is to choose a column that provides retention factors between 1 and 5 for most analytes. As an example, reversed-phase chromatography stationary phases include C8, C18, and other alkyl chain lengths. The longer chain length increases retention for nonpolar solutes. Resolution of analytes that have retention factors near 1 might be improved by using a C18 rather than a C8 stationary phase. Changing from a C8 to a C18 column will probably not

---

[7]For the origin of this and the following equations see Miller, J. M. *Chromatography: Concepts and Contrasts*, 2nd ed.; Wiley-Interscience: New York, 2005.

[8]α is also called the selectivity or selectivity factor in older literature.

improve the separation of poorly resolved peaks with retention factors near 10. Doing so will increase the total chromatogram time for no benefit.

You-Try-It 12.C
The resolution worksheet in you-try-it-12.xlsx contains several chromatograms. Tabulate retention factors and calculate the resolution between the two closest adjacent peaks in each chromatogram.

12.4   GAS CHROMATOGRAPHY (GC)

Gas chromatography is an instrumental separation that is applicable to any solute that can be vaporized intact. This requirement eliminates salts and many biologically relevant molecules. In some cases, thermally labile molecules can be derivatized to a different form that is stable when vaporized. Due to the inherent greater efficiency of capillary GC columns, separations are performed by GC rather than LC when possible.

A gas chromatograph consists of a compressed gas supply for the mobile phase, an injection port, the column supporting the stationary phase, thermostated oven, a detector, data acquisition electronics, and computer (Figure 12.7). The injection port is a rubber septum through which the needle of a syringe is inserted to inject the test portion into the flowing mobile phase. The injection port is maintained at a temperature higher than the boiling point of the least volatile component in the sample mixture. All analytes are flash volatilized and enter the column. The components in the test portion separate as they pass through the column owing to differences in their partitioning between the mobile gas phase and the stationary phase. The mobile phase, or *carrier gas*, is an inert gas such as helium, hydrogen, or nitrogen. There is no chemical interaction with the carrier gas and solutes separate based on their varying interactions with the stationary phase. A stationary phase should be selected that has a strong interaction with the analytes, i.e., use a stationary phase that matches the polarity of the analytes.

The schematic in Figure 12.7 is simplified. The dark circle on the flow control represents a knob to adjust the gas flow rate to the column. Other controls, not shown in the figure, adjust the oven temperature and the detector sensitivity. Modern instruments control these parameters via computer, along with a data acquisition system and a robotic autosampler to inject test samples. Actual instruments can also contain more than one column and different types of detectors. This flexibility makes it easy to run different types of solutes on different stationary phases with one instrument.

There is a large variety of GC stationary phases and different types of detectors, and these instrument components are discussed in turn below. For a stationary phase that matches the polarity of the analytes, the analytes will

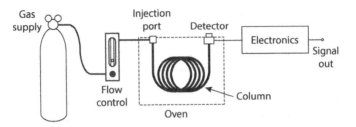

Figure 12.7   Schematic of a gas chromatograph.

usually elute in the order of lowest to highest boiling points. As a reminder, the relative polarity of solutes from nonpolar to polar is

aliphatic hydrocarbons (alkanes) < olefins (alkenes) < aromatic hydrocarbons < ethers < esters, ketones, aldehydes < alcohols, amines < organic acids

Since the partitioning is dependent on temperature, the separation column is contained in a thermostat-controlled oven. A good chromatogram will resolve all peaks in the minimum total time. Simple mixtures can be separated at a constant temperature, which is called *isothermal* chromatography. When a mixture contains both weakly and strongly retained compounds, it is not possible to achieve a good separation of all components. This condition is known as the *general elution problem*. It occurs so often that almost all modern gas chromatographs use a programmable oven so that the temperature can be ramped higher as a run proceeds. The separation is started with the oven at a temperature that achieves good separation of the early eluting compounds, that is, those with lower boiling point. The temperature is then ramped higher at some rate to speed the elution of the higher boiling point compounds. This *temperature programming* procedure allows complete separation of solutes with a wide range of boiling points in a reasonable amount of time. An additional advantage is that the increasing temperature ramp tends to sharpen the later peaks in a chromatogram.

12.4.1   GC Columns

GC columns are of two designs: packed or capillary.[9] Due to the higher efficiency, almost all analytical measurements are made with capillary columns. The exception is analysis of room-temperature gases and hydrocarbons containing one to five carbon atoms. Separating these small molecules requires a stationary phase with high surface area to achieve suitable retention. Molecular sieves are the most common column packing for room-temperature gases and carbon absorbents are used for hydrocarbons. Molecular sieves are zeolite (aluminosilicate) materials with extensive pores of ≤1 nm. The gaseous species are separated by both size and adsorption on the packing material, and

---

[9]Packed columns are an older technology that is still in use.

these separations are known as *gas–solid chromatography*. Due to the pressure drop caused by the packing material, packed columns are relatively short, 0.5–2 m total length. They are typically a glass or stainless-steel coil with an inside diameter of 1–5 mm.

Capillary columns are fused silica (purified glass) capillaries that have an inside diameter small enough for gas-phase species to equilibrate rapidly with the stationary phase coated on the inner walls of the capillary.[10] Since they are open, the pressure drop is less severe than for packed columns and lengths can be 10–100 m length, with 30 m being typical. Capillary columns provide much higher separation efficiency than packed columns but are more easily over-loaded by too much sample. Too much analyte totally saturates the stationary phase so that the plug of solute spreads, resulting in skewed peak shapes. Since syringes have a capacity on the order of 1 μl, the injector has a built-in *split* to reduce the amount of sample that enter the column. The split typically directs only 1:10 to 1:100 of the injected solution to the capillary column. The majority of solution is diverted through a vent to a waste line containing an activated carbon trap.

Table 12.2 lists typical inner diameters for commercially available GC columns. Decreasing the column diameter provides a distinct advantage in separation efficiency by reducing the $C \cdot u$ term in the van Deemter expression. This term is also affected by the thickness of the stationary phase, which will be scaled so the equilibration time in the stationary phase is similar to that in the mobile phase. The thickness of the stationary phase is generally 500–1000 times thinner than the column diameter due to the much slower diffusion of solutes in a stationary phase compared to the gas phase. For a given column diameter, thicker coatings are used for more volatile solutes and thinner coatings work better for higher molecular weight analytes. The disadvantage of the lower analyte capacity in the more efficient columns is the need for a more sensitive detector.

TABLE 12.2   Internal Diameter of Commercial GC Columns

Internal Diameter (mm)	Plate Number, N*	Analyte Capacity (μg)
0.53	39,000	1.0
0.32	69,000	0.5
0.25	87,800	0.05
0.20 or less	110,000	<0.05 or less ("Fast GC")

* For a 30-m column.

[10]Simple capillary columns are properly referred to as wall-coated open-tubular (WCOT). There are other capillary designs with additional support material on the inner surface to provide greater surface area and higher analyte capacity that we will not discuss.

Although I discussed physical characteristics first, the most important factor in selecting a GC column is to match the polarity of the stationary phase to the analytes. Since different compounds separate based on small differences in their interaction with the stationary phase, a strong overall interaction is desirable. The stationary phases are generally a polymer coating on the inner wall of the capillary. After the polymer coats the column wall or packing material, it is usually cross-linked to increase stability and prevent it from gradually bleeding out of the column at elevated temperatures.[11]

Table 12.3 lists a small sampling of stationary phases. Most of these coatings consist of a polysiloxane backbone with hydrocarbon functional groups. The polarity of the stationary phase is increased by increasing the fraction of the more polar functional groups. The most nonpolar stationary phase consists of polysiloxane with only methyl groups. More polar stationary phases increase the fraction of phenyl or other groups on the polymer. For very polar analytes, polyethylene glycol (a.k.a. carbowax) is commonly used as the stationary phase.

To summarize, a capillary column is selected based on the following factors, in order of importance:

- Select stationary phase to match polarity of analytes for strong interaction.
- Select column i.d. (inner diameter) depending on analyte capacity.
- Select film thickness (thinner for high-molecular weight analytes).
- Select length based on the efficiency that is required. A greater efficiency is needed for:
  - a difficult critical pair, or
  - a large number of analytes.

Very often, manufacturers have columns tailored to different types of analytes. In a manufacturer catalog you will find columns and methods for room-temperature gases, petroleum products, polyaromatic hydrocarbons, fragrances,

TABLE 12.3 GC Stationary Phases

Stationary Phase	Functional Groups	Polarity
Dimethyl polysiloxane	$-CH_3$	Nonpolar
5%Diphenyl, 95%dimethyl polysiloxane	$-CH_3$ and $-C_6H_5$	Nonpolar
35%Diphenyl, 65%dimethyl polysiloxane	$-CH_3$ and $-C_6H_5$	Moderate polarity
50%Cyanopropyl/phenyl, 50%dimethyl polysiloxane	$-C_3H_6CN/-C_6H_5$, and $-CH_3$	Moderate polarity
Polyethylene glycol (carbowax)	$-OH, -O-$	Polar

[11]Cross linking is the formation of bonds between polymer chains.

fatty acid methyl esters (FAME), chlorinated pesticides, and many other specialized applications.

You-Try-It 12.D
The GC worksheet in you-try-it-12.xlsx lists several sets of analytes. Use manufacturer literature to determine suitable stationary phases for these different separation problems.

12.4.2 GC Detectors

Most gas chromatographs are modular and the detector at the end of the column is easily interchanged. Different detectors will be used depending on the sensitivity and selectivity that is necessary for a given analysis. The most common detectors for gas chromatographs are the following:

- Thermal conductivity detector (TCD)
- Flame ionization detector (FID)
- Electron capture detector (ECD)
- Mass spectrometric detector (MSD)

These detectors will also have different linear ranges, which are determined with calibration standards. The following descriptions provide the characteristics that differentiate the different detector designs.

A thermal conductivity detector (TCD) consists of an electrically heated wire or thermistor, which is a resistor that is sensitive to temperature. The temperature of the sensing element depends on the thermal conductivity of the gas flowing around it. When an eluting molecule exits the column, it displaces some of the carrier gas and changes the thermal conductivity of the mobile phase. The temperature of the detector element changes, causing a change in resistance that create the signal. Being based on a bulk physical property, the TCD is not as sensitive as other detectors. However, it is nonspecific and detects any analyte. It is a common detector when using packed GC columns. The TCD is also nondestructive, allowing the collection of analytes for further analysis by condensing them as they exit the column.

A flame ionization detector (FID) consists of a small jet burner head, hydrogen and air gas supplies, and a cylindrical collector electrode positioned above the burner. The hydrogen and air support a flame through which the GC column effluent passes. The collector electrode has a negative voltage bias of 200–300 V and collects any positive ions that occur in the flame. The hydrogen/air flame produces very few ions, resulting in a low background. When organic solute molecules elute from the column, they create ions such as CHO^+ in the flame. These ions are collected on the collector electrode, creating an electrical current. The current signal is directly proportional to the number of ions created in the flame, and thus on the amount of organic molecules exiting the column. The sensitivity of the FID will vary depending on the number of carbon and heteroatoms atoms in a molecule. As always, quantitation requires a pure reference of each analyte for calibration. The FID is extremely sensitive with a large dynamic range. It is nearly universal for organic molecules except for small oxidized compounds such as formic acid and formaldehyde. Its main disadvantage is that it destroys the analytes as they elute from the column.

The ECD uses a radioactive beta emitter, typically ^{63}Ni on the inside surface of a metal cylinder. The beta particles ionize some of the carrier gas or an additional make-up gas such as N_2.[12] This process produces a plasma containing free electrons within the volume of the detector, which provides a steady current on a voltage-biased electrode. When molecules that contain electronegative functional groups, such as halogens, phosphorous, and nitro groups pass through the detector, they capture some of the plasma electrons. These negative ions tend to recombine with positive ions in the plasma, reducing the current measured at the electrode. The ECD can be much more sensitive than the FID for electronegative analytes, but it has a limited dynamic range. It finds its greatest application in environmental analysis for chlorinated and brominated pollutants. In these applications, the lack of sensitivity for other organic compounds makes it selective for the analytes of interest.

There are numerous other specialized detectors that are not discussed here, such as detectors for molecules containing N, P, or S. As with the ECD, this selectivity can be very important to try to quantitate selected components in samples that can contain hundreds of different volatile molecules. The detector essentially makes other compounds invisible, greatly improving the ability to quantitate the analytes of interest. The mass spectrometer detector discussed in the next section also has this capability to select only certain molecular fragments to display in a chromatogram. It is counterintuitive that reducing the analyte signal can improve the sensitivity of a measurement. The key aspect is that a more selective detector can decrease the baseline and overlapping peaks by orders of magnitude more than the compound of interest, so analyte peaks are easier to quantitate.

12.5 GAS CHROMATOGRAPHY MASS SPECTROMETRY (GC-MS)

A mass spectrometric or mass-selective detector (MSD) is a mass spectrometer interfaced to the effluent of a separation column. Since the analytes are already in the gas phase, the interface is simply a heated transfer line from the GC column to the mass spectrometer ion source. A key advantage of an MSD is that it can provide a mass spectrum to help confirm the identity of each peak in the chromatogram. Unlike the high-energy ICP that totally atomizes organic

---

[12]Beta particles are electrons emitted at high energy from a nucleus.

molecules, a GC ion source retains significant molecular fragments, including the unfragmented molecular ion.

As a reminder, a mass spectrometer ionizes analyte species and separates the ions based on their mass-to-charge ratio, *m/z*. The general process of obtaining a mass spectrum involves

- creating gas-phase ions (ion source);
- separating the ions in space or time on the basis of their mass-to-charge ratio (mass-selective analyzer);
- measuring the quantity of ions of each mass-to-charge ratio (ion detector).

Figure 12.8 shows a simple block diagram of a mass spectrometer detector. The vacuum system is needed to prevent collisions that could change the energy or trajectory of ions and to protect the detector and other components from oxidation that would degrade performance. The ion source for GC-MS is inside the vacuum system, making the GC-to-MS interface simpler than in ICP- and LC-MS sources that operate at atmospheric pressure. Since the light carrier gas molecules have a very high molecular speed, they are quickly pumped out by the vacuum system. Figure 12.8 omits a great deal of instrumental detail. Not shown are extraction and acceleration ion optics to transfer ions from the source region into the mass analyzer and the associated electronics for data acquisition and control.

12.5.1 Ionization Methods

The primary ionization method in GC-MS is electron ionization (EI), which was described in the previous chapter. Most instruments also have the capacity to perform *chemical ionization* (CI), which causes less fragmentation and ensures that the molecular ion is observed. One drawback is that EI efficiency drops significantly for molecules with mass greater than ≈500 Da. Since GC solutes must be volatile, this issue affects only a small subset of analytes. These heavier analytes can be ionized by CI or a recently developed GC-to-MS interface called cold EI.

CI uses a reagent ion to react with the analyte molecules and convert them to ions, usually via a proton transfer to form the $[M + H]^+$ ion. This ionization

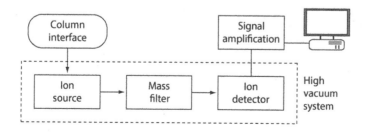

Figure 12.8 Schematic of mass spectrometer components.

process is "softer" than direct EI and causes much less fragmentation. The reagent ions are produced by introducing an approximately 100-fold excess of auxiliary gas such as methane, ammonia, or isobutane into the EI ion source. For methane, interaction with the electron beam produces relatively stable ions such as CH_5^+ and $C_2H_5^+$. These ions can react with analyte molecules to form a protonated molecular ion:

$$CH_4 + e^- (70\,eV) \rightarrow CH_4^+ + 2e^-$$
$$CH_4^+ + CH_4 \rightarrow CH_5^+ + CH_3$$
$$M + CH_5^+ \rightarrow [M + H]^+ + CH_4$$

There will be some fragmentation of the $[M + H]^+$ ion, but it is much less than in EI because very little energy is transferred to the molecule in the proton transfer process.

Negative chemical ionization (NCI) will occur for solutes with electronegative constituents such as halogen atoms or nitro groups. NCI can be more sensitive for these types of solutes and it provides greater selectivity since many other solutes do not form negative ions. The selectivity advantage of working in negative-ion mode can be extended to other classes of solutes by derivatizing analyte molecules with an appropriate agent.

Numerous examples of mass spectra are in the next You-Try-It exercise. Figure 12.9 shows an example of selectivity in *extracted-ion chromatograms*. These chromatograms are generated from one data set of a cleaning solvent

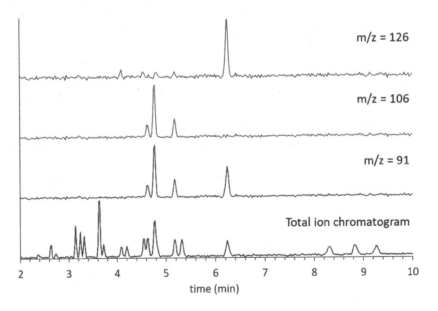

Figure 12.9 Total-ion (bottom) and extracted-ion chromatograms.

that contained chlorotoluene, xylenes, and other organic compounds. The control software records the full mass spectrum at each point in the chromatogram. The bottom chromatogram is the total ion chromatogram, which is a display of the total ion count from all peaks in the mass spectrum at a given time. It shows all solute peaks. The upper chromatograms are extracted-ion chromatograms where the signal is due only to the number of ions at the listed m/z. These signals are much lower and the extracted-ion chromatograms have been offset and scaled for visibility. The top chromatogram ($m/z = 126$) identifies the chlorotoluene peak and the $m/z = 106$ chromatogram show peaks that are xylene or contain a xylene fragment. The xylenes and chlorotoluene also appears in the $m/z = 91$ chromatogram due to the toluene fragment (tropylium ion). As noted above, using more selective detection can make it easier to quantitate a peak by reducing overlap with adjacent peaks.

You-Try-It 12.E
The mass-spec worksheet in you-try-it-12.xlsx contains sample mass spectra. Determine the major fragment ions and predict the analyte identity.

12.6   HIGH PERFORMANCE LIQUID CHROMATOGRAPHY

The principles and many of the stationary phases for liquid-phase analytical separations are the same as in SPE and column chromatography (refer to Section 2.8). The most common implementation is called *high performance liquid chromatography* (HPLC), which uses a pump to push mobile phase through the column at pressures of 3000–4000 psi. Variations include low pressure liquid chromatography (LPLC) and medium pressure liquid chromatography (MPLC). These systems tend to use large diameter columns for preparative-scale separations. There is substantial overlap with flash chromatography, which uses air pressure to force liquid through a column primarily for product cleanup. Ultra-performance liquid chromatography (UPLC) is an extension of HPLC where higher pressure, up to 10,000 psi, allows the use of smaller column packing to achieve greater efficiency.[13]

By using high pressure, the mobile phase can be forced through columns containing packing material of small diameter. HPLC packing particles are typically of 3–10 μm diameter, compared to approximately 100 μm for gravity-based or open column chromatography. As discussed in terms of the van Deemter equation, smaller particles lead to smaller A and C terms and a more efficient separation (see Figure 12.6). The enhanced efficiency of UPLC results

[13]Different manufacturers will use different terms with ultra-high performance liquid chromatography (UHPLC) also being common.

from higher pump pressures, allowing the use of even smaller diameter packing. The practical result of more efficient separations is to allow more components to be separated or a separation to be completed on a shorter column in less time.

Complete HPLC instruments generally consist of two reservoirs of mobile phase, a high-pressure pump for each mobile phase, an injector, a separation column, a detector, and a data collection system. The mobile phase solvents are degassed to eliminate formation of bubbles and the pumps provide a steady high pressure with no pulsation. The output from the two pumps are combined so that the composition of the mobile phase entering the column can be varied. This process, called gradient elution, will be explained after discussion stationary phases.

To withstand the high pressure, the hardware consists of stainless-steel tubing, fittings, and columns. The liquid test sample is introduced into a loop of tubing in the injector with a syringe. Once this loop of tubing is filled with sample solution, the injector is rotated to place the sample loop in line with the flowing mobile phase. Turning the injector handle to inject the test solution also sends a signal to the data collection system that a new run has begun. Figure 12.10 shows a simplified schematic of an HPLC instrument. Not shown are the solvent degassers or accessories such as an autoinjector, guard column, multiple detectors, or computer data acquisition. The electrical signal is converted to a digital value by data acquisition electronics and recorded on a computer as a function of time.

The software in commercial chromatographs returns both the height and integrated areas of each peak. When peaks are not baseline separated, these integrated areas must be checked closely to ensure that the software selects reasonable cutoffs between peaks. For complicated cases of overlap and peak asymmetry, the peaks must be deconvoluted for integration. Although the integrated peak areas are usually more reliable for quantitative measurements, the peak heights can be a better choice for moderate overlap. Figure 12.5(b) showed an example of overlap in two of the later peaks where using the peak heights could be more accurate than trying to integrate the peak areas. The other case where peak heights can be more accurate for quantitation is when

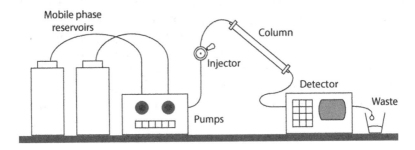

Figure 12.10   Schematic of an HPLC.

the amounts of analytes are near the detection limit. In such cases, the noise on a chromatogram can produce errors in the integrated areas, making peak heights a better choice for quantitation.

12.6.1 HPLC Columns and Stationary Phases

The stainless steel HPLC columns are typically 10–30 cm in length and 3–5 mm inner diameter. Short columns of 3–10 cm in length are used for fast analytical separations and as guard columns. Guard columns are placed before a full-length analytical column to trap irreversibly binding junk and extend the lifetime of the more expensive analytical column. Just as in open column chromatography, the different components in a mixture pass through the column at different rates because of the differences in their partitioning behavior.

HPLC stationary phases include many of the same materials that are used in SPE and open column chromatography, but as noted above, supported on much smaller packing particles. Table 12.4 categorizes different types of HPLC stationary phases with their target analytes. The first step in developing an HPLC method is to choose a suitable approach based on the nature of the solutes. For many types of mixtures, multiple approaches can separate the target analytes. Section 2.6 described the underlying interactions for most of these stationary phases, and that information is not repeated here. Compared to the materials discussed in Section 2.6, Table 12.4 drops affinity chromatography and adds size-exclusion, chiral, and hydrophobic interaction chromatography. Despite the name, affinity chromatography is used in a stepwise SPE mode to isolate one specific biomolecule. The following paragraphs briefly describe each type of stationary phase.

Reversed-phase chromatography is the most common form of HPLC due to the wide range of analytes that can dissolve in the mobile phase. Reversed-phase partitioning uses a relatively nonpolar stationary phase and a polar mobile phase, such as a mixture of methanol or acetonitrile with water. The most common stationary phases are n-octyldecyl (C18) chains, n-octyl (C8) chains, and phenyl groups. The stationary phase is bonded to support particles and manufacturers use "capping" methods to cover any exposed silica or alumina surface. The capping prevents adsorption in addition to the desired interaction

TABLE 12.4 Types of Stationary Phases

Type	Analytes	Mobile Phases
Reversed-phase	Nonpolar to moderate organics	Polar solvents
Normal-phase	Moderate to polar organics	Nonpolar solvents
Adsorption	Isomers	Any
Chiral	Enantiomers	Any
Hydrophobic-interaction	Proteins	Aqueous buffer
Ion-exchange	Cations and anions	Aqueous with pH buffer
Size-exclusion	Polymers and biomolecules	Any

with the stationary phase. The choice of stationary phase is dictated by matching the polarity of the solutes; nonpolar solutes separate best with C18 and moderate polarity solutes separate well with C8 or other stationary phases. The solutes will elute in the order of most polar to least polar.

Normal-phase partitioning uses a polar stationary phase and a nonpolar organic solvent, such as *n*-hexane, methylene chloride, or chloroform, as the mobile phase. The stationary phase is a bonded siloxane with a polar functional group. The most common functional groups in order of increasing polarity are

- cyano: $-C_2H_4CN$
- diol: $-C_3H_6-O-CH_2CH(OH)CH_2(OH)$
- amine: $-C_3H_6NH_2$
- dimethylamine: $-C_3H_6N(CH_3)_2$.

Normal-phase methods are used less frequently than the reversed-phase columns, but they are necessary for solutes that are only soluble in nonpolar solvents. Using HPLC for these types of solutes is less common because many of these analytes can be separated by GC. Elution order is the opposite compared to reversed-phase chromatography. With a normal-phase column, solutes elute in the order of least polar to most polar. The next two methods are sometimes included as normal-phase methods, but the separation mechanism depends on molecular geometry in addition to polarity.

Adsorption chromatography is a more common approach in column chromatography rather than in HPLC. The main advantage of adsorption chromatography in HPLC is in separating isomers. Isomers have different molecular shapes, leading to different degrees of physisorption on a surface. Shape-based differences are called *steric effects* and provide a means to differentiate two isomers. Owing to their very similar chemical properties, isomers can be difficult to separate by other methods.

Chiral or enantioselective chromatography separates chiral compounds, which, in the simplest case, are stereoisomers or enantiomers that are nonsuperimposable mirror images of each other. Separation of enantiomeric mixtures is challenging because the stereoisomers are identical chemically, and usually requires the high efficiency of HPLC to separate well. Chiral analytes are separated by using a chiral selector in the separation process. The chiral selector can be either a chiral stationary phase or a chiral additive in the mobile phase. One stereoisomer of a chiral molecule is usually bonded to a polymer, which is then coated onto a silica packing material. Separation occurs because two chiral analytes will interact with the one isomer on the stationary phase differently.

Hydrophobic-interaction chromatography (HIC) uses a similar stationary phase as in reversed-phase chromatography, usually a phenyl group or a short alkyl chain. The mobile phase contains a high concentration of ammonium sulfate, which is decreased during the separation. Ammonium sulfate causes the protein surface to reorient and expose hydrophobic regions that are usually buried in the interior of the protein. Decreasing the ammonium sulfate concentration in the mobile phase allows the proteins to convert back to their normal configuration. As the hydrophobic regions of the protein return to the interior of the biomolecule, the proteins elute. HIC is usually better than reversed-phase chromatography at maintaining the activity of the proteins after separation. Because hydrophobic interaction chromatography varies the mobile phase composition, HIC is much easier to use with programmable pumps rather than manually creating a gradient with an open column.

The ion-exchange interaction and the variety of stationary phases were discussed in detail for SPE applications in Section 2.7. Here we detail how these stationary phases are applied for a continuous chromatographic separation. Using a strong anion-exchange stationary phase with a quaternary ammonium group as an example, the distribution of anion solutes, A^{a-}, between the stationary and mobile phases can be written as

$$-N(CH_3)_3^+ \, OH^- \, (s) + A^{a-} \, (aq) \rightleftharpoons -N(CH_3)_3^+ \, A^{a-} \, (s) + OH^- \, (aq)$$

where the leading dash on the quaternary ammonium group represents a linkage to the support particle. The (s) indicates an ion immobilized on the stationary phase, and (aq) indicates that the ion is in the mobile phase. A distribution constant, K_D', for this equilibrium can be written as

$$K_D' = \frac{\left[-N(CH_3)_3^+ \, A^{a-} \, (s) \right] \left[OH^- \, (aq) \right]}{\left[-N(CH_3)_3^+ \, OH^- \, (s) \right] \left[A^{a-} \, (aq) \right]}$$

In loading analytes in an SPE cleanup, conditions are set to retain all anions. In an equilibrium description, we say the reaction is far to the right. By changing the conditions, in this case by rinsing the SPE cartridge with a high concentration of OH^-(aq), the equilibrium shifts to the left and all anions are displaced from the stationary phase and elute. Ion chromatography will use the same stationary phase, but conditions are set so the analyte ions are only partially retained. That is, there is a significant fraction of any given analyte ion in both the retained and the mobile phases. For the equilibrium shown above, different anions elute through the column at different rates because of their varying interaction with the stationary phase.

Intuitively, we can guess that an A^{3-} ion will have a stronger interaction than A^{2-}, which will interact more strongly than A^-. Elution order will be A^-, followed by A^{2-}, followed by A^{3-}. Now the question arises: why does the strength of the electrostatic interaction differ for ions of the same charge? Coulomb's law gives the force, F, between two point charges, q_1 and q_2, to be inversely proportional to the square of their separation, r_{12}^2:

$$F \propto \frac{q_1 q_2}{r_{12}^2}$$

As we predicted, a higher charge has a stronger interaction. Ions of the same charge vary because they have varying separation distances. Any ion in solution will be hydrated. That is, it will have a sphere of tightly bound water molecules that prevent other species from getting closer. The degree of hydration will vary for different ions, and each ion will have a different distance-of-closest approach (see Tables 5.9 and 5.10). In terms of equilibrium, we can also say that each different ion has a different value of K_D'. The net result is that ions of the same charge will be retained on the column for different lengths of time.

One of the most significant applications of ion-exchange chromatography is in the analysis of anions for which there are no other rapid analytical methods. It maintains speciation information, such as NO_2^-, NO_3^-, SO_3^-, and SO_4^{2-}, which can be lost in other methods. Ion chromatography is also used for amino acids, protonated amines, proteins, and other biochemical species.

Size-exclusion stationary phases have pores of varying sizes to separate macromolecules based on physical size. The pore-size distribution is selected to match the expected range of the analytes. Other common names for this approach are *gel-permeation chromatography* (GPC) for synthetic polymers and *gel-filtration chromatography* (GFC) for biomolecules. The mobile phase for GPC is usually an organic solvent and for GFC it is water or a pH buffer. The stationary phase is chemically inert, but it has a distribution of pore sizes. The retention time for a macromolecule of a particular size depends on the fraction of pores that it can enter. The size range where macromolecules can enter some but not all pores is called the *selective permeation* region. Smaller macromolecules that can enter more pores take a longer path to transit the column and elute later. Larger solutes cannot enter as many pores and they pass through the column more quickly. Macromolecules that are too large to enter any of the pores are not retained and they pass through the column first. These unretained molecules produce the first peak in the chromatogram, known as the *exclusion limit*. Likewise, molecules that are smaller than the smallest pore can enter all pores. They elute as the last peak in the chromatogram, known as the *permeation limit*. There is no separation among the molecules at either of these limits. The most common application of GPC for synthetic polymers is to determine the distribution of molecular weights in a sample. For biological polymers, GFC is used to fractionate different biomolecules or to separate macromolecules from salts and small molecules.

12.6.2   Gradient Elution

As noted for gas-chromatographic separations, we have the general elution problem when a test sample contains both weakly and strongly retained components. We can have this same issue in HPLC when using *isocratic* conditions, that is, at a constant mobile-phase composition. A mobile-phase composition that separates the early eluting species can result in a long time to elute the strongly retained components. We solved this problem in GC by using temperature programming, but the temperature of an LC column filled with solvent cannot be changed quickly. The better approach is to vary the mobile-phase composition as the separation progresses to speed the elution of the strongly retained components. We already introduced this concept of *gradient elution* in decreasing the ammonium sulfate concentration during a separation in hydrophobic-interaction chromatography.

When using two solvent reservoirs and two pumps, the pumps can be programmed to vary the composition of the mobile phase that enters the column as the run proceeds. The initial conditions should separate the early eluting compounds. The mobile phase composition is then varied at a rate that separates all components in a reasonable time. The programmed variation depends on the type of separation, but the mobile phase composition is changed to shift the distribution of solutes from retention on the stationary phase to being in the mobile phase. Using the anion-exchange equilibrium discussed above, the initial mobile-phase pH is selected so the earliest peaks separate. The programmable gradient will then vary the mobile-phase composition from two reservoirs that contain buffers of different pH. After the weakly retained anions elute, the mobile-phase pH is raised, increasing $[OH^-]$, to speed up the elution of the more strongly retained anions.

As another example, reversed-phase chromatography uses a nonpolar stationary phase. The mobile phase is a mixture of water and a miscible organic solvent such as methanol or acetonitrile. After allowing the weakly retained solutes to elute, the fraction of the organic component in the mobile phase is increased to disrupt the interaction between the retained solutes and the stationary phase. Doing so decreases the retention time of solutes that are strongly retained at the initial mobile phase composition.

The following list outlines the method development for an HPLC analysis. This list assumes that any interferences have been removed in prior sample cleanup procedures. Method development can be very time consuming, but luckily many procedures are published in the primary and manufacturer literature.

- Choose the separation approach based on the nature of the analytes and their solubility in mobile phases.
- Choose a column, usually based on manufacturer application notes.
- Select a column temperature. Temperature-sensitive proteins are often separated in a cold room to maintain activity. Small molecule separations are usually done at room temperature, but elevated temperatures can speed the separation ($\approx$3-fold faster 50°C).
- Run isocratic experiments to determine a starting composition that separates weakly retained solutes.
- Develop a gradient that separates all solutes in the shortest time.[14]

---

[14]Note that gradient elution is not used in size-exclusion separations. The pore-size distribution is selected to match the size of the macromolecules and there are nominally no chemical interactions.

As an illustrative example, parabens are organic compounds that are used in cosmetics, lotions, and other consumer products as preservatives to kill bacteria. Figure 12.11 shows the structure of the series of methyl through butyl paraben (systematic name: methyl 4-hydroxybenzoate, etc.). As is typical for a real sample containing many components, the parabens must be extracted from these consumer products. After extraction and SPE cleanup, a mixture of methyl, ethyl, propyl, and butyl parabens can be separated and quantitated by liquid chromatography.

Figure 12.12 shows the chromatogram of a standard solution that contained 5.0 ppm of each paraben. The column contained a reversed-phase C18 stationary phase, the mobile phase was 50:50 methanol/water, and detection was by absorbance at 254 nm. Before reading further, look at Figure 12.11 and predict the elution order, from short to longer retention time, for these paraben molecules. The first very small peak at t_R = 0.77 min in the figure is a solvent peak due to the sample solvent being slightly different in composition from the mobile phase.

The nonpolar character of these compounds increases as the alkyl chain length increases. Remember that in reversed-phase chromatography, the mobile phase is polar and the stationary phase is nonpolar. As alkyl chain length

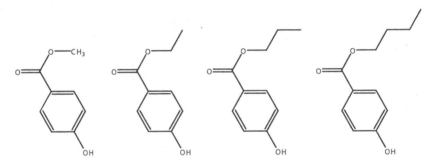

Figure 12.11   Structures of common parabens.

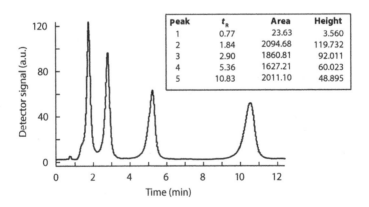

peak	t_R	Area	Height
1	0.77	23.63	3.560
2	1.84	2094.68	119.732
3	2.90	1860.81	92.011
4	5.36	1627.21	60.023
5	10.83	2011.10	48.895

Figure 12.12   HPLC chromatogram of 5.0-ppm paraben mixture.

increases in the parabens, the molecule becomes more nonpolar, increasing the interaction with the C18 stationary phase. A stronger interaction results in a longer retention time. This series of parabens will therefore elute in the order of methyl paraben (t_R = 1.84 min), ethyl paraben (t_R = 2.90 min), and so on.

The data table shown in Example 12.4 is a printout from the computer data acquisition system, which provides peak heights and integrated peak areas for each peak. Since the peaks in this chromatogram are well separated, the integrated peak areas are preferred for quantitation. For both peak areas and peak heights, the four parabens vary slightly in their absorbance at 254 nm. This chromatogram of a standard solution provides the calibration data to quantitate the amount of each paraben in an unknown sample, as illustrated in the following example.

Example 12.4 Chromatogram Calibration. A hand lotion sample was extracted with methanol and cleaned up on a C18 SPE column by washing with water and eluting with methanol. The following peaks were reported in an HPLC separation of this test portion, collected under the same conditions as in Figure 12.12. What parabens are present and at what concentrations?

Retention Time, s	Peak Area
1.89	1082.9
5.40	488.2

The 1.89-s peak matches the 1.84-s peak in the standard chromatogram and is due to methyl paraben. The 5.40-s peak matches the 5.36-s peak in the standard chromatogram and is due to propyl paraben.

Figure 12.12 provides the calibration data to determine instrument sensitivity for each paraben. In this example with no units given on the y-axis, the peak area has units of minutes. Since the standard and unknown test portions were measured under the same conditions, I will use "unit area" to avoid confusion. For methyl paraben the proportionality factor is

$$\frac{5.0 \text{ ppm}}{2094.68 \text{ unit area}} = 0.00239 \text{ ppm/unit area}$$

and for propyl paraben the proportionality factor is

$$\frac{5.0 \text{ ppm}}{2011.10 \text{ unit area}} = 0.00249 \text{ ppm/unit area}$$

Using these proportionality factors, we can determine the analyte concentrations in the lotion test portion:

$$c_{\text{unk}} = \text{signal} \times \text{proportionality factor}$$

For methyl paraben

$$c = (1082.9 \text{ unit area})(0.00239 \text{ ppm/unit area}) = 2.6 \text{ ppm}$$

and for propyl paraben

$$c = (488.2 \text{ unit area})(0.00249 \text{ ppm/unit area}) = 1.2 \text{ ppm}$$

Note that I have corrected the number of significant figures for the precision in the 5.0 ppm standard solution. Keep in mind that these calculations give us the concentrations in the processed test portion. To determine the concentrations in the original lotion, we must factor in extraction efficiency, dilutions, and any other sample processing factors.

12.6.3 HPLC Detectors

As in GC, a variety of HPLC detectors are available that offer different sensitivity, selectivity, linear range, and spectral information. In general, a detector is a transducer at the end of a separation column that produces an electrical signal when a solute elutes. Optical and electroanalytical detector consist of a cell that the HPLC effluent flows through. Other detectors, such as evaporative light-scattering and mass-spectrometer detectors, require an interface to remove the liquid mobile phase. More than one detector can be used for a separation, with the condition that a detector that destroys the sample is placed last. Likewise, a fraction collector can be positioned after a nondestructive detector to recover the analytes.

Table 12.5 lists common HPLC detection mechanisms. Each type of detector will have a different sensitivity and linear range, which can be determined by running standards. The refractive index (RI) and the evaporative light scattering detectors (ELSD) will respond to almost any analyte, making them the most generic HPLC detectors. The RI detector measures the refractive index of the mobile phase. Solutes displace some of the mobile phase and

TABLE 12.5 HPLC Detectors

Detection Method	Selectivity	Characteristics
Refractive index	Universal	Less sensitive
ELSD	Universal	Nonvolatile solutes
UV/Vis absorption	Variable, wavelength-dependent	Common for organic solutes, full spectrum possible with array detector
Fluorescence	Fluorescent solutes, wavelength-dependent	Can be very sensitive
Electrochemical	Redox active solutes	Redox active
Conductivity	Ions only	Ions
Mass spectrometer	Universal or highly selective	Can provide structural confirmation

changes the RI. Since this detector measures a bulk physical property, it is not as sensitive as the other detectors. It also has the disadvantage that gradient elution will produce a sloping baseline as the run progresses.

The main requirement for the ELSD is that the solutes be less volatile than the mobile phase. The ELSD nebulizes the column effluent in a flow of nitrogen gas to produce a spray of small droplets. These droplets pass through a heated tube to evaporate the mobile phase. The temperature of this drying tube is optimized for rapid evaporation of the mobile phase without vaporizing solute molecules. The result is a "mist" of solute particles. These particles are detected when they transit a laser beam and scatter light to an optical detector. The ELSD has a limited dynamic range, but it fills an important niche for analytes that do not absorb light in the UV/Vis.

Many organic compounds, including molecular ions such as proteins, have absorption bands in the ultraviolet or visible spectral region and can be detected by absorption spectroscopy. A mobile phase must be chosen that does not absorb at the wavelength where the analytes are detected. (Table 4.5 listed UV cutoffs of common solvents.) The effluent flows through a quartz flow cell, which serves the same purpose as the sample cuvette in an absorption spectrophotometer. The path length is quite small, and weakly absorbing species are detectable only at higher concentrations. Given that a in the Beer-Lambert relationship ($A = abc$) is wavelength dependent, the sensitivity for each individual analyte must be determined with a pure reference. Likewise, changing the detection wavelength requires recalibration. Absorbance detectors that utilize an array detector can provide the full UV/Vis absorption spectrum for each peak that elutes. This capability is useful for peak identification and can help to identify chromatographic peaks that have more than one component.

For fluorescent analytes, a fluorescence-based detector can provide much higher sensitivity than other detection methods. Analytes that have no intrinsic fluorescence will sometimes be derivatized with a fluorophore to increase sensitivity. A common example is in the analysis of amino acids. The analytes have a common chemical structure so that one type of reaction will work for all amino acids. In addition to increasing the sensitivity, addition of a fluorescent component gives each amino acid a similar sensitivity. Fluorescence detection also provides a level of selectivity since nonfluorescent components are invisible to the detector. The detector consists of a flow cell and an excitation source, which is usually tunable. After selecting an appropriate excitation wavelength, fluorescence intensity is monitored at one or multiple wavelengths.

Various electroanalytical processes can be used for HPLC detection. Potentiometric detection is possible and research groups have developed miniature detectors and arrays for various applications. Commercially, amperometric methods are more common where the signal is an electrical current due to an oxidation or reduction of an analyte as it elutes through an electrochemical cell. They have the advantage of being able to select or scan the applied potential to provide some selectivity in analyte detection. Amperometric methods are especially useful for small biomolecules such as carbohydrates and amino acids, where other detectors have lower sensitivity.

Many of the detectors described above can be used in ion chromatography. Inorganic and some small organic ions however are not electroactive at a suitable potential or do not have absorbing chromophores in the UV/Vis region. Any ion can be detected by measuring the electrical conductivity of the mobile phase as it exits the column. The presence of the ionic solutes in the mobile phase increases the solution conductivity and appears as a measurable signal. Ion chromatographic mobile phases already contain ions that create a high background for electrical conductivity, making it more difficult to measure the conductivity due only to the analyte ions as they exit the column. This problem is greatly reduced by selectively removing the mobile phase eluent ions after the ion-exchange separation column and before the detector. This is done by converting the mobile phase ions to a neutral form or removing them with an eluent suppressor. Most eluent suppressors consist of an ion-exchange membrane with acid or base to neutralize OH^- or H^+, respectively.

As in GC, interfacing an HPLC column to a mass spectrometer can provide structural information to help confirm the identity of each chromatographic peak. Unlike GC-MS, the interface and ionization mechanisms are more complicated due to the difficulty in isolating the analyte molecules from the liquid mobile phase. The LC-to-MS interface usually consists of a nebulizing spray outside of the vacuum system. Drying gas and heated regions increase evaporation of the mobile phase. Commercial systems can be configured to switch between several ionization methods, which are listed in Table 12.6. Ions are extracted into the mass spectrometer through one or more stages of differential pumping to reach the vacuum system. The LC-MS ionization methods cause little fragmentation. The vacuum interface will often include settings to induce collisions to fragment the molecular ion to gain more structural information. The following discussion describes the general ionization mechanisms, but many details of these interfaces are omitted.

TABLE 12.6 LC-MS Ionization Methods

Ionization Method	Analytes	Characteristics
Atmospheric pressure photoionization (APPI)	Small molecules	Nonpolar
Atmospheric pressure chemical ionization (APCI)	Small molecules	Moderate polarity
Electrospray ionization (ESI)	Small and large molecules	Polar molecules

Atmospheric pressure photoionization (APPI) uses a UV discharge lamp to ionize the analyte molecules. The ionization mechanism can be direct or indirect via a dopant such as toluene or acetone. Direct ionization occurs when the analyte molecule absorbs a UV photon to create $M^{\cdot+}$. Indirect ionization via a dopant can occur via charge transfer to create $M^{\cdot+}$, or by a proton-transfer reaction that forms $[M + H]^+$. The APPI method is less common than the other two methods, but it is used for nonpolar molecules that are not ionized by APCI or ESI.

In *atmospheric pressure chemical ionization* (APCI), the column effluent mixes with nitrogen nebulizer gas and passes through a heated tube to desolvate the solute molecules. A needle in the ion source is biased at 2000–4000 V to create a corona discharge that ionizes nitrogen and water molecules in the nebulizer gas. The net result is production of water clusters containing a proton, for example, H_3O^+ and $(H_2O)_2H^+$. These clusters provide a source of protons that can transfer to analyte molecules to create $[M + H]^+$.

A notable advance in mass spectrometry in the past few decades is the development of ionization methods for large biomolecules such as proteins, DNA, and RNA. *Electrospray ionization* (ESI) can ionize these large biomolecules and couples well with HPLC. An ESI source consists of a very fine capillary that is held at several thousand volts relative to a counter electrode. As mobile phase sprays out of the capillary, the droplets pick up charge. The droplets become smaller as neutral solvent molecules evaporate. The charge density increases until the repulsive force causes a droplet to break into even smaller droplets, which is known as a coulombic explosion. For small molecules, the ionization mechanism is thought to be ion ejection, where a singly charged analyte molecule is ejected from the droplet due to the electrostatic repulsion. For large molecules, all the solvent evaporates, leaving an analyte molecule with multiple charges. This mechanism is called the charge residue model. An advantage of creating ions with multiple charges, on the order of 10–20, is that it lowers the m/z ratio so that large molecules can be analyzed in a quadrupole mass analyzer. For example, a protein of 30,000 Da that has charges of +15–20 will show mass spectral peaks at $m/z = 2000$, 1875, 1765, 1667, 1579, and 1500. All of these peaks are within the typical instrument limit of $m/z = 2000$.

You-Try-It 12.F
The LC worksheet in you-try-it-12.xlsx lists several sets of analytes. Use manufacturer literature to determine suitable stationary phases and detectors for these different separation problems.

12.7 ELECTROPHORESIS

Electrophoresis is a separations technique that is based on the mobility of ions in an electric field. It finds its greatest utility in separating large biomolecules such as proteins, DNA, RNA, and fragments of these species. Very simply, positively charged ions migrate toward a negative electrode and negatively charged ions migrate toward a positive electrode. This movement is called *electrophoretic mobility*. For safety reasons, one electrode is usually held at ground potential and the other is biased positively or negatively. Ions separate because they have different migration rates depending on their total charge, size, and shape.

12.7.1 Instrumentation

The instrumentation for an electrophoretic separation consists of a high voltage supply, electrodes, buffer solution, buffer reservoirs, and a support for the "running" buffer. There are three types of supports, flat porous substrates such as filter paper or cellulose acetate strips, polymeric gels, and glass capillary columns. Open capillary tubes are used for many types of samples including small molecules. Capillary-based instruments will include some type of detector to measure the separated components, similar to the in-line detection of HPLC.

The two most common flat supports are polyacrylamide gel for protein mixtures or smaller DNA fragments and agarose gels for larger nucleic acids. These two materials differ in pore size, with agarose being larger to accommodate the larger biomolecules. The pore size of polyacrylamide can be varied by using different amounts of a cross-linker when casting the gel.[15] Also, when preparing the gel, a comb is placed at one end to create wells for sample loading. The gel is supported on either a horizontal or a vertical mount, and multiple samples are run in separate lanes. Applying a typical voltage of 100 V results in a runtime of 30–60 min. After the separation is completed, the paper or gel is stained to visualize the separated components.

12.7.2 SDS-PAGE

SDS-PAGE stands for sodium dodecyl sulfate (SDS) polyacrylamide gel electrophoresis (PAGE). It is the most widespread type of electrophoresis and finds its greatest use in molecular weight analysis of proteins and other biomolecules. SDS is a surfactant (detergent) that dissociates oligomeric proteins into its subunits and unfolds the individual polypeptides. The SDS binds to most polypeptides in a ratio of one SDS molecule per two amino acids. Since

---

[15]Cross linking a GC stationary phase improves the temperature stability. Varying the degree of cross linking in a gel provides a means of varying the size of the open pores.

SDS is anionic, it gives the resulting complexes a negative charge with a fairly constant charge-to-mass ratio. The electrophoretic migration rate is therefore determined only by the size of the SDS–protein complex. The polyacrylamide gel support has a pore size distribution that allows all proteins to migrate, but not at equal rates. Smaller proteins migrate faster and larger proteins migrate more slowly because they are unable to pass through the smaller pores in the gel. Molecular weights are determined by running marker proteins of known molecular weight simultaneously with samples. Figure 12.13 shows a sample gel that has been run and stained. The molecular weight markers are in the far left lane and the labels give the number of base pairs. These standards of known molecular weight provide the calibration to determine the molecular weight and identity of the separated analytes in the other lanes.

There are two common variations to improve resolution in gel electrophoresis, discontinuous electrophoresis, and isoelectric focusing. Discontinuous (disc) electrophoresis uses two or more gels that are buffered at different pH. Figure 12.14 shows a schematic of a disk electrophoresis setup. The gel is held on a support so that the two ends are in contact with the buffer solution. Samples are loaded in the wells at the top and then the voltage is applied. The first gel is called the *stacking gel* and the longer gel is called the *running* or *resolving gel*. The stacking gel is very open and proteins migrate through it very quickly. When proteins reach the resolving gel their migration slows, causing them to "stack up." The net result is that as they migrate from one gel to the other they become concentrated into sharp bands, which produces much higher resolution than with conventional gel electrophoresis.

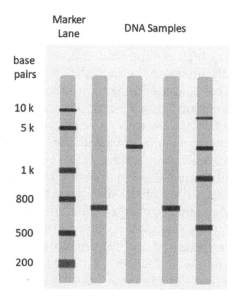

Figure 12.13   Schematic of a gel electropherogram.

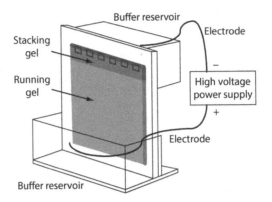

Figure 12.14 Schematic of a discontinuous electrophoresis apparatus.

Proteins have a large number of charged functional groups and will have a specific pH at which they have an overall charge of zero. This pH is called the *isoelectric point*. At lower pH, they will be charged positively and at high pH they will be charged negatively. Since the isoelectric point depends on the amino acid composition of the protein, it is different for most proteins. Resolution can also be improved using isoelectric focusing. In this technique, the support gel maintains a pH gradient. As a protein migrates down the gel, it reaches a pH that is equal to its isoelectric point. At this pH, the protein is neutral and no longer migrates, that is, it is focused into a sharp band on the gel.

12.7.3 Capillary Electrophoresis (CE)

CE, or capillary zone electrophoresis (CZE), performs the electrophoretic separation in a small-diameter glass capillary. The main advantage of this approach is that a very high electric field can be applied because the small capillary can efficiently dissipate the heat that is generated. Increasing the electric field reduces the separation time to produce very efficient separation. Another advantage compared to gel electrophoresis is that an instrumental detector, similar to an HPLC detector, can be placed in-line with the capillary. The analyte quantitation is simultaneous with the separation and subsequent staining and reading of a gel is not necessary.

The migration of solutes in CE is due to *electroosmotic flow*. Electroosmotic flow occurs because the surface of the silica glass capillary contains negative oxygen ions that attract positively charged counterions from the buffer (Figure 12.15). When the high voltage is applied, these positively charged ions migrate toward the negative electrode and drag solvent molecules in the same direction (the negative charges on the surface are fixed in place). During a separation, uncharged molecules move at the same velocity as the electroosmotic flow with minimal separation. Positively charged ions move faster and negatively charged ions move more slowly. The ions separate because they migrate at different rates because of differences in charge, size, and shape.

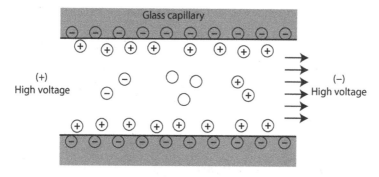

Figure 12.15 Schematic of the double layer on a capillary surface.

Unlike in HPLC, there is no mass transfer or eddy diffusion broadening in CE, resulting in very efficient separations. The main line-broadening mechanism is diffusion, and the rapid separation at high voltage minimizes this problem. In addition, electroosmotic flow produces a relatively flat profile of the solvent movement across the diameter of the capillary. Owing to friction at the column wall in HPLC, there is a difference in the mobile phase flow between the center and the walls of a column, which contributes to peak broadening. The net result is that CE will have much higher resolution than HPLC, making it the preferred separation method in a number of applications. However, the greater complexity and sample-dependent variability in results with CE means that HPLC remains a more common analytical method for less demanding separations in analytical laboratories.

Capillaries are typically of 50 μm inner diameter and 0.5–1 m length, and the applied potential is 20–30 kV. Figure 12.16 shows a schematic of CE separation. A small volume of sample, ≈10 nl, is injected at the positive end of the capillary. Commercial instruments have a robotic autosampler to perform this step. Owing to electroosmotic flow, all sample components migrate toward the negative electrode where the separated components are detected. CE detection is similar to detectors in HPLC, and include absorbance, fluorescence, electrochemical, and mass spectrometry. Owing to the small amount of analyte, a sensitive detector is required. Note that data acquisition electronics and computer data storage are not shown in the figure.

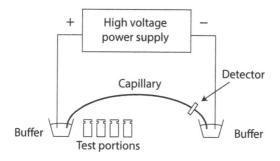

Figure 12.16 Schematic of capillary electrophoresis.

The capillary can also be filled with a gel, which eliminates the electroosmotic flow. Separation is accomplished as in conventional gel electrophoresis but the capillary allows higher resolution, greater sensitivity, and on-line detection. Another variation, micellar electrokinetic capillary chromatography (MEKC), uses micelles in the buffer to introduce partitioning into the migration time for specialized analyses. A number of other CE variations are in constant development to extend the high resolution of this method to more solutes and applications.

Chapter 12. What Was the Point? This chapter described a variety of instrumental methods for separating and quantitating analytes in a mixture. The underlying principles for the chromatographic separations are the same as discussed for extractions and column chromatography. As we saw for spectrometric methods, a number of different techniques can often be used for a given analytical problem. The choice of a given method is often based on subtle or practical factors. In addition to the examples discussed in this chapter, there are numerous other approaches that are tailored for specific applications.

PRACTICE EXERCISES

1. Consider the following chromatographic detectors:
   - ECD
   - FID
   - Fluorescence detector
   - Mass spectrometer
   - RI detector
   - TCD
   - UV/Vis absorption detector
   (a) For each detector, indicate if it is a feasible detector for only GC, only HPLC, or both GC and HPLC.
   (b) Which detector will be most appropriate in a separation of the major products from a fermentation reactor, alcohols, aldehydes, and ketones, if the components require identification of their molecular structure? Explain your rationale.
   (c) Which detector will be most appropriate in a separation of the major products from a fermentation reactor, alcohols, aldehydes, and ketones, if the components must be collected at the end of the column for IR absorption spectroscopy? Explain your rationale.

2. Make a table of GC detectors and list the advantages and disadvantages of each. Specify the types of analytes that are measurable by each detector.

3. Predict the elution order of benzene, *n*-hexane, and *n*-hexanol in
   (a) reversed-phase partition liquid chromatography

(b) normal-phase partition liquid chromatography
(c) capillary GC with a nonpolar stationary phase
Give the advantages and disadvantages of these three methods and suggest factors to consider in choosing one over the others.

4. Explain the mechanism of interaction in ion-exchange chromatography.

5. Choose from this list of chromatographic columns to answer the next questions. Note if the sample requires processing, such as extraction, before analysis. Also suggest suitable detectors for the analytes.
   - bonded C18 ($-C_{18}H_{37}$) reversed-phase HPLC column
   - cation-exchange HPLC column
   - anion-exchange HPLC column
   - capillary GC column containing polyethylene glycol
   - capillary GC column containing polydimethyl siloxane
   (a) Which column will be the most appropriate to separate and quantitate halides, bromate, nitrate, nitrite, and phosphate in an extract of livers from rats that were fed a diet of hot dogs?
   (b) Which column will be the most appropriate to separate ammonium, calcium, magnesium, morpholine ($C_4H_8ONH_2^+$), potassium, and sodium in blood samples?
   (c) Which column will be the most appropriate to separate the alkanes in samples of gasoline?
   (d) Which column will be the most appropriate to separate and identify barbiturates in blood samples?

6. A series of similar acids, including acetyl salicylic acid (aspirin, $pK_a = 3.49$), benzoic acid ($pK_a = 4.20$), salicylic acid (2-hydroxybenzoic acid, $pK_a = 2.97$), and nicotinic acid (niacin, $pK_a = 4.85$) are to be separated by anion-exchange chromatography.
   (a) What is the appropriate pH for the mobile phase? Explain.
   (b) How should the buffer pH be changed in a gradient elution to speed up the elution of the more tightly held ions? Explain.

7. Why is HPLC rather than GC the most common chromatographic method for analyzing mixtures of proteins?

8. Use the following data, taken from the chromatograms in Figure 12.5, for the questions below. Table 12.7 contains peak data for the reference chromatogram and Table 12.8 contains data from a test portion of the cocoa extract. The reference solution contained 1.0 μg/ml of each of the analyte.
   (a) What is the concentration of each component in the cocoa extract?
   (b) The sample preparation was to dissolve 50 mg of cocoa powder in hot water and then extract with three 10-ml portions of diethyl ether. What is the concentration, as a weight percent, of each component in the cocoa powder?

TABLE 12.7   Chromatogram Peaks of Standard Solution

t_R, min	Substance	Peak Height, arb. units
1.0	Theobromine	7.2
1.7	Theophylline	12.4
4.3	Caffeine	3.9

TABLE 12.8   Chromatogram Peaks of Cocoa Powder Extract

t_R, min	Peak Height, arb. units
0.8	17.9
1.0	12.2
1.7	7.0
2.0	6.2
3.2	3.6
4.3	2.9
5.0	2.6
5.8	2.3
6.6	2.1
7.8	1.8

9. What are the relative merits and tradeoffs of SIM and TIC when using mass spectrometric detection?

10. Make a table listing the mass spectrometer designs discussed in Chapter 10 with their general characteristics and performance measures such as resolution and limit of detection. Use an Internet search to find as many specifics in manufacturer information as you can for GC-MS and LC-MS instruments.

11. Describe the basic operation of gel electrophoresis. Explain the polarity of the electrodes, the need for the buffer, and the relative migration rate of large and small proteins. Explain why at least one lane of molecular weight standards is run with the test portions.

12. List several methods to visualize the separated biomolecules after running a gel. Specify any selectivity afforded by the different methods.

13. Use an Internet search to find one or more detailed protocols for creating gels and performing SDS-PAGE. Explain why different amounts of crosslinker are used for different protein analytes. The description in this chapter was brief and skipped many details. List the additives used to prepare the buffers and the test portions in SDS-PAGE and describe their purpose.

14. Use an Internet search to find suitable separation methods for amino acids. Choose one each using GC, HPLC, and electrophoresis and compare their relative performance.

15. The speciation of arsenic can be as important as total concentration in terms of health effects. Look up common forms of arsenic in groundwater and suggest a chromatographic method to couple to an ICP-MS to measure the different forms.

16. Use an Internet search to find examples of separations using CE. Tabulate specifications such as type and number of solutes, detection limits, and speed of analysis. Include any special sample preparation or derivatization steps in the analyses. Categorize the types of analytes as environmental, biomedical, forensic, etc. Describe any common factors in the examples that you find.

17. What are total-ion, selected-ion, and extracted-ion chromatograms? What are the advantages of these different data acquisition and display modes?

Part III. What Was the Point?

Part III of this text introduced the concepts and details of instrumental methods that are common in chemical measurements. These methods, based on spectroscopy or the measurement of an electrical charge or current, have been developed for the very high sensitivity that they can achieve. The instrumental separations are especially powerful by combining a sensitive detector to a separations column to provide separation and quantitation in one process.

Standard operating procedures for instrumental methods require the proper use of QA and QC control samples, blanks, and standards to achieve accurate results. Using a new method often requires extensive method validation to develop procedures that minimize interferences and bias. For many types of analytes, more than one type of instrumental method can provide accurate results. Selecting a method is often determined by practical matters, that is, equipment availability, speed of analysis, or cost. There are many more instruments than could be described in this introductory text. When you encounter new methods, do try to connect the underlying principles to what you know. You will find that many methods share a common basis for operation.

Epilogue. What Is Next?

Congratulations! You made it to the end of the textbook. I do not know what is next for you, but I wish you well. I hope that this text has helped prepare you for your future challenges.

INDEX

Basics of Analytical Chemistry and Chemical Equilibria: A Quantitative Approach, Second Edition. Brian M. Tissue.
© 2023 John Wiley & Sons, Inc. Published 2023 by John Wiley & Sons, Inc.
Companion Website: www.wiley.com/go/tissue/analyticalchemistry2e